Verification and Validation in Computational Science and Engineering

by

Patrick J. Roache

hermosa publishers

PO Box 9110
Albuquerque, New Mexico 87119-9110
USA
http://www.hermosa-pub.com/hermosa

International Standard Book Number 0-913478-08-3
Library of Congress Catalog Card Number 98-96214
Printed in the United States of America

Cover Illustrations

The cover background illustrations are from pages 332 and 343 of the classic paper "The Approximate Arithmetical Solution by Finite Differences of Physical Problems Involving Differential Equations, with an Application to the Stresses in a Masonry Dam", *Transactions of the Royal Society of London, Series A*, Vol. 210, pp. 307–357, by Lewis Fry Richardson (1910).

Warnings - Disclaimer

This book is intended to provide information in regard to the subject matter covered. Reprinted material is quoted with permission where required, and sources are indicated. It is sold with the understanding that the publisher and author are not engaged in rendering legal, accounting, professional engineering, or other professional services. If legal or other expert assistance is required, the services of a competent professional should be sought. Efforts have been made to make this book as complete and accurate as possible. However, there may be mistakes both typographical and in content. Therefore, the text should be used only as a general guide and not as the ultimate source of information. The purpose of this book is to educate. The publisher and author have neither liability nor responsibility to any person or entity with respect to any loss or damage caused, or alleged to be caused, directly or indirectly by the information contained in this book.

If you do not wish to be bound by the above, you may return this book to the publisher for a full refund. Direct all inquiries to Hermosa Publishers, P. O. Box 9110, Albuquerque, New Mexico 87119-9110, USA.

PREFACE

The title of this book, like most titles, is a compromise. Most of my experience, and therefore most of the examples in this book, are taken from that branch of Computational Science (or Computational Engineering, or Computational Physics, or Computational Mathematics) known as Computational Fluid Dynamics or CFD. Often, this term CFD is taken to apply to only computational aerodynamics, but I use it here more broadly to include computational aeroacoustics, groundwater flow and transport, ocean modeling, meteorological modeling, combustion modeling, free convection heat transfer, neutron transport, magnetohydrodynamics, etc. I also have some experience in computational electrodynamics, examples from which appear in this book as well. However, the principles covered in the book are much more general, applying to virtually all numerical solutions of partial differential equations, including solid mechanics, structural dynamics, chemistry, etc., even some finance models. Use of the term "partial differential equations" in the title would be descriptive but would have led to an overly long title, and perhaps would have seemed too mathematical. Much of the book would be properly covered in the title "Computational Mathematics", but "Validation" (as used herein - see Chapter 2) strictly speaking is not a mathematical issue (as is "Verification") but is an issue of physical science. Also, the related issue of "Certification" is most obviously an engineering term, related to somewhat specific engineering projects.

Within the broader realm of Computational Science and Engineering, I do not treat "real time" problems, e.g. on-line nuclear reactor simulations, which need robust input error checking. This is not a Verification and Validation issue in Computational Physics or Mathematics. (It certainly is a consideration for code quality, but not for Verification and Validation, as defined herein.) The term "Computer Science" as used in university departments would sometimes cover this material as well, but more often refers to issues of hardware and system software, rather than applications codes for partial differential equations as considered herein. The somewhat amorphous term "Computational Science and Engineering" includes applications codes but also (in one definition) is concerned with providing "total problem-solving environments" (i.e., computer systems), and this book would not qualify for such an inclusive definition. In the end, I chose the more inclusive title, although the text will often use "CFD" as the descriptive term covering partial differential equations in science and engineering.

The examples used herein obviously cannot cover the entire range of computational solutions of physically important partial differential equations. Some notable omissions are radiation equations, Kortweg-de Vries equations, molecular modeling, etc. But most readers of the *Journal of Computational Physics*, or the SIAM Numerical journals, or the AIAA and ASME and AIChE Journals, should have an interest in the book's subject.

Topics other than Verification and Validation are treated, such as Confirmation, Certification, Benchmarks, and Quality Assurance. These are related to but distinct from Verification and Validation. The terms are not as universally defined nor as centrally important as Verification and Validation. "Quantification of Uncertainty" is an excellent inclusive term; frankly, the main reason I did not use it is that I already have two other publications with that title. (Also, some authorities now suggest that it should not be used synonymously for errors that are not due to lack of physical information.)

The background required for a reader of this book includes a first level course in partial differential equations and basic numerical methods including some linear algebra and some numerical solutions for partial differential equations. It should not be necessary to be an expert in CFD. The methods and issues discussed are applicable to general purpose commercial codes, as well as specialized scientific codes. For the specific examples, it would be advantageous to have some background in *either* aerodynamics, fluid dynamics, groundwater flow and transport, or electrodynamics. I do not by any means intend to discourage applied mathematicians, but the discussions of validation would benefit if the reader had some minimal experience with *physical* experiments. I do not consider "CFD Lab" courses, as valuable as they may be, to be an adequate substitute for all physical experimental courses. (I understand that it is now possible at some universities to earn a bachelor's degree in engineering without ever taking a physical laboratory course; it is my opinion that the recipients of these degrees have been cheated in their education.)

Many readers will feel that I have not done justice to the extensive literature on error estimation in Finite Element methods. As I describe in Chapters 4 and 7, my opinion is that these methods are useful for guiding grid adaptation but less useful for the title subject of this book. More basically, I hope to excuse myself by quoting the legendary mathematical physicist S. Chandrasekhar, explaining why, in his classic *Hydrodynamic and Hydromagnetic Stability* (Oxford, 1961), he had no reference to viscous shear flow. "[T]his is a large and highly specialized field and I have not felt myself competent to write about it. This argument is indeed a general one: in the last analysis an author chooses to write about only those matters in which he has some confidence of his understanding."

I have taken two minor liberties when quoting sources. In this book, I have capitalized terms like Validation, Verification, Calibration, etc. in order to emphasize that the terms are used in a limited *technical* (vs. general) context - a major theme, especially in Chapter 2. To maintain consistent emphasis, I have capitalized these terms in quotations even when the original authors did not. Also, I have replaced underlines appearing in the original sources with italics, just because it looks better in type. The emphasis of the original authors is maintained.

In my reading of papers by respected colleagues in the title area, my experience has invariably been mixed. Even when reading what I consider to be the very best papers, those which provide significant insights and which I have quoted extensively in this book, I find many points of disagreement. Sometimes these points are minor or semantic, but often they are major and substantive. It has not seemed to be worthwhile, or even feasible, to discuss each of these points individually with citations of the original papers, as would be done in a review (e.g., for *AIAA Journal* or *Applied Mechanics Reviews*). I know for a fact that some of the disagreement is due simply to time-lag in the evolution of the concepts in the communities, and that many of the authors would now change their position (and terminology) upon further thought. This has certainly been my own experience. Extrapolating, I am confident that I will change my opinion on some of my ideas expressed herein. As a corollary, I am sure that this book contains internal inconsistencies. I ask the reader's tolerance.

ACKNOWLEDGMENTS

It may surprise (and, if so, hopefully disappoint) some readers to discover that not everyone *cares* about numerical accuracy. I have been most fortunate in receiving research funding support over years from a minority in the applied mathematics, science, and engineering communities who believed it to be an important issue, not by giving just lip service, but also money and work. Foremost have been Larry Rapagnani, formerly of the U. S. Air Force Phillips Laboratory, Rip Anderson, Mel Marietta, and Mert Fewell of Sandia National Laboratories, Jim McCroskey of U. S. Army–Ames, and Tom Stauffer of the U. S. Air Force Armstrong Laboratories. Thanks to them for the years they supported my approach to numerical accuracy, i.e. they bought what I was selling. Likewise, Dr. Jack L. Melchor, through his Melchor Foundation, and the University of Notre Dame provided me with a Visiting Professorship in the Department of Civil Engineering and Geological Sciences for the fall of 1996, during which much of this manuscript was written. My thanks also go to Notre Dame Professors Joannes Westerink, William Gray, and Jerry Marley for setting up the visit.

For moral support and intellectual stimulation, I am deeply indebted to my colleagues at Ecodynamics, Sandia, Notre Dame, and throughout the CFD community, and to my wife, Dr. Catharine Stewart-Roache. I apologize for omissions, but these come to mind (listed alphabetically): Dan Aeschliman, John Bergstrom, Rebecca Blaine, Fred Blottner, Peter Bradshaw, Ismail Celik, Ray Chin, the late Tom Davis, Dave Dietrich, Roger Eaton, Kathy Economy, Dick Ewing, Mert Fewell, Joe Ferziger, Chris Freitas, Mohammed Gad-El-Hak, Kirti Ghia, Urmila Ghia, Bill Gray, Phil Gresho, Fred Habashi, Scott Hagen, Hank Happ, Daniel Haworth, Eric Jumper, Patrick Knupp, Benny Leonard, Eddie Leonardi, Clark Lewis, Robert

MacKinnon, Mel Marietta, Steve McCormick, Jim McCroskey, Unmeel Mehta, Tom Mueller, Bill Oberkampf, Sam Paolucci, Dominique Pelletier, Steve Piacsek, Andrew Pollack, Joe Powers, Michael Rabbitt, the late John Rakich, Helen Reed, Art Rizzi, John Ruge, Tom Russell, Patrick Ryan, Kambiz Salari, Bill Saric, Tom Stauffer, the late Joe Steger, Stan Steinberg, Catharine Stewart-Roache, Demetri Telionis, Joe Thompson, Alex Treadway, Randy Truman, Graham de Vahl Davis, Paul Van Gulick, David Walker, Bob Warming, Joannes Westerink, Frank White, Dave Wilcox, David W. Zingg.

This work was partially supported by Sandia National Laboratories and the United States Department of Energy under Contract DE-AC04-76DP00789, Dr. M. G. Marietta and Dr. M. E. Fewell, Technical Monitors; by the United States Army Aviation Systems Command, Aeroflightdynamics Directorate under Contract NAS2-13618, Dr. W. J. McCroskey, Technical Monitor; by United States Air Force contract "Adaptive Grids for Raven", Dr. N. L. Rapagnani, Technical Monitor , and United States Air Force contract F08635-93-C-0085, Dr. T. Stauffer, Technical Monitor.

TABLE OF CONTENTS

Dedicated to
Orestes Augustus Brownson, 1803–1876.

"He lived a life of integrity."

PART I
OVERVIEW

Part I will present an overview of the subject, approached by way of historical development. The history will include not only ideas but attitudes, and involves several controversies. Considerations of terminology and semantics have proven to be essential, as attested by confusion in the literature. However unraveling the semantics can be tiresome, and the reader could proceed directly to the more substantive material in Part II if preferred.

1

INTRODUCTION

1.1 SKETCH OF HISTORICAL DEVELOPMENT OF CFD

As noted in the Preface, the subject of numerical solution of partial differential equations (PDE's) is more broad than Computational Fluid Dynamics (CFD), but I will often use the acronym CFD to serve for the entire computational PDE area, including computational aeroacoustics, groundwater flow, electrodynamics, plasma dynamics, solid mechanics, structural dynamics, heat conduction, neutron transport, groundwater flow and transport, ocean modeling, meteorological modeling, combustion modeling, free convection heat transfer, neutron transport, magnetohydrodynamics, chemistry, enhanced oil recovery processes, etc., even some finance models. The methods and issues to be discussed are equally applicable to general purpose commercial codes and to specialized scientific codes.

The historical roots of CFD can be traced to Richardson's 1910 paper on the computation of stresses in a masonry dam. Although performed with hand calculations by "computing boys," paid piecework, the work was truly multidimensional, and the paper itself is highly recommended reading for any practitioners interested in their own cultural heritage. See also the brief history in Roache (1972). The paper by Emmons (1970) was a landmark study for its systematic approach. The popular science article by Harlow and Fromm (1965) alerted many in the scientific community to the potential of the computer simulation approach, presenting at least qualitatively accurate solutions of previously intractable problems without the constricting assumptions used by theoreticians. By the time of the publication of the first book with "CFD" in the title (Roache, 1972) it was beginning to be realized that we were in the midst of a revolution, in fluid dynamics and indeed in all areas of science and engineering. The traditional categories of "theoretical" and "experimental" were being forced to make way for a third category: "computational." For a while, people tried to fit the computational work into the old "theoretical" category, but it did not work. As I wrote in 1972, "it is this author's contention that computational fluid dynamics is a separate discipline, distinct from experimental fluid dynamics and from theoretical fluid dynamics, with its own techniques, its own difficulties, and its own utility."

The revolution was enabled by the phenomenal progress in computing power, which continues today. Yet it cannot be emphasized too much that progress was and is *equally* achieved by improvements in algorithms. For example, Cramer's method for solving linear systems of equations, which I was taught in high school and undergraduate school, is so inefficient for large systems that its operation count is literally astronomical. For modern problems with $O(10^6)$ grid points, the algorithmic improvement over Cramer's rule (and other primitive algorithms, e.g. the naive discrete Fourier transform replaced by the modern Fast Fourier Transform) can overshadow the hardware increases spanning the age of the abacus to parallel architecture supercomputers.

Nevertheless, we are now living with the historical heritage of marginal computing power. As computing power increased exponentially, perhaps doubling every two to three years, the vision of the practitioners increased even faster, abetted by the increase in complexity of the physics modeled and by the dimensionality. If one could solve (to at least qualitative accuracy) a two-dimensional incompressible steady laminar separated flow problem on a grid of $O(100 \times 100)$ nodes at some calendar date (say 1970), then one could swallow up a two-order of magnitude increase in computing power by going to three-dimensional flows, another two orders of magnitude by going to time-dependent flows, another one order of magnitude by going to turbulent flow with Reynolds-averaged Navier-Stokes equations and a two-equation model of turbulence, another order of magnitude with a Large Eddy Simulation of turbulence, another one or two orders of magnitude with brute-force calculation of turbulence by Direct Numerical Simulations (DNS) in square channels at low Reynolds numbers. This is roughly where we are in aerodynamics analysis at the time of this writing (1998). Further generations of computing power can be eaten up, several orders of magnitude at a time, by increases in Reynolds numbers and by engineering geometries (easily one order of magnitude, two for full aircraft). Grid resolution also has a healthy appetite; for a three-dimensional time-dependent problem, a factor of 2 increase in computer speed is eaten up by merely a 19% increase in resolution, even assuming optimally efficient methods. An order of magnitude increase in computer speed does not allow even a doubling of resolution, but only a 78% increase. There is more. Going from *analysis* to *design optimization* is good for a minimum of one order of magnitude for even the simplest system with one or two design parameters; two or even three orders of magnitude are more realistic. In groundwater flow and transport, a geologic disposal site for radioactive waste originally designed with back-of-the-envelope calculations is now analyzed with *thousands* of geostatistical realizations of property fields and scenarios, each calculation involving two-dimensional and three-dimensional time dependent calculations with three different CFD codes (WIPP PA Dept., 1992). Detailed geochemistry calculations in groundwater remediation studies, coupled with groundwater flow, could saturate any conceivable computer system. As is now well recognized, the post-processing of data, especially from time-dependent simulations, demands more computing power than previous levels of simulations. That is, the computer resources devoted in 1998 to visual post-processing probably exceeds the power devoted to performing the simulations themselves in 1980. In many installations, "antique" computers purchased 3 to 5 years earlier as computing engines are kept around for post-processing tasks.

The point is that the universe of possible problems is so extensive, and the power of simulations is so great, that CFD practitioners have often focused on *qualitative* simulation of the *next* more difficult problem class, rather than on achieving *quantitative* accuracy on the *previous* problem class. The result of this situation sometimes has been to cause a *decrease* in the quality of simulations published.

It is easy to be sympathetic to the wave of virtually unbridled optimism accompanying this revolution. In the aerodynamics community, the touchstone was the 1975 article in *Aeronautics and Astronautics* by Chapman, Mark and Pirtle (1975), which predicted (and called for) CFD to "supplant" wind tunnel experiments. The lack of restraint was perhaps due in part to the natural necessity of "selling" the CFD potential to management, composed as always of an older generation of technocrats who were not directly involved in the newer technical field of CFD, and who were more inclined to skepticism than to enthusiasm, who held the purse-strings, and who were to make the decisions about relative funding of CFD vs. experimental facilities.

To the credit of the CFD community, the ensuing debate was not simply characterized by "us vs. them," i.e. by CFD practitioners vs. experimentalists. From the CFD community, there were two immediate and critical responses to the Chapman et al article, one by Bradshaw (1975) and one by myself (Roache, 1975). Bradshaw pointed out the naiveté of the "supplanting" philosophy especially in regards to turbulence modeling, while I emphasized the limitations inherent in the mathematical approximations. Interestingly, we now recognize these as issues of the two title categories of this book, i.e. Validation and Verification, respectively. (See Chapter 2 for definitions.) This more cautious and realistic philosophy was reiterated by other conscientious CFD practitioners in a division of opinion that persisted for years, arguably to this day. Especially noteworthy were the editorial comments of three successive editors of the Proceedings of the AIAA Computational Fluid Dynamics Conferences. J. Rakich (1973), R. T. Davis (1975), and F. G. Blottner (1977) each took unequivocal stands against the overly optimistic view of Chapman, Mark and Pirtle (1975). (Rakich, 1973 anticipated the latter article.) Each explicitly stated that wind tunnels would not be replaced by computers, invited participation by experimentalists in the CFD conferences, and called for collaboration between computational, experimental, and analytical fluid dynamicists. Implicitly, they took stands against the even more aggressively optimistic salesmanship of others during this time. (An anonymously authored parody of the situation appeared in the mid 1980's, and is reproduced here in Appendix D with additional comments.)

At the round table discussion of the 1988 AGARD conference on Validation in CFD (AGARD, 1988), H. Hornung noted approvingly that numerical performance had improved over previous years, yet the claims were more modest, and that "reduction of the extreme claims is a very healthy sign." In the same round table discussion, P. Sacher noted that "Nobody talks anymore of replacement of wind tunnels by the computer. I am really happy to find that."

The salesmanship persists today in the arena of CFD commercial codes. There exists a deleterious Darwinistic effect here. Most potential customers want to hear only good news. The more conscientious salespeople will point out limitations of their products, and often lose sales to less scrupulous salespeople. The best support I can give to the conscientious salespeople is to encourage enlightened customers, which is a thrust of this book.

The division of opinion persisted but not so much in a spirit of debate. Unfortunately, as is often the case in human affairs, the opinions solidified and there arose separate camps that did not much talk to each other. The effects were very significant, with the optimists largely winning the day in the management struggles. While arguably good for computing power and computors like myself, the result for fluid dynamics as a whole was deleterious. Aeschliman et al (1995) also attribute to the Chapman, Mark and Pirtle (1975) article the widely recognized historical trend (especially in the U.S.A.) of reduced funding for experimental work, to the point where irreplaceable facilities were scrapped, graduate students went unfunded, and cultural skills were almost lost. As they note, it is now possible at some colleges for someone to obtain a Bachelors Degree in engineering or physics and "never witness, let alone actively participate in an actual experiment," the genuine physical laboratory courses having been totally "supplanted" with "CFD Labs." I obviously believe in CFD education and CFD Labs, but I agree with Aeschliman, Oberkampf, and Blottner (1995): "Modelers who have no laboratory experience are unlikely to appreciate the true complexity of the real world."

Aeschliman, Oberkampf, and Blottner (1995) also note that the optimistic predictions of Chapman, Mark and Pirtle (1975) have failed to materialize, even though the actual increase in computing power exceeded their requirements. Likewise, Rizzi and Vos (1996), note "the fundamental question of how well available RANS [Reynolds-Averaged Navier-Stokes] codes and turbulence models predict the maximum lift of a simple transport aircraft remains largely unanswered." Also, Aeschliman et al (1995) note that the continued closing of experimental facilities will result in "increasing dependence on new and unvalidated CFD codes for solutions to the most difficult remaining flow problems. The National Aerospace Plane, which was to be designed and developed with a very heavy reliance on CFD, and the ensuing programmatic fiasco surrounding NASP after the truth became known, was not necessarily a unique event."

1.2 THE NEW IMPETUS TOWARD HIGHER QUALITY SOLUTIONS

The Stanford Turbulence "Olympics" (Kline et al, 1981) made a tremendous effort to establish high standards of turbulence simulations, and the organizational aspects still stand as a paradigm for this type of workshop or community endeavor (e.g., see Rizzi and Vos, 1996). Unfortunately, the first conclusion of the Review Committee (chaired by Howard Emmons) was that the numerical quality of the solutions was so poor that it was impossible to draw meaningful conclusions about the relative merits of various turbulence models. This same story has been repeated many times in conferences, symposia and workshops since then.

Needless to say, there have always been some individuals and groups who produced high quality, reliable CFD calculations. Two decades before the present book was written, Blottner (1977) stated the present issue with clarity. After distinguishing the two distinct sources of error from "either inappropriate governing equations or inaccurate numerical solution procedure" (i.e., the issues now named Validation and Verification; see Chapter 2), Blottner introduced the term "credibility" and noted its necessary requirement: "Computational fluid dynamicists must perform grid refinement studies whenever they present numerical studies as experimentalists are expected to put error bounds on their results. Only if this is done can the needed credibility of computational fluid dynamics be established."

However, individuals like Blottner were the exception, and the general level of quality even as late as the mid-1980's was unacceptable, in my opinion. This evaluation is based not only on reading of final published articles, but on my experience on the Review Committee of the Stanford Turbulence "Olympics" (Kline et al, 1981) and as an editor and reviewer. In 1985, Frank White, the Editor of the American Society of Mechanical Engineers (ASME) *Journal of Fluids Engineering* created the position of Associate Editor for Numerical Methods, which formally recognized the special needs of this discipline. In 1986, we published (Roache, Ghia and White, 1986) in JFE an "Editorial Policy Statement on the Control of Numerical Accuracy" which instituted "higher standards on the control of numerical accuracy." Since the rationale and needs (in 1986) for the policy statement are explained in the announcement, it is repeated below. (See also Appendix B.)

1.2.1 ASME Journal of Fluids Engineering 1986—"Editorial Policy Statement on the Control of Numerical Accuracy"

A professional problem exists in the computational fluid dynamics community and in the broader area of computational physics. Namely, there is a need for higher standards on the control of numerical accuracy.

The numerical fluid dynamics community is aware of this problem but, although individual researchers strive to control accuracy, the issue has not to our knowledge been addressed collectively and formally by any professional society or journal editorial board. The problem is certainly not unique to the JFE and came into even sharper focus at the 1980–81 AFOSR HTTM-Stanford Conference on Complex Turbulent Flows. It was a conclusion of that conference's Evaluation Committee that, in most of the submissions to the conference, it was impossible to evaluate and compare the accuracy of different turbulence models, since one could not distinguish physical modeling errors from numerical errors related to the algorithm and grid. This is especially the case for 1st-order accurate methods and hybrid methods.

The practice of publishing comparisons based on coarse grid solutions, without systematic truncation error testing, may have been acceptable in the past. Certainly 10–15 years ago any calculation was of interest, and much of the exploratory work deserved publication, as many researchers lacked the computational power or funds to do a thorough and systematic error estimation. We are of the opinion that this practice, however understandable in the past, is outmoded and that, with powerful computers becoming more common, standards should be raised. Consequently, this journal hereby announces the following policy:

The Journal of Fluids Engineering will not accept for publication any paper reporting the numerical solution of a fluids engineering problem that fails to address the task of systematic truncation error testing and accuracy estimation.

Although the formal announcement of this journal policy is new, it has been the practice of many of our conscientious reviewers. Thus the present announcement is not a change in policy so much as a clarification and standardization.

Methods are available to accomplish this task, such as Richardson Extrapolation (when applicable), calculations with a high- and low-order method on the same grid, and straightforward repeat calculations with finer or coarser grids. As in the case of experimental uncertainty analysis, "...any appropriate analysis is far better than none as long as the procedure is explained." Whatever the authors use will be considered in the review process, but we must make it clear that *a single calculation in a fixed grid will not be acceptable,* since it is impossible to infer an accuracy estimate from such a calculation. Also, the editors will not consider a reasonable agreement with experimental data to be sufficient proof of accuracy, especially if any adjustable parameters are involved, as in turbulence modeling.

We recognize that it can be costly to do a thorough study, and that many practical engineering calculations will continue to be performed on a single fixed grid. However, this practice is insufficient for publication in an archival journal.

1.2.2 Later Policy Statements and Other Inititatives

Although apparently innocuous, this policy was not universally welcomed nor was it implemented without difficulty. See Appendix A, a reprint of Roache (1990), for my experiences with the implementation of this policy, including the objections put forth on its practicality. The *ASME Journal of Heat Transfer* implemented a similar policy in 1994 (see Appendix B). In 1988 and 1989, there were special sessions at the ASME Winter Annual Meetings on numerical accuracy. In June 1989, Clark Lewis organized a session of invited papers at the AIAA Thermophysics Conference on CFD "code validation/verification/certification." The papers from this well attended session were published in an issue of the *AIAA Journal of Spacecraft and Rockets* (Vol. 27, No. 2, March-April 1990). In his introductory comments as *JSR* editor-in-chief, Lewis (1990) noted that "computational fluid dynamics has reached the age where those who practice it and publish the results of code development and applications must demonstrate the accuracy and value of their work." In October 1989, Tinsley Oden organized a workshop at the Texas Institute for Computational Mechanics on the somewhat broader issue of reliability. In the early 1990's, the ASME Coordinating Group on Computational Fluid Dynamics sponsored several Symposia on the general subjects of "Quantification of Uncertainty in CFD" and benchmarking, with motivation principally provided by Ismael Celik and Chris Freitas; e.g., see Celik and Freitas (1990), Celik et al (1993), Freitas (1993b, 1995b), Johnson and Hughes (1995). In 1994, the *International Journal of Numerical Methods in Fluids* published its statement (Gresho and Taylor, 1994). In 1993 the AIAA Fluid Dynamics Technical Committee, with the encouragement of the Chairman David Walker, passed (almost universally) its recommendation for a similar policy, which was adopted for *all* AIAA publications and first published in January 1994. See Appendix B for reprints.

The AIAA Policy Statement is admirable on at least four accounts. First, it includes requirements for both computational and experimental uncertainty in a single statement, thus indicating that nothing special is being required of the computational work beyond simple good engineering practice. Second, the policy statement is to be re-published yearly. Third, the statement is included with instructions to journal manuscript reviewers. Fourth, the core statement (see Appendix B) is now included in the "Information for Contributors to Journals of the AIAA" published in every issue.

In 1994, the impetus for quality work reached a new plateau. The ASME *Journal of Fluids Engineering* editor, D. P. Telionis, accepted the recommendation of the ASME Coordinating Group for CFD and its Chairman, C. J. Freitas, and published a much expanded and more specific Policy Statement (supplanting Roache et al, 1986) which included the categorical rejection from publication of 2-point first-order upstream methods. That policy statement and the introduction by Freitas are reproduced in Appendix B.

Once again, and more understandably, the new policy caused considerable discussion; see Leonard (1995a,b), Leonard and Drummond (1995), Freitas (1995a), Roache (1995).

At the time of this writing (1998), the AIAA CFD Committee on Standards is developing a detailed guide covering the issues of numerical accuracy (Verification, Validation and related issues).

In a visionary editorial in the ASME *Journal of Fluids Engineering*, Telionis (1995) described how the revolution in Internet information technology can be used to favorably affect all these issues and broader ones of the review process and publication lag, critical comments, author revisions, etc. For example, gory details of Verification (see the succeeding chapters) could be relegated to a longer electronic version of a paper, with only a short summary statement in the printed version. Besides shortening the printed version and removing clutter for readers interested primarily in the results, the electronic mode provides another tremendous advantage. No longer will authors have to struggle to reduce the electronic paper to 6 pages, so no longer will they have that excuse for not providing convincing evidence of grid convergence. Reviewers can, in good conscience, be more demanding of numerical quality. Likewise for the expository aspects of the papers. Electronic versions can provide a multi-level publication, and the longer, more leisurely versions can avoid what G. H. Douglas has called the "cobblestone writing" style, in which the requirement for brevity fosters staccato statements of fact–subject, verb, object; subject, verb, object; ...– one after another, like an old wagon bouncing along on a cobblestone street from one hard unyielding fact to the next. Wouldn't it be wonderfully ironic if electronic on-line publishing became responsible for a return to Victorian prose style?

1.2.3 Commercial Codes and Users

New aspects of the problem are caused by the now extensive use of general purpose commercial codes. In 1972, when my first CFD book was published, there were no CFD commercial codes. A few researchers had distributed their codes (including source code) to a select audience. Usually, the authors had negative experiences with misuse and mostly became discouraged with the idea of wide use of their codes. The codes were used without sufficient knowledge of either the sensitivity to iterative convergence parameters or the fluid dynamics, applied without conscientious grid resolution studies, and applied outside the regime of physical parameters intended. In the worst cases, the users modified the source code, reducing the accuracy, and then published inaccurate results attributed to the original code author. In fairness to the users, it must be noted that the code documentation (both internal and external, and both theory and users manuals) was usually atrocious; only slowly did the CFD community develop an appreciation for the significant work required to produce adequate documentation. In the groundwater flow discipline, in 1988 the U. S. Geological Survey published and released into the public domain its code which became known universally as MODFLOW (McDonald and Harbaugh, 1988). Its clear documentation, both internal and external, became the standard of performance and contributed to the code's immense popularity and use.

The difficulties change and compound as one considers in turn

a) research codes for a limited class of problems, developed and used only by the authors and their close colleagues,
b) public domain codes for a limited class of problems and for which source code is released,
c) commercial codes for a limited class of problems and for which source code is not accessible, and,
d) the toughest case, general purpose commercial codes for which source code is not accessible.

The methods for Verification and Validation to be described in this book are applicable to general purpose commercial codes, but the user must have confidence that the numerical methods as described in the manuals are actually those implemented in the code, with no undocumented special case "switches," "fixes," etc. This is especially true in the difficult area of turbulence modeling. Even for thoroughly documented codes, the most conservative Verification techniques may be required because of the unfortunate prevalence in general purpose commercial codes of using "hybrid" methods. The details will be discussed in later chapters; for now, suffice it to say that CFD algorithm developers have long known that there is a trade-off between code robustness and accuracy, and CFD code marketers know that there is little market for numerical accuracy but much demand for bullet-proof code robustness. General purpose CFD codes must be treated with skepticism in any new application by any conscientious user. See the results in the "CFD Triathlons" organized by C. J. Freitas (Freitas, 1993b, 1995b) for an indication of the state of use of CFD general purpose commercial codes on laminar problems. Also, note that it is very difficult to devise test cases for non-trivial industrial problems and to obtain quality data for Validation comparisons; e.g., see the CFD World Users Association publications (Muller, 1994; de Vahl Davis, 1995; Muller and Loffler, 1996).

In the same vein, the issue of user education must be addressed—not education in code use, but in fluid dynamics (or whatever the discipline involved). It is asking too much of a code vendor to make up for lack of technical training (or plain sloppy thinking) in a user. A CFD code is not an "aerodynamicist on a chip" or a "groundwater analyst on a chip." The folklore of both communities abound with horror stories of technically unqualified users who care little for the accuracy of the results, and who obtain from CFD codes answers that are either patently absurd, or that are correct but could have been obtained with elementary calculations by hand or handbook.

It is apparently the case that user education is more of a practical concern in fluid dynamics than in some other disciplines, notably structural analysis. It is obvious that there is no analogue of fluid turbulence in structures, but the more significant distinction may be that the simplest level of analysis for structures (involving linear stress-strain, small deformations, steady state, etc.) happens to be very useful, whereas the analogous level in fluid dynamics (potential flow) by itself has few practical applications.

Likewise, user (and management) expectations are often unrealistic, and should be corrected whenever possible. In this regard, let me state the bad news right now.

It is totally unrealistic to expect that one can approach a new problem (not close to a previously calculated problem) that is difficult (not potential flow, etc.) and expect to achieve reliably accurate answers with a single calculation.

1.3 A PERSONAL ANECDOTE ILLUSTRATING THE IMPORTANCE OF SYSTEMATIC CODE VERIFICATION

I would like to introduce and hopefully motivate the subject of code Verification (and more generally, Reliability) with a personal anecdotal example (Roache, 1982). The subject area is not fluid dynamics, but electrostatics.

In the mid 1980's, working with colleagues at the Tetra Corporation, I developed the ELF (for **EL**ectric **F**ield) codes for the design of laser electrodes and high power switches (Roache et al, 1984). Later (after the experience described below) commercial versions known as the TetraELF codes were marketed. The problems involved boundary fitted coordinate generation in two-dimensional and three-dimensional geometries using elliptic grid generation methods

(Thompson et al, 1974), and variants by Steger and Sorenson (1979) and Thomas and Middlecoff (1980), solved by semidirect elliptic marching methods (Roache, 1995). The codes ultimately included time-dependent strongly nonlinear source terms, solution adaptive grid generation, interior boundaries of dielectric materials, and other complications (Roache et al, 1984).

An early production version of the two-dimensional ELF code certainly produced plausible, intuitively appealing solutions, both for the nonorthogonal boundary-fitted grid generation and for the solution of the "hosted" equations for the electric potential ϕ. The code had been exercised over a variety of problems for several years. It had failed to converge in the boundary-fitted grid generation task only on one class of electrodes with a slit-like geometry, which failure was inexplicable. Other than this failure to achieve a solution (which is not the same, nor as dangerous, as achieving a false solution) the codes were used with much confidence; the answers looked good, even compared to an exact solution.

I had ostensibly benchmarked the code by comparison to the classical solution for the Rogowski family electrode case (Lorrain andCorson, 1962) based on traditional complex variable theory, and restricted to linear steady two-dimensional problems (vacuum solution). The code solved the (transformed) Laplacian equation for potential ϕ with an error of 0.4% using only a 13×13 cell grid. This I considered excellent performance for a relatively coarse grid, and I believe that most practitioners would be confident in such a code. Certainly, many codes are in use that have not been benchmarked as well and are used with confidence.

My colleague at Tetra Corporation, Henry J. Happ, was not so confident. I patiently explained how the code could not be expected to produce the exact results, that 0.4% was to be expected for this coarse level of discretization, and was certainly adequate for engineering design and analysis. Ignoring my explanation and greater experience, Happ proceeded to perform a systematic grid convergence study of the old fashioned kind, described for one-dimensional problems in any old numerical analysis textbook. This required considerable work for a two-dimensional problem using the computers of the day. Starting with a more coarse grid of 5×5 cells, as the grid was refined, the solution converged *at the expected rate* for a 2nd-order numerical method. Unfortunately, it did not converge to the correct answer. The error from the Rogowski solution was reduced, as expected, but the error did not approach zero. Rather, the asymptote for the error appeared to be about 0.2%. The code accuracy was zeroth order, i.e., wrong.

Chastened by this numerical experiment, I investigated the source code thoroughly, looking for some subtle error, perhaps in a boundary condition implementation near the logical corners of the grid, that might cause this small error in the results. What I found instead was a *gross* error, namely, a factor of 2 error in the stencil coefficient for the cross derivative term in the transformed equations. The error appeared in both the equations for the grid generation, and in the discretization of the hosted equation. (At least I had been consistent!)

When the code was corrected, the solution error did asymptote to zero. The factor of 2 error in cross-derivatives affected the grid generation by less than 0.01% in the location of any x and y of the nodes, and the electric field by 0.016%. Note that the electric field, being the gradient of the solution for the potential ϕ, is a difficult quantity to predict since the maximum occurs at the curved boundary.

This insensitivity of the solution and its gradient to a gross error in coding was an affront to intuition. However, intuition was salved by re-running the previously failed grid generation for slit-like geometries. With the error, iterative convergence of this nonlinear problem (the nonlinearities are introduced by the coordinate transformation) was obtained only with extreme measures of 20 continuation steps plus the use of extensive under-relaxation. The resulting "mesh" was a "mess," with coordinate lines that crossed and extended outside the physical domain. When the coding error was corrected, the code converged to a good grid with only 2 continuation steps.

How could such a gross error have escaped detection in so many previous exercises? The answer is, the same way that such a gross error could make so little difference (maximum 0.016% in the gradient) in the solution; *the test problems had not sufficiently exercised the terms in the equations*.

The insensitivity originated in the mathematics of the grid generation procedure, which was based on the homogeneous Thompson-Thames-Mastin (Thompson et al, 1974) or the original Winslow (1967) method. For this generator, the continuum interior equations are identical to those of an orthogonal and indeed *conformal*, angle-preserving transformation. The discrete coordinate transformation generated is close to the conformal transformation, differing only because of truncation error and the distribution of boundary values of the grid. (The boundary nodes were located by equidistribution in arc length, but this distribution is not far from that of a conformal transformation grid.) *For a conformal transformation, the cross derivative terms in the transformed Laplacian operator are identically zero*. Thus, the gross error of a factor of 2 occurred in terms that were very small for the problem studied, and thus escaped detection when only absolute size of the error for a single grid solution was examined. However, when a *systematic grid convergence study* was performed, the presence of an error was quickly indicated. My experience has been that such systematic grid convergence testing is remarkably sensitive; other examples will be given in Chapter 3.

Why is not this kind of systematic grid convergence study used more often? Certainly, it has been used since the early 20th century in ordinary differential equations (ODE's). Virtually anyone who has taught numerical methods for ODE's has had the experience of students obtaining a good answer with their own Runge-Kutta integration code, but observing that the method does not converge with 4th-order rate on their problem, but only at 2nd-order rate. (Their are half a dozen errors in understanding and coding the RK4 algorithm that yield a consistent 2nd-order method but are still in error.) I believe that the bad habit of not performing systematic grid convergence studies to Verify the convergence rate of a multidimensional code developed historically because of limited computer resources. We certainly now have adequate resources available to do the job, but practitioners have lowered the standards from the old days of ODE solutions and are often content with superficial Verification exercises, or none at all. (Another reason was the lack of general exact solutions for difficult PDE's like the Navier-Stokes equations. This reason is now obsolete; see the Method of Manufactured Solutions in Chapter 3.)

What was frequently done was to appeal directly to agreement with physical experiments. (In the terms to be described more specifically in the next chapter, practitioners passed over Verification and went directly to Validation.) The general inadequacy of this practice is clear, considering the previous example on an important class of problems with only a 0.4% error arising from a gross coding mistake. In many experiments (certainly most aerodynamics and all groundwater flow experiments) this level of error would not be detectable with confidence. It is now the dominant opinion (see citations in Chapter 10) that there is a "continuing need for high quality experiments that are designed specifically for CFD code [Validation]" (Melnick et al, 1996).

Also, it is clear to all who have thought and wrote seriously that

- Verification and Validation are separate activities,
- Verification should precede Validation, and
- These two terms had best be defined with some specificity in a technical context.

Verification and Validation must be defined, distinguished, and contrasted to other activities in the arena generally described as "confidence building" or "Quantification of Uncertainty." These necessary semantic distinctions are the subject of the next chapter.

REFERENCES FOR CHAPTER 1

Aeschliman, D. P., Oberkampf, W. L., and Blottner, F. G. (1995), "A Proposed Methodology for Computational Fluid Dynamics Code Verification, Calibration, and Validation," *Proceedings 16th International Congress on Instrumentation in Aerospace Simulation Facilities*, WPAFB, Ohio, 95–CH3482–7, July 1995, pp. 27.1–27.13.18–21.

AGARD (1988), *Validation of Computational Fluid Dynamics*, Lisbon, Portugal, 2–5 May 1988, NATO Advisory Group for Aeronautical Research and Development, AGARD CP 437, Volume 1: Symposium Papers and Round Table Discussion. Volume 2: Poster Papers. December 1988.

AIAA (1994), "Editorial Policy Statement on Numerical Accuracy and Experimental Uncertainty," *AIAA Journal*, Vol. 32, No. 1, January 1994, p. 3.

ASME (1994), Editorial Board, "Journal of Heat Transfer Editorial Policy Statement on Numerical Accuracy," *ASME Journal of Heat Transfer*, Vol. 116, November 1994, pp. 797–798.

Blottner, F. G. (1977), "Introduction to the Proceedings of the 3rd AIAA Computational Fluid Dynamics Conference," *Proceeding of the 3rd AIAA Computational Fluid Dynamics Conference*, Albuquerque, New Mexico, June 27–28, 1977, pp. 1–2.

Bradshaw, P. (1975), "Computers vs. Wind Tunnels," in Letters, *Astronautics and Aeronautics*, September 1975, p. 6.

Celik, I. and Freitas, C. J. (1990), Eds., *Benchmark Test Cases for Computational Fluid Dynamics*, ASME FED Vol. 93, Book No. H00598–1990.

Celik, I., Chen, C. J., Roache, P. J., and Scheurer, G. (1993), *Quantification of Uncertainty in Computational Fluid Dynamics*, ASME FED Vol. 158. ASME Fluids Engineering Division Summer Meeting, Washington, D.C., June 20–24, 1993.

Chapman, D. R., Mark, H., and Pirtle, M. W. (1975), "Computers vs. Wind Tunnels for Aerodynamic Flow Simulations," *Aerospace America*, April 1975, pp. 22–35.

Davis, R. T. (1975), "Introduction to the Proceedings of the 2nd AIAA Computational Fluid Dynamics Conference," *Proceedings of the 2nd AIAA Computational Fluid Dynamics Conference*, Hartford, Connecticut, June 19–20, 1975, p. 1.

de Vahl Davis, G. (1994), "Going for Quality in CFD Codes," *Proceedings 2nd World Conference in Applied Computational Fluid Dynamics*, World User Association–CFD, Basel, Switzerland, 1–5 May 1994, pp. 18.1–18.6.

Emmons, H. (1970), "Critique of Numerical Modeling of Fluid–Mechanics Phenomena," *Annual Review of Fluid Mechanics*, Vol. 2, 1970, pp. 15–36.

Freitas, C. J. (1993a), "Editorial Policy Statement on the Control of Numerical Accuracy," *ASME Journal of Fluids Engineering*, Vol. 115, No. 3, September 1993, pp. 339–340.

Freitas, C. J. (1993b), Ed., *The CFD Triathlon: Three Laminar Flow Simulations by Commercial CFD Codes*, FED Vol. 160, ASME Fluids Engineering Conference, Washington, DC, June 20–24, 1993.

Freitas, C. J. (1995a), "The End of Justification, the Beginning of Implementation," *ASME Journal of Fluids Engineering*, Vol. 117, No. 1, March 1995, p. 9.

Freitas, C. J. (1995b), "Perspective: Seclected Benchmarks for Commercial CFD Codes," *ASME Journal of Fluids Engineering*, Vol. 117, June 1995, pp. 208–218.

Gresho, P. M. and Taylor, C. (1994), "Editorial," *International Journal for Numerical Methods in Fluids*, Vol. 19, 1994, p. iii.

Harlow, F. H. and Fromm, J. E. (1965), "Computer Experiments in Fluid Dynamics," *Scientific American*, Vol. 212, No. 3, pp. 104–110.

Johnson, R. W. and Hughes, E. D. (1995), *Quantification of Uncertainty in Computational Fluid Dynamics–1995*, ASME FED Vol. 213. ASME/JSME Fluids Engineering and Laser Anemometry Conference, Hilton Head, South Carolina, August 13–18, 1995.

Kline, S. J., Cantwell, B. J., and Lilley, G. M. (1981), Eds., *Objectives, Evaluation of Data, Specifications of Test Cases, Discussion, and Position Papers. The 1980–81 AFOSR/HTTM–Stanford Conference on Complex Turbulent Flows*, Vol. 1, ThermoSciences Division, Mechanical Engineering Department, Stanford University.

Leonard, B. P. (1995a), "Comments on the Policy Statement on Numerical Accuracy," *ASME Journal of Fluids Engineering*, Vol. 117, No. 1, March 1995, pp. 5–6.

Leonard, B. P. (1995b), "Uncertainty in Future of CFD," ICOMP News, Ohio Aerospace Institute, Cleveland, Ohio, Vol. 2, No. 1, January 1995, pp. 1–2.

Leonard, B. P. and Drummond, J. E. (1995), "Why You Should Not Use 'Hybrid', 'Power Law' or Related Exponential Schemes for Convective Modelling–There Are Much Better Alternatives," *International Journal for Numerical Methods in Fluids*, Vol. 20, 1995, pp. 421–442.

Lewis, C. H. (1990), "Comments on the Need for CFD Code Validation," *AIAA Journal of Spacecraft and Rockets*, Vol. 27, No. 2, March–April 1990, p. 97.

Lorrain, P. and Corson, D. P. (1962), *Introduction to Electromagnetic Fields and Waves*, W. H. Freeman Company, San Francisco, 1962.

McDonald, M. G. and Harbaugh, A. W. (1988), *A Modular Three–Dimensional Finite–Difference Ground–Water Flow Model*, U. S. Geological Survey Book 6, Modeling Techniques. Available from Scientific Software Group, P. O. Box 23041, Washington, D. C. 20226–3041.

Melnik, R. E., Siclari, M. J., Marconi, F., Barber, T., and Verhoff, A. (1995), "An Overview of a Recent Industry Effort at CFD Code Certification," AIAA Paper 95–2229, AIAA 26th Fluid Dynamics Conference, San Diego, California, 19–22 June 1995.

Muller, A. (1994), Ed., *Proceedings of the Basel World CFD User Days 1994, Second World Conference in Applied Computational Fluid Dynamics*, Basel, Switzerland, 1–5 May 1994.

Muller, A. and Loffler, B. (1996), Eds., *Proceedings of the Basel World CFD User Days 1996, Third World Conference in Applied Computational Fluid Dynamics*, Basel, Switzerland, May 1996.

Rakich, J. V. (1973), "Introduction to the Proceedings of the Symposium on Computational Fluid Dynamics," Proceedings of the AIAA Computational Fluid Dynamics Conference, Palm Springs, California, July 19–20, 1973, p.1.

Rizzi, A. and Vos, J. (1996), "Towards Establishing Credibility in CFD Simulations," AIAA Paper 96–2029, 27th AIAA Fluid Dynamics Conference, New Orleans, Louisiana, June 17–20, 1996.

Roache, Patrick J. (1972), *Computational Fluid Dynamics*, First Edition, Hermosa Publishers, P. O. Box 9110, Albuquerque, New Mexico 87119-9110, USA. http://www.hermosa-pub.com/hermosa. Revised Printing 1976. See also Roache (1998).

Roache, P. J. (1975), "Computers vs. Wind Tunnels," in Letters, *Astronautics and Aeronautics*, September 1975, pp. 4–5.

Roache, P. J. (1982), "Semidirect/Marching Methods and Elliptic Grid Generation," *Symposium on the Numerical Generation of Curvilinear Coordinate Systems and Use in the Numerical Solution of Partial Differential Equations*, J. F. Thompson, Ed., Nashville, Tennessee, Elsevier Science Publishing Company, April 1982, pp. 729–737.

Roache, P. J., Steinberg, S., and Moeny, W. M. (1984), "Interactive Electric Field Calculations for Lasers," AIAA Paper 84–1655, AIAA 17th Fluid Dynamics, Plasma Physics, and Lasers Conference, Snowmass, Colorado, 25–27 June 1984.

Roache, P. J., Ghia, K. and White, F. (1986), "Editorial Policy Statement on the Control of Numerical Accuracy," *ASME Journal of Fluids Engineering*, Vol. 108, No. 1, March 1986, p. 2.

Roache, P. J. (1990), "Need for Control of Numerical Accuracy," *AIAA Journal of Spacecraft and Rockets*, Vol. 27, No. 2, March–April 1990, pp. 98–102.

Roache, P. J. (1995), *Elliptic Marching Methods for Domain Decomposition*, CRC Press, Boca Raton, Florida.

Roache, P. J. (1998), *Fundamentals of Computational Fluid Dynamics*, Hermosa Publishers, P. O. Box 9110, Albuquerque, New Mexico 87119-9110, USA. http://www.hermosa-pub.com/hermosa. To appear.

Steger, J. L. and Sorenson, R. L. (1979), "Automatic Mesh–Point Clustering Near a Boundary in Grid Generation with Elliptic Partial Differential Equations," *Journal of Computational Physics*, Vol. 33, No. 3, December 1979, pp. 405–410.

Telionis, D. P. (1995), "Editorial," *ASME Journal of Fluids Engineering*, Vol. 117, September 1995, pp. 329–330.

Thomas, P. D. and Middlecoff, J. F. (1980), "Direct Control of the Grid Point Distribution in Meshes Generated by Elliptic Equations," *AIAA Journal*, Vol. 18, No. 6, June 1980, pp. 652–656.

Thompson, J. F., Thames, F. C., and Mastin, C. W. (1974), "Automatic Numerical Generation of Body–Fitted Curvilinear Coordinate System for Field Containing Any Number of Arbitrary Two–Dimensional Bodies," *Journal of Computational Physics*, Vol. 15, No. 3, July 1974, pp. 299–319.

Winslow, A. (1967), "Numerical Solution of the Quasilinear Poisson Equations in a Nonuniform Triangle Mesh," *Journal of Computational Physics*, Vol. 2, pp. 149–172.

WIPP PA Dept. (1992), *Annual Performance Assessment for the Waste Isolation Pilot Plant, Vol. 1, Preliminary Comparison with 40 CFR 191, Subpart B, Vol. 2: Technical Basis*, Performance Assessment Department, Sandia National Laboratories, Albuquerque, New Mexico December 1992.

2

SEMANTICS: TERMINOLOGY, TAXONOMIES, AND DEFINITIONS

The purpose of a fish trap is to catch fish, and when the fish are caught, the trap is forgotten.
The purpose of a rabbit snare is to catch rabbits. When the rabbits are caught, the snare is forgotten.
The purpose of words is to convey ideas. When the ideas are grasped, the words are forgotten.
Where can I find a man who has forgotten words? He is the one I would like to talk to.

"Means and Ends," *The Way of Chuang Tzu*, Thomas Merton
(New Directions Publishing Corporation, New York, 1965)

2.1 INTRODUCTION

This chapter is based primarily on Roache (1995), "Verification of Codes and Calculations" and Roache (1997), "Quantification of Uncertainty in CFD."

Background discussion, definitions and descriptions will be given for some terms related to confidence-building in CFD. Examples will be given of worthwhile semantics vs. worthless semantics, and practical definitions vs. effete philosophizing. The *critical* distinction is made between Verification vs. Validation. Distinctions will be made between numerical errors vs.

conceptual modeling errors; iterative convergence vs. grid convergence (or residual "accuracy" vs. discretization accuracy); adequate and inadequate error taxonomies; Confirmation, Calibration, Tuning, and Certification; Verification of numerical accuracy of codes vs. Verification of individual calculations; truncation error vs. discretization error; customer illusions vs. customer care; and Quality Assurance vs. quality work.

The editorial "we" or "our opinion" will be used herein to refer to an opinion held not just by myself but by a significant number of practitioners in the field, although not necessarily a majority. The more restricted "I" will be used to indicate more individual opinion, although not necessarily unique. The definitions and distinctions have developed over years, with less standardization in the earlier years. The definitions did not flow inexorably from the etymology or the common use (as some philosophers would have us believe) but rather developed as the need arose to make the distinctions. None of the terms has enjoyed a pristine use in the literature, i.e., all have been used by authors (including myself) in ways inconsistent with the present definitions, and some continue to be so. Nevertheless, the definitions and usage proposed herein are becoming standardized, enabling communications to become more efficient and precise. Also, the ideas behind the words are important.

"Verification and Validation" and the broader area of confidence-building in CFD is a curious subject. The growing recognition of its importance is attested to by the policy statements given by ASME (Roache et al, 1986; Freitas, 1993a; Editorial Board, 1994) and AIAA (1994) and other journals (Gresho and Taylor, 1994), as noted in Chapter 1. See Roache (1990) reproduced herein as Appendix A for my experience in implementing the policy in Roache et al (1986). The tone of articles can be fairly legalistic, yet the area is quite subjective, dependent on opinions, world-view, philosophy of science, philosophy of engineering, and appeals based on common sense as much as mathematics. This is partly due to the semantics involved.

2.2 SEMANTICS

I fear that many readers will not be interested in this discussion, but will dismiss it as "mere semantics." But it is important, and in my experience can have major consequences to projects. "Semantics," after all, is "the study of meaning, especially in language," and would seem to be a worthwhile activity if people want to know *what it is* they are arguing about. The negative connotation of "mere semantics" arises when people argue uselessly about words without looking beyond, to the ideas behind the words. For example, the choice of "Verification" or "Validation" was originally arbitrary (in my opinion), and is now recommended solely because of common developing use. (I have published articles using the opposite definition.) In a common English thesaurus, "verify," "validate," and "confirm" are all synonyms, but the words are used herein, and generally in code Quality Assurance (QA), as *technical terms* with more context-specific meaning.

Such technical terms are preferably related to common use, although regrettable exceptions occur. (For example, I have always found the term "moment" as used in mechanics to be annoying because its technical meaning is orthogonal to common usage, whereas "torque" is adequate and commonly understood.) But the term's *technical meaning* is defined independent of common use, and in a *specific technical context.* The same word can have different technical meanings in different technical contexts. Biologists use the word "model" for a laboratory rat with a human-like disease, dentists speak of a "calculus" that Leibnitz would not claim, and "Confirmation" is a religious rite. Even the word "error," which a non-technical person would never suspect of ambiguity, has multiple meanings depending on context. Programming errors are *mistakes* but discretization errors are not. In common terminology, "error" and "mistake" are virtually synonymous.

This is not a universally accepted attitude toward semantics. In a widely quoted paper which has been recently described as "brilliant" in an otherwise excellent *Scientific American* article (Horgan, 1995), Oreskes et al (1994) think that we can find the real meaning of a technical term by inquiring about its common meaning. They make much of supposed *intrinsic* meaning in "verify" and "validate" and, like a Greek morality play, agonize over "truth." They come to the remarkable conclusion that it is impossible to verify or validate a numerical model of a natural system.

Now most of their concern is with groundwater flow codes, and indeed, in geophysics problems, Validation is very difficult. But the authors extend this to all physical sciences. They even state the following: "We are left with the conclusion [from their flawed syllogism] that we can never verify a scientific hypothesis of any kind." They clearly have no intuitive concept of error tolerance, or of range of applicability, or of common sense. My impression is that they, like most lay readers, actually think Newton's Law of gravity was proven wrong by Einstein, rather than that Einstein defined the limits of applicability of Newton. But Oreskes et al (1994) go much further, quoting with approval (in their footnote 36) various modern philosophers who question not only whether we can prove any hypothesis true, but also "whether we can in fact prove a hypothesis false." They are talking about physical laws (not just codes, but any physical law). Specifically, we can neither validate nor invalidate Newton's Law of Gravity. One might expect that even a philosopher who would not accept "proof" of Newton's inverse-square law could at least accept the common experience that would disprove a "*direct*-square law" hypothesis, i.e. that gravitational attraction *increases* with the square of distance. But not so; to these philosophers, neither Newton nor anti-Newton can be "proven." See also Konikow and Bredehoeft (1992) for a similar outrageous claim. (What shall we do? No hazardous waste disposals, no bridges, no airplanes, no construction codes for earthquake areas, no analysis or predictions of atmospheric pollution, no...)

If one insisted on reserving an overly rarefied concept of "proof" only for mathematics, one could still reasonably use the word "validate" with a common-sense (or more precise) concept of specified accuracy uncertainty. One could say with Hull (1997) (in the context of creationism arguments against the "unproved theory" of evolution) that "Newton's celestial mechanics...is a well-validated theory (in the sense of explaining or predicting the result of an observation) that is unproved and incomplete. We even know its successor–general relativity. Nevertheless, we launch Mars probes, among other space voyagers, confidently using Newton's mechanics.

Unproved (but validated) theories are quite useful! The electric power industry is based on Maxwell's theory of electricity and magnetism, unproved and incomplete as it is."

It is important to recognize that these controversies are not just the stuff of academic debates, but have serious consequences, especially on public policy. For example, it can (and has) happened that ill-conceived legalistic (regulatory) definitions of Validation could, if applied rigidly and without the leaven of common sense, categorically eliminate a geologic repository as a possibility for hazardous or radioactive waste disposal. It is not necessary to argue the relative merits of different system categories here, but we can at least allow the possibility that such a geologic repository *could*, at least in principal, be the best (safest and cheapest) method of disposal. The problem here with regard to overly legalistic definitions of Validation is *time* and *spatial data*. The time scale legislated for low-level radioactive waste containment is 10,000 years and clearly no tests can be run to "Validate" a model at this time scale. Likewise, the spatial data regime for one candidate site covers hundreds of square kilometers to a depth of at least 700 meters, and clearly pumping tests at length scales of 10's of meters to completely cover the area are not possible (WIPP PA, 1992; Helton et al, 1995,1996). Unless common sense and good engineering practice prevail, the legalistic definition of "Validation" could *categorically* eliminate this potentially best category of waste disposal. For the benefit of those who would prefer to eliminate it, with a feeling of moral superiority, I note two additional facts. (1) The pseudo-intellectual agonizing over "truth" and impossibly rarefied definitions of "Validation" will ultimately eliminate *all* conceivable disposal schemes, not just geologic systems. (2) The dangerous waste already exists, it will not go away, and in fact is presently stored in systems which are fragile and dangerous even by common-sense standards. Also, this sophomoric attitude towards semantics likewise would eliminate bridges (built, after all, on ground, with unknown data) and aircraft; the atmosphere provides the "open system" which Oreskes et al (1994) claim eliminates any possibility of "Validation."

For a mild but worthwhile criticism of Oreskes et al (1994), see Rykiel (1994). For my complete criticism of Oreskes et al (1994), rejected for publication by *Science*, see Appendix C. For criticisms of Konikow and Bredehoeft (1992), see de Marsily et al (1992) and Leijnse and Hassanizadeh (1994). This latter paper is important for unraveling the semantics associated with the word "model," in particular, whether "model" includes parameter values. They make the worthwhile distinction of "strong" and "weak" definitions of "model" (based on whether or not it includes parameters or is only a conceptual model), and corresponding strong and weak senses of Validation. See also related discussion of modeling philosophy in Tsang (1991) and in the entire Special Issue of *Advances in Water Resources* (Gray, 1993) devoted to "Research Perspectives in Hydrology."

The Konikow and Bredehoeft (1992) paper, entitled "Groundwater Models Cannot be Validated," asserts that "In the end, action concerning waste disposal will be a judgment; a professional judgment by the scientific community and a judgment by society." I assert that questions like "Can groundwater models be Validated?" and "Do we or don't we model?" are phony questions. The fundamental and meaningful question is, "Do we make decisions qualitatively or quantitatively?" All the geological information and insight possible (age of rocks, chemical composition, color, ...) cannot provide the quantitative information on groundwater flows that are necessary for decisions on important subjects like hazardous and/or radioactive waste disposal. Once we decide to make decisions quantitatively (and we must), then we *do* model. The next question is one of scale of the model; a "back of the envelope" calculation, based on geological insight etc., *is* a model. It is a model using the limit of crude resolution, a single cell or control volume, or what chemical engineers refer to as a "batch model." Whether such a batch model is adequate, or high-resolution computer solutions of partial differential equations are required, depends on the accuracy required. In any case, the models must be Validated in some sense, at some level of accuracy, no matter how crude, in order to be quantitatively useful.

Clearly, we are interested in normal speech and practical definitions, applied in the context of engineering and science accuracy, not in such worthless semantics and effete philosophizing.

Below, definitions and descriptions are given for some terms related to confidence-building in CFD. (I will not try to define "confidence building" with any precision.) To reiterate, these are technical terms defined in a technical context, not just common language.

2.3 CODE VERIFICATION AND VALIDATION: NUMERICAL VS. CONCEPTUAL MODELING

First and foremost, we must make the essential distinction between Code Verification and Validation. Following Boehm (1981) and Blottner (1990),we adopt the succinct description of "Verification" as "solving the equations right," and of "Validation" as "solving the right equations."

"Verification" ~ solving the equations right.

"Validation" ~ solving the right equations.

The code author defines precisely what continuum partial differential equations and continuum boundary conditions are being solved, and convincingly demonstrates that they are solved correctly, i.e. usually with some order of accuracy, and always consistently, so that as some measure of discretization (e.g., the mesh increments) $\Delta \to 0$, the code produces *a* solution to the continuum equations; this is Verification. Whether or not those equations and that solution bear any relation to a physical problem of interest to the code user is the subject of Validation.

In a meaningful though perhaps overly scrupulous sense, a "code" cannot be Validated, but only a calculation (or range of calculations with a code, for a specific class of problems) can be validated. In my experience, dealing with other than algorithm developers, this is a difficult concept and requires continual reiteration.

Another way to make the distinction (i.e., to get to the idea behind the words, beyond "mere" semantics) is to speak of numerical errors vs. conceptual modeling errors. An example of conceptual modeling vs. numerical modeling is the assumption of incompressibility. This is clearly a conceptual modeling assumption. Is it the code builder's fault, or any criticism of a commericial code itself, if the user incorrectly applies it? For example, dynamic stall of helicopter rotor blades involves compressibility at a surprisingly low free-stream Mach number. Results from an incompressible code may not agree with experiment very well, but we cannot say that the code fails Verification because it was applied to compressible flow, although we may have some sympathy for the user who is fooled by dynamic stall. But no one would have sympathy for a user who applied an incompressible flow code to a reentry vehicle at Mach 20. In this example, and in many practical cases, the lack of agreement with experiment is not a *code* problem, but a *modeling* problem

"Model" includes more than the code. "Model" includes conceptual modeling assumptions (e.g., incompressibility, symmetry, etc.). "Model" (in this strong sense, see Leijnse and Hassanizadeh, 1994) also includes data input to the code, e.g., geometry data (which are not so easy to determine accurately as many people assume) and boundary conditions and initial conditions). These can lead to failure of Validation of a *model*, with possibly no criticism of the *code*.

Another way to make the distinction between Verification and Validation is to follow the classical distinction between mathematics and science. Mathematics is a tool of science, often the predominant language of science. But mathematics exists by itself. It would be "true" regardless of any correspondence to the natural world. Verification is seen to be essentially and strictly an activity in mathematics, the mathematics of numerical analysis. Validation is essentially and strictly an activity in science (and engineering science): physics, chemistry, fluid dynamics, even the "soft" sciences of economics, sociology, etc.

"Verification" ~ mathematics

"Validation" ~ science/engineering

The typical Computer Science view of "code Verification" is not mine. "Verification" in my view does not include all aspects of code QA. For example, it does not include the important and nagging concerns of version control, or archiving of input data, or documentation (external and internal). Less obviously, it does not (in my view) include *reading* of source code. Blottner (1990) described his verification of a Navier-Stokes code and included the fact that the Fortran source code was examined. Although perhaps useful, and contributes to confidence building, I do not consider it to be part of Code Verification *per se* (nor of "Confirmation"; see below). Even as part of Code QA, the practice is of less value than it might appear at first glance. Consider the ridiculousness of regulatory agency personnel reading source code when they cannot comprehend the algorithms. Or consider how difficult it is, even if one comprehends the algorithm, to read someone else's source code for a three-dimensional semicoarsening multigrid method, or a full spectral element code, or a block-7 implicit multi-equation turbulence model. We can read source code, either our own or someone else's, interminably and still not be able to claim Verification. On the other hand, if the grid convergence tests prove that the code is (say)

2nd-order accurate, it would take some strange coding to nullify this test. (Perhaps some in-line limit on the parameter range that was not "hit" in the grid convergence test is possible.) Except for these unusual cases, reading of the code appears to be neither necessary nor sufficient for Verification.

Surely few would claim that reading of code is any substitute for Verification via grid convergence testing (although some of my clients have tried). If we included reading of the source code as part of Code Verification, then the code author's reading could be claimed as a "partial Verification." We know it does not prove a thing. In my view, "Verification" (in this restricted, technical meaning) should treat the code as a "black box," without reference to its internal organization. Note, then, that the present "Verification" has nothing in common with the type of software systems that perform internal consistency checks, look for "dead code" (i.e., un-exercised code segments, which might be left-over artifacts from early stages of code development), and otherwise perform computer-science-type of code "Verification" by consideration of the internal code.

As reading source code does not constitute partial Verification, neither does extensive use of a code, nor publication, as some authors have claimed. (Especially in groundwater flow codes, where the physical parameters and initial conditions are so poorly known, extensive experience with so-called "real world" calculations (even hundreds of cases calculated by many different users) would be an inefficient and, more importantly, *inconclusive* way to approach Verification. Likewise, journal publication and wide code distribution do not constitute Verification. Code Verification is not some kind of exercise in democracy, anymore than proving a theorem is. *First* one verifies a code's accuracy, *then* one publishes. (Note that, when codes with many options combinations are considered, no journal would publish a complete set of Verifications, nor would most referees consider the test in sufficient detail.) Extensive code use and journal publication certainly have value. They add to confidence building, and are worthwhile for general code QA, and are helpful to improve code documentation, garner suggestions for input/output improvements, obtain data on performance measures for a wide range of problem parameters, demonstrate robustness and portability, etc. They are just not part of Verification as technically defined herein.

On the issue of robustness, note I do not consider a code bomb (i.e., a divide by zero, or instability) to be a failure of Code Verification in the present context. As long as the code does not lie to the user, Code Verification has not been breached. Code robustness is certainly desirable, and is properly part of QA and overall code evaluation. From a QA "customer" point of view, robustness is often paramount, unfortunately sometimes to the point that it trumps all other considerations. (Frankly, users often do not care about accuracy, so long as the computational results are smooth, and are obtained with no intelligent interaction required.) But in my use of the terms, robustness is not an issue in Verification. (Besides, it is well known that robustness is often achieved at the expense of numerical accuracy, as noted in Chapter 1.)

Since Validation involves comparisons with experiment, there is necessarily an error tolerance involved, and what level is acceptable depends on the use intended. Such considerations are incorporated into Mehta's (1995) more expanded definition.

"*Validation* is defined as the process of assessing the credibility of the simulation model, within its domain of applicability, by determining whether the right simulation model is developed and by estimating the degree to which this model is an accurate representation of reality from the perspective of its intended uses." (Mehta, 1995)

This definition accepted by Mehta in the aerospace industry is in agreement with that accepted by Tsang (1991) in the groundwater modeling community, as given originally by Schlesinger (1979).

Validation: "substantiation that a computerized model within its domain of applicability possesses a satisfactory range of accuracy consistent with the intended application of the model." Schlesinger (1979)

Tsang (1991) also explicitly includes the specification of performance metrics in the definition of (groundwater flow) model Validation, and notes that "almost by definition, one can never have a Validated computer model without further qualifying phrases." Tsang also notes the different aspects of Validation with respect to (a) a process (i.e., governing equations) and (b) a "site-specific system," which includes geometric structure, stratigraphy, fractures and other geologic characteristics, plus boundary and initial conditions [both of which are notoriously vague in groundwater modeling]. Tsang also takes the ostensibly precise approach of only speaking of Validation of computer *models* rather than *codes*. (As we noted earlier, it is also a meaningful though overly scrupulous distinction to claim that neither a code nor a model can be Validated, but only an individual calculation.) Tsang states that the term "code Validation" is illogical, but close reading shows that his problem arises from not distinguishing Validation from Verification (a term he does not use at all). Likewise, some of his "other Validation methods" (other than the basic approach of comparing model predictions with laboratory or field data) would be classified as Verification or model sensitivity studies in present terminology, and the rest are simply variants of the basic approach. (Given this relatively minor caveat on semantics, Tsang's paper is highly recommended reading for modelers of groundwater flow and transport.)

Code *Verification* is defined rather tersely by the IEEE (Jay, 1984): "Formal proof of program correctness." I agree with Oberkampf (1994) in his evaluation. "Although very brief, this definition brings unprecedented clarity to the meaning of the term, and it adds a new perspective to the issue. Specifically, this definition bluntly requires correctness or veracity of the prediction, without bringing in supporting topics such as what is being predicted or how it is done." Although more general than the definition used herein, this IEEE definition is actually compatible with ours, is more general, and is still compatible with the distinction between Verification and Validation. That is, "program correctness" for a PDE code would naturally include "solving the equations right," and of course a definition of what those (continuum and discretized) equations are, without getting into the question of whether certain problems are appropriate for those equations and that code, i.e. Validation.

The definition of Verification given by Golub and Ortega (1992), quoted approvingly by Jameson and Martinelli (1996), is somewhat muddled, since it uses the word "verification" (apparently in a common, non-technical sense) within the definition of Validation (in a technical sense). Likewise, Jameson and Martinelli use "validation" as we do here for the physical model correctness, but only for the" overall approach" or the "final validation." (I find it difficult to distinguish the "overall approach" from the submodels.) They also state "verification is needed for every single claim made in constructing the model, or submodels (both physical and numerical)," i.e. they use "verification" in regard to the physical model as well as the numerical process. Likewise, they also use "Validation" for the "numerical solution process" requiring that it be "separately verified and validated." Other attempted distinctions made are also unclear, e.g. whether they intend "numerical scheme" to include the grid resolution. Nevertheless, the paper contains excellent practice and results, and is recommended reading. (See also excerpts in the present Chapter 8.)

The question of "solution uniqueness" always arises with nonlinear equations, and its position in the "Verification" or "Validation" distinction is important. If, as stated carefully above, we take "Verification" to mean simply that "*a* solution" to the continuum PDE's is obtained, then the problem of PDE non-uniqueness is avoided. This is probably being too easy on the code developer in some cases. Physically inadmissible solutions should be eliminated by the code, e.g., a shock-tube simulator should be required to eliminate expansion shock solutions. But nature abounds with physically non-unique solutions in fluid dynamics, e.g., hysteresis of airfoil stall and recovery, and bi-stable fluid amplifiers. If nature cannot decide which solution to produce, we cannot expect more of a code. Note that obtaining "*a* solution to the continuum PDE's" may involve sorting through multiple numerical solutions for nonlinear problems; e.g., see discussion in Chapter 8. See also especially Stephens and Shubin (1981) whose study of Euler solutions indicates that the multiple numerical solutions converge to the same solution as the grid is refined. Note that numerical artifacts may select one multiple solution over another; for example, Oberkampf et al (1995) cite the example of Levy et al (1995) showing that the manner in which the left-hand (iterative driver) side of each difference equation is formulated impacts whether the symmetric or asymmetric solution of steady state vortex flows over slender bodies at angle of attack is obtained. Oberkampf et al (1995) label the asymmetric solutions for this class of flows as "spurious," but in general we cannot be categorically dismissive. For example, computations of both laminar and turbulent flows through symmetric expansion channels can produce either symmetric or asymmetric solutions (provided that the entire channel flow is calculated, rather than having symmetry enforced through the *fiat* of computing only half the flow). Contrary to wide-spread intuition, nature prefers the asymmetric solution; this is the basis of bi-stable fluid amplifiers (e.g., see IAHR, 1982). Likewise, flow over a geometrically symmetric cavity (rectangular or spherical) does not produce a symmetric flow, but one in which the axis of rotation of the primary recirculating vortex is skewed. In general, it is asking too much of simulations to always select preferred solutions (preferred by nature, or by intuition of the modeler), in my opinion. Sometimes, the "spurious" (or at least, unanticipated) solutions are demonstrably artifacts of discretization, as when the asymmetry tends to disappear as the grid is refined (Sengupta et al, 1995). Other times, the unanticipated solution persists as the grid is refined, and is therefore "Verified." Stable or not, observable in nature or not, unique or not, such Verified simulations are mathematical solutions. Sorting out the other aspects is not an aspect of Verification; it is an aspect of Validation, but it should be approached with great respect for the subtlety of nature.

A closely related question is the *stability* of a solution. Although people usually think of artificial (i.e., numerical artifact) instability as a code difficulty, the equally common situation is artificial stability, e.g., a 1st-order time-dependent code whose artificial viscosity damps out disturbances, so that accurate calculations of physical fluid instability would require impractically high resolution. In my opinion, it is asking too much of a code to exactly mimic stability boundaries except in the limit of $\Delta \to 0$. For example, it is not a failure of Verification for a full Navier-Stokes code to produce steady shear layer solutions at Reynolds numbers known to be unstable; an unstable solution is still a genuine solution. (It is certainly not necessary, as claimed in a recent SIAM article by Johnson et al [1995] to analytically and *a priori* solve the hydrodynamics stability problem in order to estimate CFD numerical errors!)

Also, I believe strongly that Code Verification can and *should* be completed without appeal to physical experiments. The first part of this statement claims that Code Verification should be *completed*, i.e., Code Verification is *not* an ongoing exercise. Verification, as we have said, is an exercise in mathematics, not science. When one proves a theorem, the work is completed. Proving the formula for solution of a quadratic equation is not ongoing work. This is not to say that one could not have made an error in the proof of a theorem, nor that Confirmation exercises (see below) are not valuable in Confidence-building. It is to say that Code Verification is a mathematical activity that in principle comes to a conclusion, e.g. a code is or is not 2nd-order accurate. The second part of the statement claims that Verification can and should be achieved without using physical experiments. That it *can* is unarguable–very general methods are available (see the Method of Manufactured Solutions in Chapter 3). That it *should* is perhaps debatable; Oberkampf (1994) and Aeschliman et al (1995) allow Verification for carefully performed experiments in parameter regimes wherein confidence is high between the correspondence of the governing equations and the physical science. I suppose the physical experiment could be regarded as a sort of continuum analog computer that produces "solutions" to the governing equations, but at best such a definition would be strained and unnecessary. Experimental agreement is part of Validation, not Verification, and the concepts are distinct enough to be worth preserving. Also, as Roberts (1988) noted, justification of a purely mathematical approximation does not need physical experiments, e.g. if you want to know if the irrotational approximation is good, then compare potential code results with Euler results. Not only are physical experiments not needed, they may just be confusing, because one now has to sort out two sources of error, the rational approximation error and the physical experimental error. The other extreme is to consider high-order DNS (Direct Numerical Simulations) of turbulence to be of such high quality that simpler turbulence models can be "Validated" by comparison with this mathematical realization. Again, I would prefer to restrict the terms

Verification and Validation. (Alternately, one can use the somewhat less precise terms of "Confirmation" and/or "Benchmark"; see below.)

Validation has highest priority to engineers and scientists because "nature" is the final jury. But any experience with laboratory experiments will quickly disavow the absolute quality of experimental data. (It is asking too much of a CFD code to agree with wind tunnel test data when these data do not agree with other wind tunnel data, nor even from one run to another in the same wind tunnel.) Thus, I strongly believe that complete Verification of a code (and a calculation–see below) should *precede* any comparisons with experimental data, i.e., Verification first, then Validation. This is not so much a matter of definition as it is recommended engineering practice, but its wisdom is attested by the voice of experience (often using differing semantics), e.g., Bradley (1988), Marvin (1995), Aeschliman et al (1995), Porter (1996), Melnik et al (1995, p. 23), Sindir et al (1996), Jameson (1992), Jameson and Martinelli (1996). To quote Jameson (1992): "Simply comparing experimental data with numerical results provides no way to distinguish the source of the discrepancies, whether they are due to faulty numerical approximation or programming, or to deviations between the math model and the true physics." The aerospace consortium experience in Verification, Validation and Certification (see definitions below) reported in Melnik et al (1995) was forcefully clear, as follows.

"In the past constraints on computer resources have led to shortcuts using inadequate coarse grids leading to inadequate Verification. In this project, because we were able to use grids that were sufficiently fine, we were able to observe first-hand the pitfalls of working with inadequately Verified codes [calculations]. From our experience in the project it is clear that proper Verification [of calculations] is an absolutely necessary first step in the code Certification process. This basically involves the determination of the grid required to achieve a specified level of numerical accuracy." (Melnik et al, 1995)

Note another contrast between Verification and Validation: Verification is *completed* (at least in principle, first for the code, then for a particular calculation) whereas Validation is *ongoing* (as experiments are improved and/or parameter ranges are extended).

This brings up another aspect of experiments. One sometimes hears complaints that a "CFD code needs additional data" that experimenters typically do not measure. For example, data are published on turbulent flow in channel expansions which do not characterize the details of the boundary layer ahead of the expansion (nor sometimes elementary quantities like boundary layer thickness). This complaint neglects a fundamental fact of life:

CFD codes require no more information than the physics.

In fact, a CFD code may be expected to require *less* information than the full physics, due to simplifying approximations in the turbulence theory. If experimenters have not measured these quantities, then they have an uncontrolled experiment, regardless of whether CFD codes will be used. The question, of course, is whether these unmeasured quantities are important to the physics of interest. [Marvin (1995) uses the apt term "well-posed validation experiment."] This seems to me to be a major opportunity for CFD to contribute to experimental work; CFD can be used to predict the importance of difficult-to-measure quantities like wind tunnel flow angularity, non-uniform stagnation enthalpy, surface waviness, etc. as discussed lucidly by Aeschliman et al (1995). A premier example given in the study by Haynes et al (1996) is the sensitivity of boundary layer transition to free stream vorticity, which is "an unusually difficult experiment" but which can be modeled by CFD. (See further discussion in Chapter 10.)

When I say that Verification is completed, rather than ongoing, I am addressing only a completed code, or code version, and perhaps only a limited set of option combinations. Large scientific and commercial codes develop over years, and have multitudinous combinations of options. In this sense, we can say that Code Verification is an ongoing process, but only because the word "Code" as used is amorphous, really referring to many codes (all with the same name, but hopefully different version numbers), each of which in turn must be Verified.

For guidelines on Verification and Validation in the nuclear industry, see American Nuclear Society (1987).

2.4 CODE CONFIRMATION

Some Computer Science or QA people would have Code Verification necessarily performed by someone other than the code developer, sort of an "arm's length transaction" philosophy. In my view, it is ridiculous to expect code builders to not Verify their codes! Verification is a necessary part of code development. Code authors would be remiss in their duty if they released a code without Verification. I would trust a code builder's Verification results if presented in full; fraud is not usually the issue. But if it is, and if further tests are required (or repeats of the original Verifications to check for fraud or oversight) then I would suggest the use of the term "Confirmation" for calculations independently run by other than the code builder.

Also, the suite of problems can be used for re-Verification (e.g. after porting to a new computer or a new compiler, or to be run after addition of new options) and for user training. Code use independent of the code builder is probably more of a genuine issue in Validation than in Verification, especially when judicious "tweaking" or Tuning of code parameters can be involved; e.g., see recommendations in Porter (1996), and the distinction between "prediction" and "postdiction" made in the 1980–81 Stanford Turbulence "Olympics" (Kline et al, 1981).

As in Roache et al (1990), we recognize five distinct regimes where errors can be made in Verifying (or in Roache et al, 1990, "Benchmarking") a CFD code, even without considering the Validation question of whether the right equations are being solved for the target problem:

1. in code generation (either by hand or using computer Symbolic Manipulation; see below);
2. in code instructions (e.g., in a user manual or comment cards);
3. in problem set-up;
4. in defining and coding a test case (analytical solutions are often more difficult to code than numerical solutions); and

5. in the interpretation of code results.

The first two are errors of the code author. The last three are errors of the code user, although ambiguous or scant code documentation can put some of the responsibility back onto the code author. "Verification" of a code removes (1) and, if done thoroughly, (2), but (3–5) still contain the potential for errors in any new application. We reluctantly conclude that there will be a continuing need for users to construct and exercise test cases *even when using Verified codes*.

Obviously, it is good common sense to build more confidence with more problems, even if the code is Verified in the sense of a theorem. In the same way, a high school student will "confirm" the general solution of the quadratic equation for a particular case by back-substitution of the solutions. This exercise does not mean that the student does not "believe" the formula, but it is prudent and pragmatic to do so, and it builds confidence, not only in the theorem but in one's understanding of the theorem. Thus, although I consider that Code Verification should be completed and therefore *is not* on ongoing activity, Code Confirmation *is* naturally an ongoing activity (as is Validation), even for well-exercised commercial codes.

Also, as a practical political consideration, not everyone can appreciate the generality of a mathematical proof or a thoroughly executed Code Verification. This is particularly true when a physically unrealistic solution is used to Verify a code (see "Method of Manufactured Solutions" in Chapter 3). Although mathematically sound, the exercise will not inspire confidence in the mathematically unsophisticated. For such people, an extensive suite of realistic-*looking* Confirmation problems, even if they alone fail to constitute a rigorous Code Verification, may build more user confidence than a single rigorous but unreal-appearing Code Verification problem.

Indeed, this skeptical attitude is not only forgivable but often wise, prudent, and pragmatic. I have the same attitude in regard to "proofs" of grid generation algorithms, single-grid error estimators, iterative convergence rate proofs, efficiency comparisons based on operation counts, gains from parallelization, and other topics. [For example, in Roache and Steinberg (1984), I gave a counter-example to a widely-accepted and important theorem on grid generation; see also Knupp and Luczak (1995).] Still, the skeptic should keep in mind the possibility, in principle and often in practice, of rigorous Code Verification in the sense of a mathematical theorem.

2.5 BENCHMARKS AND INTER-CODE COMPARISONS

"Benchmarking" of codes is a less specific term than Verification or Validation. Williams and Baker (1996) prefer to reserve the term "benchmarking" for code-to-code comparisons; this has the advantage of clearly distinguishing it from Verification and Validation, which they define as we do herein, i.e. comparisons with closed form mathematical solutions and with experiments, respectively. Unfortunately, the term is commonly used more inclusively. It commonly refers to comparisons of results of the code in question to some other standard or "benchmark," which might be an analytical solution, a physical experiment (usually on simple configurations involving a single dominant physical phenomenon (Melnik et al, 1995; Rizzi and

Vos, 1996), a rigorously performed numerical solution, or just another code solution at comparable grid resolution, which has been termed "inter-code comparison." Often, the term "benchmarking" has been used in meeting papers wherein rigorous grid convergence testing has not been performed, and comparisons have not been precise; such "Benchmarking" exercises could be classified with "Confirmation" exercises above. (See Hutchings and Iannuzzelli, 1987 for popular level examples of such non-rigorous benchmarking in fluid dynamics, and Rizzo, 1991 for a popular level description of convergence studies for Finite Element structures calculations.) The more rigorous Benchmarking exercises could be classified with Verification or Validation exercises above, depending on whether the Benchmark is a rigorous numerical solution (e.g., see Ghia et al, 1982; de Vahl Davis, 1983; Celik and Freitas, 1990; Celik et al, 1993) or a physical experiment. The "CFD Triathlons" of Freitas (1993b, 1995) were based on "Benchmarks" that are experimental data; thus, these exercises could all be described as Validation.

The term Benchmarking is also widely used for the distinct activity of comparisons not of accuracy, but execution speed, comparing different codes on the same computer, or different computers running the same code. On old computers (until the 1980's) such comparisons were relatively straightforward, dependable, scaleable, and repeatable. This is no longer the case with modern computer architectures, flexible system configurations and memory management, and compiler options. Particularly for purchasers of Workstations and PC's, speed Benchmarking on non-CFD codes from different manufacturers do not correlate well with CFD code performance (Dudebout and Fahs, 1996).

Precise inter-code comparisons could be used in principle to rigorously Verify a code. If Code A has been rigorously Verified against analytical solutions, it could be used to approximately (in an ordered sense) reproduce another analytical solution that could be used legitimately as the basis of comparison for Code B, a kind of indirect Verification of Code B. This would be straight-forward if both codes produced essentially the same discrete solution at all grid refinements; e.g., both used simple linear FEM and differed only in the solver, so that the answers from each agreed to within the iterative convergence tolerance. Less straight-forward would be the Verification of a 4th-order accurate Code B by comparison with a previously Verified 2nd-order accurate Code A, though it could be done convincingly.

Unfortunately, in practice, inter-code comparisons are usually rather loose; e.g., see the PSACOIN exercises (Nies et al, 1990) and the international projects INTRACOIN, HYDROCOIN and INTRAVAL (Larsson, 1992). In the HYDROCOIN exercise (OECD, 1988), "Level 1" was titled "Code Verification," yet *no* participating group Verified the theoretical rate of convergence. Surely some agreement of (say) a new code with an old and widely used code could legitimately build some confidence. Note, however, that there is nothing to be gained by demonstrating agreement with an inaccurate code! This occurs surprisingly often. Also note that a good code (say a 2nd- or higher order accuracy code) may appear to the naive to be inaccurate when its results are compared with results of four or five 1st-order or hybrid codes, all of which are in rough agreement. (Furthermore, the 1st-order solutions are smoother, and often appeal more to the naive user's intuition.) In such a case, the statistical "outlier" is more credible.

Mathematics is not an exercise in democracy! Also, note that group exercises in inter-code comparisons invariably are not only comparisons of code accuracy *per se* but also exercises in conceptual modeling and code user skills as well. (See e.g., Nies et al, 1990; Larsson, 1992; Freitas, 1993b, 1995; de Vahl Davis, 1994; Muller, 1994) Furthermore, the exercises challenge the ability of organizers to create meaningful comparison problems.

2.6 CODE CERTIFICATION, QUALITY ASSURANCE, AND ACCREDITATION

"Code Certification" goes beyond Code Verification and Validation, and is used with less uniformity; see definitions in Mehta (1989, 1991, 1992a,b, 1995, 1996), Melnik et al (1995), Rizzi and Vos 1996). Following Mehta (1988), Aeschliman et al (1995) define Code Certification as "the entire process of establishing the credibility of a code, i.e., a Certified code has been Verified, Calibrated, and Validated." (See the section on "Calibration and Tuning" below.) In a later publication, Mehta (1996) described code Certification as consisting of Verification and Validation of a code, plus rules for its use (on a project) and complete documentation. He also speaks of certifying the simulations as well as the code. In the field of groundwater modeling, Tsang (1991) simply equates Certification of a code with proper Verification plus proper documentation. The aerospace consortium experience in Verification, Validation and Certification reported in Melnik et al (1995) use the term Certification similarly but with an additional distinction. They associate Certification with a different set of variables than Validation. While both involve experimental data comparisons, typically (at least in the aerospace community), Validation has focused on simple geometries usually involving a single dominant physical phenomenon with data comparisons to fundamental flow quantities (velocities, pressures, etc.). In contrast, Certification must focus on the needs of the aerospace design process, which involve more complex geometries or complete configurations involving multiple physical phenomena, with data comparisons to performance quantities like lift and drag coefficients, nozzle efficiencies, etc. In my opinion, this distinction seems to be non-essential. Surely one could compare to experiment and Validate or invalidate any quantity. My position agrees with Porter (1996), who calls for the Validation process to "address at least three levels of flow problems: fundamental physics, unit or component flow problems, [and] overall systems."

Many authors in aerospace (e.g., Melnik et al, 1995) speak of three levels of codes, namely research, pilot, and production codes, and associate Certification only with the production level. However, Validation (in the general sense of comparison with experiments, perhaps for a more limited range of parameters than Certification) is not so restricted. That research and pilot codes should also be compared to experiments is the opinion of Marvin (1995), Melnik et al (1995) and Rizzi and Vos (1996), all of whom also adopt the same three-level description code development.

Rizzi and Vos (1996) also describe Certification somewhat differently than Melnik et al (1995), linking it more with Verification and mathematics rather than with Validation and science. "Verification and Certification address the accuracy of the computational model, and...Calibration and Validation are concerned with the suitability of the physical model." Also, "Certification is concerned with programming issues, e.g. logic checks, programming style, documentation, and Quality Assurance issues, e.g. rerunning the suite of Certification test cases after each new version release in order to be certain that no new errors have been introduced into the previously Certified version." They also include in Certification the issues of model options and robustness. They quote the definition of Certification (above) from Mehta; however, Mehta includes Validation in Certification, whereas Rizzi and Vos do not. (They do not note the inconsistency.) Thus we see that "Certification" can and has been taken to refer only to Verification-type activities, or to include project-oriented Validation activities as well, by different respected authorities in the same sector of computational engineering (aerodynamics). Sindir et al (1996) describe their four-phase "Validation" procedure (outlined in Chapter 10) which would perhaps fit the description of "Certification" as used by the previously cited authors.

Generally, I find Certification and Code Quality Assurance (QA) to be practically indistinguishable, with the different terms being preferred in different engineering activities. As contrasted to Code Verification and Validation, Code Certification or QA is usually not associated with mathematics or science so much as engineering and engineering management. Certification is more of a *programmatic* concept than a scientific concept. "Code Certification" appears to be the preferred term in the Aerospace field, (Mehta, 1989, 1991, 1995; Melnik et al, 1995; Barber, 1996; Porter, 1996) while Code QA is preferred by regulatory agencies involved with groundwater remediation (WIPP PA, 1992). "Code Accreditation" appears to be simply the process of some authority (perhaps legal or regulatory) officially declaring a code to be useable for a specific project (e.g., see Mehta, 1996); no general guidelines are discernible.

"Verification" ~ mathematics

"Validation" ~ science/engineering

"Certification" ~ engineering management

Contributing to the lack of uniformity in usage is the fact that there are two extremes of "Code Certification," the smallest at the scale of the isolated code, and the largest at the scale of the engineering project. Suppose we have a code that is rigorously Verified, and then acceptably Validated for a limited set of physical problems and parameters. One could still not claim Certification or QA unless the code were documented, etc. And then, even if the code were Certified by some set of criteria, one might not be able to claim Certification for the project or *programmatic* needs.

My inclination would be to include in Code Certification the activity of Code Verification independent of the code authors, i.e., Code Confirmation, other aspects of code QA including robustness, documentation (internal, external, users manuals), version control, the QA system itself (including such clearly non-mathematical but practical considerations as the selection process for review committees, signature authority for change orders, forms for bug reports, conflict resolution procedures, etc.). For project-level Certification, one would also need Code Validation for a project-oriented range of parameters. For several authors, a principal distinction is that "Validation" involves comparisons to experiments primarily on model problems while "Certification" involves comparisons to experiments on more realistic geometries and range of operating parameters (e.g., Melnik et al, 1995; Barber, 1996). Note that "code Certification as a design tool requires specific criteria/metrics" (Melnik et al, 1995). Barber (1996) also notes that a research group (associated more with Validation) is typically interested in fundamental flow variables (streamlines, velocity profiles, etc.) whereas a design group (associated with Certification) is typically interested in performance oriented variables (lift and drag coefficients, system efficiency, etc.). Barber (1996) includes Risk Assessment and Reduction in the Certification process.

For project-oriented engineers, the activities of Code Verification/Validation/Confirmation almost form a continuum, and the three terms are often used as a suite and even an acronym, "VVC." (See Melnik et al, 1995; Porter, 1996; Aeschliman et al, 1995.) For them, it is natural to see a progression from "research codes" through "pilot codes" to "production codes" (Melnik et al, 1995). However, the qualities associated with each category by Melnik et al, while perhaps representative of aerospace practice, are sometimes arbitrary and unnecessary. For example, there would seem to be no inherent reason why a code with a "limited range of application" could not become a "production code," nor why a "research code" or a "pilot code" code not have thorough documentation.

Likewise, Melnik et al (1995) and Rizzi and Vos (1996) describe "Validation" as progressing through higher stages, from "Benchmarking" for simple flows exhibiting one single dominant flow feature and simple geometry to multi-feature flows and complex geometries. However, I would prefer to allow "Validation" to be considered thorough, i.e. at a high stage, for even a limited class of problems. The progression they describe would then not be defined as higher stages of Validation, but simply as Validation for a more general class of CFD codes.

2.7 VERIFICATION OF CALCULATIONS

Just as it is critical to distinguish Code Verification (correctness of the mathematics) from Code Validation (correctness of the science), so we must distinguish between Verification of Codes vs. Verification of (individual) Calculations.

A code may be rigorously verified to be (say) 2nd-order accurate, but when applied to a new problem, this fact provides no estimate of accuracy or confidence interval. It is still necessary to band the numerical error for the individual calculation, usually (and most reliably) by performing grid convergence tests. It would be preferable to have different words for these two "Verification" activities, but I am at a loss for a clarifying term. The very important point, independent of the semantics, is that *the use of a Verified Code is not enough.*

This point is possibly well recognized by present readers, but is not universally so. Especially in the commercial code arena, user expectations are often that the purchase and use of a "really good code" will remove from the user the obligation of "doing his homework," i.e. the tedious work of Verification of Calculations via systematic grid convergence studies. This unrealistic hope is sometimes encouraged by salespeople and by advertising.

The methodology for Verification of Calculations is a major theme of this book; see Chapters 3 and 4.

2.8 QUANTIFICATION OF UNCERTAINTY

The term "Quantification of Uncertainty," as used in the ASME Workshops (e.g., Freitas, 1993b; Celik et al, 1993; Johnson and Hughes, 1995) and elsewhere, can refer both to Verification of Codes and to Verification of (individual) Calculations as described above. The term also allows for inclusion of both error "estimation" and the more conservative error "banding," which includes a "factor of safety" as in the Grid Convergence Index (see Chapter 5) and related methods. "Quantification of Uncertainty" is distinguished from the more amorphous "Confidence Building" by the key word "Quantification," and is less project-oriented than "Code Certification" or "Quality Assurance" (see above). Etymologically, it could include Validation, of course, but whether or not this is intended must be inferred by its context. One authority (Oberkampf, 1998) prefers to restrict the word "Uncertainty" to potential deficiency of modeling due to *lack of knowledge*, e.g. uncertainty in geometry specification, contrasted to "errors" like discretization error that are not due to lack of knowledge. This usage would in fact be preferable, but unfortunately the broader use of "Uncertainty" is already widespread.

2.9 GRID CONVERGENCE VS. ITERATIVE CONVERGENCE

The literature commonly uses the term "convergence" in two completely different ways. Readers probably will know the distinction between iterative convergence vs. grid convergence (or residual "accuracy" vs. discretization accuracy). Usually, the meaning is clear from the context, but sometimes confusion occurs, e.g. when some new variant of the SIMPLE algorithm (Patankar, 1980) is presented as being "more accurate." The "accuracy" claimed here often is residual accuracy, i.e. what is better called iterative convergence accuracy or iterative speed, and has nothing to do with the order of accuracy of the discretization.

For the present subject, we note that iterative convergence can muddy the distinction between Code Verification and Calculation Verification, since iterative tuning parameters (e.g., multigrid cycles, relaxation factors, etc.) can be problem dependent. Most importantly, incomplete iterative convergence will corrupt Verification.

2.10 ERROR TAXONOMIES

2.10.1 Inadequate Error Taxonomies

Several taxonomies of errors given in the literature are inadequate and misleading, in my opinion. Not all lists are taxonomies. For example, the list "flora, fauna, mammals, dogs" is not an adequate biological taxonomy. All items are certainly part of biology, but "mammals" are not separate from fauna, rather they are part of it, as are "dogs." In a true taxonomic classification, each element of the taxonomy is counted once and only once. In the list "flora, fauna, mammals, dogs," consisting of four categories, the family pet is counted three times! Biologists (and logicians) are well aware of difficulties and sometimes arbitrariness in taxonomic classifications. There can be gray areas and elements that do not fit the supposedly complete classification system, e.g. the platypus, hermaphrodites. Also, the same set of individuals can be classified by more than one valid taxonomy, e.g. automobiles might be classified according to year of manufacture, or country of manufacture, or color, etc. But there are also clearly false taxonomies, and inappropriate or inadequate taxonomies, and we see these often in CFD.

Specifically, a listing of "sources of error" as seen often in the literature is not a taxonomy appropriate for Verification of codes or calculations. The "grid generation errors" (Ferziger, 1993; Celik and Zhang, 1995 or "grid resolution/distortion" errors (Bobbitt, 1988; Barber, 1996) or "factorization errors" (Rumsey, 1988) are not separate from discretization errors for purposes of error estimation. For the Verification of a code or a calculation, there are no such things as separate "grid generation errors," nor are there separate "errors associated with coordinate transformations" (Ferziger, 1993). Indeed, bad grids do add to discretization error *size*, but do not add new terms other than discretization errors. This does not mean that one grid is as good as another, or that a really bad grid cannot magnify errors, but only that these so-called "grid generation errors" do not have to be considered separately from other discretization errors in a grid convergence test. If the grid convergence test is performed, and the errors are shown to reduce as $O(\Delta^2)$, for example, then all discretization errors are verified. One does not need to separately estimate or band the "grid generation errors."

Such listings *are* useful for analysis of algorithms. For example, one can study the effect on (formal) accuracy of grid angle θ. We know that as two families of coordinate lines approach parallelism ($\theta = 0$) the transformed PDE's (continuum and discretizations) degenerate. For θ small, the effect is to increase the size of the coefficients in the error expansion, but not the *order*. It has often been stated in the grid generation literature that strong coordinate stretching reduces the order of centered differencing in logical space to 1st order. In fact, it can be demonstrated (confirmed by numerical experiments) that severe stretching like $\chi = a \cdot \exp(x)$ or even the ridiculous $\chi = a \cdot \exp[\exp(x)]$ greatly increase the size of the error (with 6-figure coefficients) but that the formal and observed order remains 2nd-order (Steinberg and Roache, 1985).

Likewise for the proposed numerical error bar (Karniadakis, 1995; see also Vanka, 1995) shown in Table 2.10.1. The contributions consist of separately estimated numerical errors from "boundary conditions," "computational domain size," "temporal errors," and "spatial errors." This is clearly an inadequate taxonomy.

The category "boundary errors" can include ordered terms like wall vorticity (when vorticity is a dependent variable) standard forms for which (Roache, 1972) can be $O(\Delta)$, $O(\Delta^2)$, $O(\Delta^4)$, etc. Such ordered errors will tend to zero as the discretization improves, so that a "boundary

- boundary conditions
- computational domain size
- temporal errors
- spatial errors

Table 2.10.1. Example of an Inadequate Taxonomy for Error Estimation.

error" from wall vorticity evaluation need not be considered separately from the grid convergence study. The category "boundary errors" would also include far-field boundary conditions, which are not ordered in Δ, i.e. the error persists even in the limit of $\Delta \to 0$. This does not mean that all far-field boundary conditions are equally bad; some are better than others, but *none* vanish as $\Delta \to 0$. On the other hand, *all* vanish as distance to the far-field boundary $L_b \to \infty$. This "boundary error" will improve as the "computational domain size" increases. The taxonomy already includes "temporal errors" and "spatial errors" (which are evaluated in the grid convergence study) and "computational domain size" errors, so that both the ordered boundary errors and the non-ordered (outflow, or "free") boundary errors are already counted elsewhere in the taxonomy. Since the intention is to provide a quantitative breakdown in the sources of numerical error in an error band, the proposed taxonomy is inappropriate for the Verification of Codes and Calculations.

Note, however, that "free" outflow boundary errors may prove to be "ordered" not in Δ but in $1 / L_b$ where L_b is the distance from the region of interest (e.g., an airfoil) to the outflow boundary. (See the data of Zingg [1992], shown in Chapter 6 to be 1st-order in $1/ L_b$.) Thus, this source of error can be evaluated by purely numerical tests, by varying the position of the outflow boundary.

In contrast, consider the commonly used approximation for near-wall density ρ of $\partial\rho / \partial n = 0$ for compressible flow, a modeling approximation that (although commonly used and accurate in some sense) cannot be evaluated easily, i.e., its error does not vanish as $\Delta \to 0$ nor with any other limit of a modeling parameter such as L_b. Thus, this error of the wall density condition is more resistant to modeling accuracy than the commonly acknowledged free outflow or far-field computational boundary. (In fact, it will get worse as Mach Number increases and $(T_w - T_{adiabatic})$ increases.) This is really a conceptual modeling error. Like incompressibility, or two-dimensionality, it cannot be evaluated numerically using only the code being Verified, but must be evaluated by comparison with some external benchmark—numerical, experimental, or theoretical.

In summary, the taxonomy of error sources shown in Table 2.10.1 is inappropriate for the Verification of Codes and Calculations.

- Physical Modeling Errors
- Discretization Errors
- Programming Errors [i.e., mistakes]
- Computer Round-Off Errors

Table 2.10.2. A Meaningful Error Taxonomy. (From Oberkampf et al, 1995.)

2.10.2 A Meaningful Error Taxonomy

By contrast, the list of error sources provided by Oberkampf et al (1995) is an appropriate and meaningful taxonomy. They give four categories of errors, shown in Table 2.10.2. A slightly expanded (though not necessarily exhaustive) taxonomy will be suggested shortly.

2.10.3 Gray Areas: "Justification"

The subject of outflow (or "open") boundary conditions does produce some fuzziness in categorization of Verification vs. Validation, in my opinion. The error can be "ordered" in $1 / L_b$, as above, and therefore can be part of Verification. That is, it is up to the user (who is doing the conceptual modeling) to estimate or band the error caused by the position of the outflow boundary. But if the code has some sophisticated free outflow condition, e.g. a simple vortex condition for Euler equations (Thomas and Salas, 1985), or the "fringe method" in DNS turbulence calculations (Marvin, 1995), then the distinction is not as immediate. Certainly the equations used are clear, and the code may "solve the equations right" (i.e., Verification), yet there exists another benchmark solely from the mathematics (the case with infinite boundary distance) which could be used to justify the outflow condition without recourse to physical experiment (which would clearly be Validation).

In any case, I agree with Oberkampf et al (1995) that there is no "free outflow" or "open" boundary condition that is "true," "correct" or "exact" in any general sense. ("What are the exact boundary conditions for the partial differential equations on an open boundary?" is a meaningless question.) And while improvements in free outflow boundary conditions are possible, they are only capable of improving efficiency by allowing computations on a smaller numerical domain for the same error tolerance as present methods. But in no way are improvements a necessity for obtaining Verified solutions. Methods presently exist that will give the correct answers, to within an arbitrary tolerance, if the position of the free outflow boundary is varied and its influence assessed, i.e. if the effect is Verified quantitatively. On the other hand, consider the use of "fuzzy boundary conditions," a term introduced by Sani and Gresho (1994) to "suggest the existence of numerical BC's that produce good numerical solutions, but if one tries to take the limit as $\Delta \rightarrow 0$ of these BC's, one obtains unacceptable BC's for the PDE's" (Oberkampf et al, 1995). In my opinion, these "fuzzy boundary conditions" should not be used, since they make convincing Verification by rigorous grid convergence testing impossible.

Another example of semantic failure or fuzzy taxonomy arises when we consider Benchmarking a code based on simplified equations against a code using more complete equations. Examples include a turbulent boundary layer code or Parabolized Navier-Stokes (PNS) code benchmarked against a Reynolds-stress averaged Full Navier-Stokes (FNS) code, and a groundwater flow code using the Dupuit approximation against a code using the full nonlinear free surface (phreatic) equations (Serrano, 1995; Knupp, 1996; Knupp et al, 1996). Presume that both codes are convincingly "Verified," i.e. they correctly solve their respective equations. Suppose that the PNS code results agree well with the FNS code results for some range of parameters (e.g., including angle of attack). This agreement is not strictly included in the term "Verification," since the Verification of the PNS code has already been completed prior to the FNS benchmarking. Then we could say that the agreement has demonstrated that the PNS code is "solving the right equations" in one sense, i.e., it justifies the use of parabolic marching. Yet to claim "Validation" would be over-reaching, since we have not demonstrated the adequacy of the turbulence model by comparison with experiment. We have "solved the right equations" only in an intermediate sense of demonstrating that the PNS equations adequately represent the FNS equations, but not in the ultimate sense of "solving the right *physical* equations."

Appealing to the other distinction between Verification and Validation based on mathematics vs. science, it is clear that such a comparison exercise should be categorized with Verification rather than Validation, but that the categorization is somewhat inadequate because the PNS code can (should) have been fully "Verified" in the more usual sense before the comparison exercise was started. But if the only choice is "Verification" or "Validation," it is safer to reserve the term "Validation" for experimental comparisons.

In some situation where nit-picking would be excusable, the solution is to introduce a finer-scale of categorization, perhaps using the word "Justification," technically defined in this specific context. (The term is in common use in theoretical papers, as when one "justifies" the assumption of some simplification to the accepted full theory, e.g. a small perturbation.) In situations dictating a legalistic distinction, (e.g., code review by a regulatory agency) one could claim a "Justification" of the simplification of mathematical models (Full Navier-Stokes to boundary layer, or compressible flow to incompressible flow, or variable properties to constant properties, or small-scale isotropy of turbulence, or Dupuit approximation, etc.) while making it clear that the physical "Validation" remains to be accomplished.

Certainly one can understand the issues in the above examples, and do good work, without introducing new terminology like "Justification." For example, in discussing the "fringe method" for inflow and outflow boundary conditions for DNS turbulence calculations, Marvin (1995) uses the overall context of "Validation," and says the comparisons "confirmed the appropriateness," obviously using "confirm" in a general sense rather than the more limited technical sense proposed herein. Fortunately, anyone with some technical knowledge of the subject area and some common sense understands the meaning behind the words without needing an overly precise, tiresome, legalistic, semantic distinction between "Validation" and "Confirm" as the terms are used by Marvin and others, and the terms "Validation," "Verification," "Confirmation," and now "Justification" as proposed here. Unfortunately, such "mere semantics" may become of vital interest when dealing with regulatory agencies such as the EPA, or with legal definitions in a NASA contract, etc.

2.10.4 An Expanded Error Taxonomy

The taxonomy of errors shown in Table 2.10.4.1 is an expanded version of that given by Oberkampf et al (1995). It still has some gray areas, but is appropriate for purposes of Verification of Codes and Calculations.

- Errors ordered in discretization measures Δ; these errors can be evaluated by grid convergence studies.
- Errors ordered in some numerical (rather than physical) parameter not associated with discretization (like L_b); these errors can be tested numerically in the code being Verified.
- Errors ordered in some physical parameter.
- Non-ordered approximations (like $\partial \rho / \partial n = 0$ or $\partial P / \partial n = 0$) that are conceptual modeling errors.
- Programming errors (unlike the other errors, these errors are *mistakes*); these can be detected by grid convergence studies for a problem with an exact solution (see "Method of Manufactured Solutions" in Chapter 3).
- Computer round-off errors; these errors can be identified by grid convergence studies or *ad hoc* approaches, but often they are simply demonstrated to not be significant.

Table 2.10.4.1. An Error Taxonomy Appropriate for Verification of Codes and Calculations.

Each category in the Taxonomy of Table 2.10.4.1 will now be elaborated upon.

- Errors ordered in discretization measures Δ; these errors can be evaluated by grid convergence studies.

The discretization measures Δ can include each space coordinate, time, or N vortices, N Fourier modes, etc. The category includes ordered artificial viscosity (e.g., Roache 1972) but this can be tricky to categorize and estimate if the parameters of the artificial viscosity (or better, artificial diffusion) change with the discretization, and with "hybrid" methods.

- Errors ordered in some numerical (rather than physical) parameter not associated with discretization (like L_b); these errors can be tested numerically in the code being Verified.

This category notably includes the location of idealized far-field boundaries, such as distance from a wing to the downstream (outflow) boundary L_b, or the cross-stream distance from a wing to the "free-stream" conditions L_∞. Note that these could be argued to be physical conditions rather than numerical parameters, since the same kind of consideration exists for wind tunnel tests (Roache, 1972). That is, one could think of numerically modeling the wind tunnel walls with no numerical error associated with the location of the walls, and relegate the conceptual modeling error (compared to the desired simulation of flight in the atmosphere) to the wind tunnel concept. In practice, the assignment to non-discretization numerical errors is probably preferable, since the effect may be evaluated by purely numerical tests in the code being Verified (i.e., it may be considered to be a mathematics problem rather than a physical science problem, and therefore Verification rather than Validation) and since the numerical simulation at the "walls" can be *better* than the physical wind tunnel simulation. For example, why model a physical no-slip wall with a boundary layer, when a slip condition better approximates the condition of free flight? Conditions at the quasi-"free stream" location (upper boundary) that are even less restrictive are available; the simple expansion wave approximation (Allen and Cheng, 1970; Roache, 1998) is not just a slip wall, but is a better model of the free-flight condition. Note that the actual order (in $1 / L_\infty$) may be unknown *a priori*, but at least we know the limit is good and the error can be investigated by straight-forward numerical experimentation (Roache and Mueller, 1970; Roache, 1972). Also, the order may be determined by numerical experiments and sometimes by theoretical considerations, e.g. of potential flow (Roache, 1972). Similar considerations would apply to the modeling in groundwater flow and transport calculations, in which the "far-field" condition may be conceptualized as a distant drainage divide and simplified in several ways.

- Errors ordered in some physical parameter.

  These are physical or conceptual modeling errors, the order of which might be determinable using perturbation methods (Van Dyke, 1975; Wilcox, 1995). Examples are $M \to 0$, the Boussinesq approximation for natural convection, number of fractures or fracture width in a dual porosity formulation, skin resistance, etc.

- Non-ordered approximations (like $\partial \rho / \partial n = 0$ or $\partial P / \partial n = 0$) that are conceptual modeling errors.

  The condition $\partial \rho / \partial n = 0$ at a wall is not quite identical to the boundary layer approximation, and is a better (less restricted) approximation than $\partial \rho / \partial y = 0$ throughout the boundary layer, but is commonly applied in situations where the boundary layer approximation is not valid (e.g., stagnation regions, shock-boundary layer regions). Another example is non-ordered artificial viscosity. The inherent artificial viscosity of nonlinear flux limiters may be difficult to classify. Errors in this non-ordered category might become graduated to the previous ordered category by later analyses. Discrete vortex methods may be difficult to classify, but the solution should get better in some sense as the number of vortices increases. The solution may "converge" in the sense of approaching a limit, but whether that limit approaches the full Navier-Stokes equations, or the inviscid Euler equations, or something else, may be open to question.

- Programming errors can be detected by grid convergence studies for a problem with an exact solution (see "Method of Manufactured Solutions" in Chapter 3).

- Computer round-off errors

Round-off error is sometimes confused with discretization error, and to emphasize the distinction people often speak of "computer round-off error" or "machine error." Round-off errors arise from the fact that computers do not work in the real number system, but only with a finite word-length subset of the real number system. They will be discussed further in Chapter 4.

2.11 TRUNCATION ERROR VS. DISCRETIZATION ERROR

"Truncation error" is an unfortunate term. Strictly speaking, it refers to the truncation at some finite value of a series, which could be analytical (e.g., the Sudicky-Frind solution, 1982) or more commonly in the present context, the Taylor-series expansion of the solution, which is the basis of developing the finite difference equations. It is a worthwhile concept because it allows one to define the "order" of the finite difference (or finite element, finite volume, etc.) method. Unfortunately, it is often used loosely in the sense of "discretization error," i.e. the error that is caused by the fact that we can only use a finite number of grid points (or another measure of discretization). In a FDM, one cannot take the limit of infinite order (i.e., limit of zero truncation error) without also taking the limit of infinite number of grid points, since high order methods require higher support. This terminology makes the limit process somewhat misleading, in my opinion. Also, it confuses the issue of solution smoothness with discretization error, since the Taylor series expansion depends on smoothness.

In the context of grid convergence tests, it is preferable to not speak of evaluating the "truncation error" of a numerical approximate solution, but rather the "discretization error" that arises because of the finite discretization of the problem. This terminology applies to every "consistent" methodology: FDM, FVM, FEM, spectral, pseudo-spectral, vortex-in-cell, etc., regardless of solution smoothness. (By "consistent" we mean, of course, that the continuum equations are recovered in the limit of infinite discretization.) The term "truncation error" is then reserved just for the examination of the order of convergence rate of the discretization. Note again the point of a taxonomy; these two "errors" are not independent. For any finite grid calculation, we do not have a "truncation error" (arising from the use of finite-ordered FDM, say) that we *add* to the "discretization error" (arising from the use of a finite number of grid points). And it is not possible to approach the limit of zero "truncation error" by arbitrarily increasing the order of the FDM or FEM without increasing the discretization. (Note in FEM we could fix a finite number of "elements" but we would still have to increase the discretization, i.e. the support within the elements.) However, the alternate is true: we can in fact approach the limit of eliminating all the discretization error by arbitrarily increasing the number of grid points without changing the "order" of the method. Thus, "discretization error" is the preferable term for speaking of the numerical error of a calculation, and "truncation error" is not separate in the taxonomy of an error estimate for a calculation. (However, the order of the truncation error is still verified in the code verification.)

Truncation error has also been defined (Ferziger, 1993) as the residual resulting from substitution of the continuum (exact) solution values into the discrete equations on some grid.

This definition can be useful for analysis of discrete methods. But again, it is not distinguishable from "discretization error" in an accuracy estimate of a real calculation.

2.12 CALIBRATION AND TUNING

The term "Calibration" is used with more latitude than Verification, Validation, or Confirmation.

I prefer to use the term "Code Calibration" to mean the adjustment of parameters needed to fit experimental data, e.g. the 6 closure coefficient values necessary for two-equation turbulence models. Some other CFD practitioners agree, e.g. Marvin (1995), who gives the following description.

"Code parameters such as turbulence models may need to be adjusted to accommodate applications for geometries and conditions outside the envelope of their original validation. Experiments intended to support this activity can be referred to as Calibration experiments." (Marvin, 1995)

As an example of Calibration, Marvin (1995) cites the work of Coakley and Huang (1992) in which basic k-ε and k-ω turbulence models are [rationally] corrected for compressibility and length scale, vastly improving the experimental agreement for surface pressures and heating rates on an ogive-cylinder-flare body at Mach number = 7. Importantly, the good agreement was maintained for other hypersonic experiments on shock interactions.

Likewise, Mehta (1996) is clear that "Calibration is not the process of determining the level of accuracy or credibility, but is the process of obtaining correction factors." Consistent with Marvin and with Mehta, Porter (1996) notes the dictionary definition of Calibration: "to standardize by determining the deviation from a standard so as to ascertain the proper correction factor." Rizzi and Vos (1996) are also consistent with this view. "Calibration is the process of Tuning or Calibrating a code with a particular fluid dynamics model to improve its prediction of global quantities for realistic geometries of design interest. This has to be done because there is no universal turbulence..."

However, other colleagues assure me that the term Calibration is used in experimental studies just as a means of ascertaining accuracy, or more properly of determining the inaccuracy, e.g. of a pressure probe or a wind tunnel test section flow. (Oberkampf, 1994 notes that the term "Validated" is never used for an experimental ground test facility because it would be a misnomer. Rather, facilities are Calibrated.) If extended to codes as in Aeschliman et al (1995), this definition, in my mind, makes "Code Calibration" almost indistinguishable from Validation, or perhaps Validation for a more restricted range of parameters. Indeed, Porter (1996) gives the definitions adopted by a NASA *ad hoc* committee.

- *Validation*: "comparison with experiment to verify [*sic*] the ability to accurately model over a range of parameters [*sic*].

- *Calibration*: comparison with experiment to provide a measure of the ability to predict specific parameters [*sic*]."

Even if we ignore the unfortunate use of term "verify" in a common- language (rather than technical context) sense, and overlook the use of "parameters" in two different senses (the first as an input parameter, the second as solution value), these NASA definitions still offer little distinction, with Validation involving a *range* of parameters but Calibration involving *specific* parameters. See also Bradley (1988).

Aeschliman et al (1995) use (as do others) the triplet "Code Verification, Calibration, and Validation" or VCV and give this description of Calibration. "We loosely interpret code 'Calibration' to mean a code's ability to reproduce valid data (not exclusively experimental) over a specified range of parameters, for some geometry, without necessarily assessing the overall correctness of all of the physical models employed. We consider Calibration to be a less-demanding element of Validation, and is addressable experimentally by the same methods." For CFD, I prefer Marvin's (1995) description above.

Bradley (1988) summarized the substance of a NASA study of a process for CFD Validation, and distinguished code Validation from Calibration as follows. (The quote is from Marvin, 1995.)

"CFD code Validation: Detailed surface and flowfield comparisons with experimental data to verify the code's ability to accurately model the critical physics of the flow. Validation can occur only when the accuracy and limitations of the experimental data are known and thoroughly understood and only when the accuracy and limitations of the code's numerical algorithms, grid density effects, and physical basis are equally known and understood over a range of specific parameters. CFD code Calibration: The comparison of CFD code results with experimental data for realistic geometries that are similar to the ones of interest, made to provide a measure of the code's ability to predict specific parameters that are of importance to the design objectives without necessarily verifying that all of the features of the flow are correctly modeled."

Note that the term "Calibration" also has traditionally been applied to the empirical adjustment of constants in a *theoretical* analysis, as in Bradley (1988). Marvin (1995) goes on to state that "During the intervening years since these definitions were formulated [1988], it has been argued that the definition of Validation is too restrictive, especially for the complex applications associated with realistic geometries. Nevertheless, NASA has maintained the definition as a goal of its Validation process."

In my opinion, definitions or descriptions should avoid attempts to enforce levels of accuracy, which descriptions are inevitably vague anyway (e.g., "detailed," "thoroughly understood," etc.) and leave these qualifications simply to the evaluation of the *thoroughness* of the Validation (or invalidation). Also, as I have noted earlier, assessment of "grid density effects" is not a code property *per se*, but rather a property of the particular code *application*, i.e., part of Verification of a particular calculation rather than Verification of a code.

Many people simply equate Calibration with the adjustment of parameters that is called "Tuning," and often "somewhat without scientific justification" (Mehta, 1996)." In addition to such "model Tuning," one may legitimately "tune" to correct for under-resolution in engineering parametric studies, especially for trends (Oberkampf, 1998). The faintly pejorative association of model Tuning is deserved if every new data set requires re-tuning, but not so if reasonable universality is obtained. Bradshaw (1992) noted that "a simple model which has been carefully calibrated" [for a particular problem] "may out-perform more advanced models on its home ground. This may be the First Law of Turbulence Modeling."

It is important to recognize that all of these activities (Verification of calculations, Calibration, Validation, and especially Certification) have associated with them error tolerances that cannot be arbitrarily defined universally but must be defined with reference to their intended uses (Marvin, 1988; Mehta, 1995; Melnik et al, 1996). Also, as noted above, Certification is a *programmatic* concept, rather than a scientific concept.

2.13 QUALITY ASSURANCE (QA) VS QUALITY WORK

Since we are making semantic distinctions in the context of confidence-building for CFD, it is worth making the distinction between the technical term "Quality Assurance" or QA vs. "quality work." QA boosters like to talk as though the two were equivalent, but formal QA is largely a system of paperwork (e.g., see ISO, 1991) run by managers. A project can meet all the formal QA requirements and still be low quality (or flatly erroneous). On the other hand, high quality work can and has been done without a modern and formal QA program. (The legendary Blackbird spy plane was test flown in 1962, long before formal QA programs.)

QA can, of course, be helpful for quality work. (If nothing else, it can be used to encourage management to support quality work.) Even the formality and paperwork are helpful in issues like version control. But if allowed to run amuck, formal QA can mire real work in forms and procedures and definitions, impeding real quality rather than fostering it. For example, formal QA procedures may require expensive re-running of an entire suite of confirmation problems whenever a code is modified. This can discourage minor code fixes and improvements. As another example, I have seen QA reviewers complain and require written justification (with committee approvals, documentation of resolution, archiving of correspondence on approved forms, etc.) for a change in input data that produced numerical value changes from 3.0 to 2.999998. And this, in a geophysical problem wherein the parameter ranges were sampled over three orders of magnitude!

2.14 CUSTOMER ILLUSIONS VS. CUSTOMER CARE

The QA dictum that "the customer is always right" is adequate for a candy store operation, but not for any professional service. Catering to customer illusions can be antithetical to true customer care. (Indeed, the prerequisite task of *identification* of the QA customer is not obvious in many cases, e.g., university teaching.) This is the great QA philosophy failure; the customer is *not* always right, and we should not always give the customers what they want. (I have not read the original QA manifestoes, so I speak only from experience and second-hand sources of what QA originally was intended to be.) I do not want my physician, lawyer, or surveyor to take orders from me.

This distinction is pertinent in code user training and education. The limits of applicability of codes must be made clear. The best commercial code cannot be an "aerodynamicist on a chip" or a "groundwater modeler on a chip," i.e., the code cannot substitute for user expertise in the technical area. Also, this distinction of true "customer care" is most important in the area of code robustness. We all know how to build a robust CFD code, following what has become a CFD joke (Roache, 1994). "The good news about first-order upstream differencing is that it always gives you an answer. The bad news is that it always gives you the same answer, no matter what the Reynolds number." It is remarkable and disheartening to see how many industrial CFD practitioners will freely admit to caring little about numerical accuracy. (A director of a very successful consulting firm, specializing in simulation of enhanced oil recovery processes, stated that in some 20 years of experience he had *never* had a request from a customer for enhanced numerical accuracy.) For those who do care, we must not gloss over the limitations of CFD and give them a false sense of security.

2.15 OTHER DISTINCTIONS: AUTHORS, USERS, MODELERS, CODE AND SOFTWARE

Another way to make distinctions is between the "code author" or "code builder" (which may of course be a team, including algorithm developers and programmers) and the "code user" or analyst. In many past situations (e.g., a graduate dissertation project) the same person performed both functions and therefore took the blame or credit for all aspects, but it is still worthwhile to make the distinction in functionality, especially with the rise of general-purpose commercial codes and software.

Another distinction is "constitutive equation developer" most typified in fluid dynamics by the "turbulence modeler." Again, the same person could and has functioned in all three capacities (code builder, code user, and constitutive equation developer) but I agree with Blottner (1990) that it is too much burden on one person.

In regard to the distinction made earlier between numerical errors and modeling errors, I cannot agree with Mehta (1996) that these may be viewed as precision and bias errors; see Chapter 9 for definitions of these terms as used in experimental error assessment.

Another distinction often made in the software engineering community is between "software" and "code." Briefly, "code" is defined as computer instructions and data definitions, whereas "software" is a more comprehensive term, including "programs, procedures, rules, and any associated documentation pertaining to the operation of a computer system ..."(Mehta, 1996). This distinguishes system software from applications software (like PDE codes) but, as readers will know, physical modelers usually overlap the terms, to no practical detriment.

2.16 SENSITIVITY, UNCERTAINTY, AND RISK

Significant practical detriment does occur in public policy discussions that fail to distinguish between terms like "Sensitivity," "Uncertainty," and "Risk." Sensitivity calculations (often called "perturbation" calculations) usually refer to determination of the change of code output (either directly calculated quantities or inferred quantities) caused by unit changes in input data (including parameters). Essentially, this is equivalent to a numerical approximation to a partial derivative, say $\partial Q / \partial d$ where Q is an output quantity of interest and d is input data or parameter. Such data cannot be transformed into Uncertainty in the results until a conceptual model is developed for the uncertainty distribution for the input parameters, which provide a value for Δd and thence uncertainty in output values $\Delta Q = \Delta d \cdot \partial Q / \partial d$. Such a concept from elementary, intuitive level calculus is simple but still confusing to many public "stakeholders" in public policy debates. The next jump, from Uncertainty to Risk, is even more difficult. The simplest definition is Risk = Uncertainty × Consequence. That is, a high-level uncertainty (say, a 1% chance of a contaminant release to the environment) may produce an acceptably low level of risk if its Consequence is small, e.g. a minor and temporary health problem.

Mehta (1996) states that "Sensitivity studies usually consider extreme values of the model parameter," but representative or mid-range values would seem to be equally significant, or more so. (Note, however, that mid-range values may not exist, or be permissible, for bi-modal distributions.) Sensitivity studies (and the consequent uncertainty and risk studies) can also be assessed not just for input data and parameters but also for differing conceptual models (also referred to as "structural uncertainty"). As Mehta (1996) notes, "The method for assessing output uncertainty is much more developed for parameter uncertainty than for structural uncertainty." Note, however, that it is always possible, in principle, to transform a structural uncertainty into an artificial parameter uncertainty by using a blending function between models. The blending function could be applied to the outputs of the different models, or in some cases to internal parameters which blend one model into another.

Such parameter sensitivity studies are, in my taxonomies, not part of either Verification or of Validation; they are simply part of the *results* of the modeling. This distinction is often lost in the (necessarily brief) reporting in journals, as when an author investigates the sensitivity of answers to some uncertainly specified physical parameter; this physical *modeling* sensitivity is usually reported in conjunction with purely numerical grid convergence tests. For example, Zha and Knight (1996) report sensitivity tests for their difficult problem of three-dimensional shock wave/turbulent boundary layer interaction. They report grid convergence tests by independent coordinate refinement (see Chapter 5), first doubling the grid in x, then in y, then in z. These are purely numerical tests. They also report (contiguously) sensitivity tests on the thickness of the upstream boundary-layer and on different isothermal wall temperatures. These are physical modeling tests. (Both categories were convincingly performed.) In most papers, this distinction will not be made, and the results may be reported under a heading like "Solution Convergence Tests" or "Verification." The distinction is perhaps worth noting, but the only important point is that all the work be done and reported clearly, as in Zha and Knight (1996).

2.17 ETYMOLOGY AND NEAR-SYNONYMS

As noted earlier, validation, verification and confirmation in common (non-technical) use are given as synonyms in a common English thesaurus. More complete definitions indicate extensive overlap of these words in some contexts, and with other related words such as substantiation, sanction, approval, authoritative, cogent, convincing, truth, authenticity, accuracy, genuineness, definite assurance, even logical. The Middle English/Latin root of "verification" is the equivalent of "truth," whereas the Latin root of "validation" is the equivalent of "strong." This etymology contributes, at best, a shade of meaning in favor of the distinction used herein. Others, e.g. Jameson and Martinelli, 1996, see a stronger etymological distinction. In any case, there is absolutely no justification for the extreme position taken by Oreskes et al (1994), arguing as though the etymology and common use have a compelling (almost magical) power over our use as technical terms.

Many other instances of near-synonyms and fuzzy usage occur. Mehta (1996) uses the words "conceptual model" and "simulation model" distinctly, yet the essence of the distinction is not stated nor is it apparent. Mehta states that credibility is established through the processes of Verification, Validation, and Certification, and he offers guidelines for various aspects of establishing code credibility, including developing and assessing models, Verification, Validation, and experimental tests. (Some of these suggested guidelines are arguable or obvious, most are useful, and all reflect the voice of experience.) "Credibility" is an easily understood term and is obviously desirable; as Mehta (1996) notes, "The significance of computer simulations depends solely on their credibility." Yet I am somewhat uncomfortable with the word, because if "credibility" becomes the stated goal of a project, then the work can devolve into image building and public relations rather than science and engineering.

Naturally, writers will use overlapping and/or inconsistent descriptions, and make distinction in their terminology only adequate for their intentions, which probably do not include philosophy of science. As stated earlier, it is best to recognize the significant arbitrariness in the original choice of technical usage, but still to recognize the widespread acceptance of a standard technical distinction where it exists, e.g. in Verification vs. Validation.

2.18 ACCURACY VS. RELIABILITY

Another worthwhile semantic distinction arises with the terms "accuracy" and "reliability," which are often used interchangeably in the engineering literature but which are, in fact, distinct. Consider the following hypothetical situation. You are solving an expensive time-dependent three-dimensional problem. Based on estimates from experience on related problems, you have a computing budget adequate to solve the problem on a $100 \times 100 \times 100$ grid. This would produce the most *accurate* solution you can afford. However, for about the same budget, you could produce two solutions on $90 \times 90 \times 90$ and $80 \times 80 \times 80$ grids. (The economics of the scaling for three-dimensional problems with $\Delta t \propto \Delta x$ indicate that the fine grid solution will cost $C \cdot 100^4 = C \cdot 10^8$ while the two grid solutions will cost $C \cdot 90^4 + C \cdot 80^4 \cong 1.07C \cdot 10^8$, about 7% more.)

The question is, how shall you spend your budget? The answer is, if you want *accuracy*, you use the 100^3 grid, because this will produce the most accurate solution. However, although you know it is the most accurate solution that you can afford, you do not know how accurate it is! If instead you want *reliability*, the answer is you opt for the 90^3 and 80^3 grid solutions. The difference between the two solutions gives you some idea of the accuracy. You have more *confidence* in the 90^3 solution; you can *rely* upon it. You have a somewhat *less accurate* solution but it is *more reliable*.

(Methods for quantifying the accuracy, and perhaps using both 90^3 and 80^3 grid solutions to achieve accuracy higher than the 100^3 grid solution, will be addressed in Chapter 5.)

2.19 ADDITIONAL REMARKS ON VERIFICATION

Finally, I would like to address some concerns raised in the literature that are not as significant as claimed, in my opinion.

Cheng (1970) and later Oberkampf et al (1995) put much emphasis on the importance of the Lax Equivalence Theorem (see Richtmyer and Morton, 1967) as being of fundamental importance to CFD. I believe the importance is exaggerated. The Equivalence Theorem may be briefly condensed to stating that *consistency + stability yields convergence*. Consistency simply means that as $\Delta \rightarrow 0$, the discrete (algebraic) equations approach the continuum partial differential equations, which is just the calculus limit. Stability is defined in the von Neumann sense (Roache, 1972). "Convergence" here means convergence of the algebraic solution to the continuum solution. Certainly this theorem is good to know, and provides some confidence. Unfortunately, it applies only to *linear* (and properly posed) initial value problems. As noted by Oberkampf et al (1995), the analyses for consistency and especially stability are "predominantly developed for very simple model problems... The model equations are always linearized equations and uncoupled from any other equations."..."Additional, but related, simplifications of consistency and stability analyses are elimination of: mixed classification partial differential equations, non-uniform grids, and boundary conditions." In the matter of mixed PDE's, they note that "in *every* supersonic flow problem modeled by the steady Navier-Stokes equations, hyperbolic and elliptic regions exist adjacent to one another." So as a practical matter, modelers cannot sit on their hands waiting for more general theorems. Also, note that numerical boundary layer solutions were obtained, Verified, and Validated many years before the Equivalence Theorem was published, so progress can be made without the rigor. The same applies to questions of existence and uniqueness.

Another concern in Code Verification is the number of user options in a code, especially general-purpose commercial codes. This is a genuine practical problem, but does not nullify claims of Verification; it just limits those claims. The exponentially expanding complexity of the option tree does not nullify the definition of Verification of Code; it simply qualifies the definition. "Code Verification" is restricted to that combination of options claimed to be Verified. There is a gray area here, as one might expect, in the judgment of the independence of options. Some knowledge of algorithm and code structure may be necessary to infer the reasonableness of simplifications of option interactions (essentially, partitioning of the full matrix of option interactions). The easy answer is the rigid one of categorical "no." This would usually appeal to the rigid SJ temperament type (see Chapter 11) that gives Quality Assurance a bad name. A more intelligent and economical approach is possible, bearing in mind that subtle and unanticipated option interactions have occurred, especially before the acceptance of structured coding and modularity.

In regard to "canned" or highly modular elliptic solvers, one can make a very strong case for independence of the options. The argument becomes unassailable if separate residual checks are made after the canned solver has obtained the solution. Essentially, the "solver" is treated as providing an initial guess for a simple point-iterative method. Different modules for the solver could be interchanged (e.g., a multigrid solver, elliptic marching method, PCG, etc.) without repeating all option-tree combinations (e.g., types of boundary conditions, time differencing schemes, well descriptions) since the algebraic solution provided by the solver is always verified (Roache et al, 1996).

2.20 CONCLUSION: LIMITATIONS OF SEMANTIC DISTINCTIONS

The semantic distinctions involved in the general area of confidence-building in CFD are important and worthwhile. While sometimes arbitrary, it is worthwhile to try to maintain uniformity of terminology, or at least to recognize the underlying conceptual distinctions and to define one's terms with appropriate precision. The most important distinction, and fortunately the most generally agreed upon, is the distinction between Verification (mathematics) and Validation (science).

The definitions used herein are sometimes at variance with the discussions in Oberkampf (1994), Sindir et al (1996), Jameson and Martinelli (1996) and others, notably in the present title subject of Verification. My position is that we *Verify a Code* by convincingly demonstrating (if not "proving" as in a mathematical theorem) that the code *can* "solve the equations right." When done properly, the exercise also Verifies the *order of convergence* of the code. Then, we *Verify a Calculation* by convincingly demonstrating that the code *has* "solved the equations right," to a rationally estimated accuracy or error band. These two exercises are purely mathematical; neither appeals to experimental data for their justification. Only in *Validation* do we demonstrate that we have "solved the right equations," with an understood context of engineering or scientific accuracy, by appealing to experimental data.

It is well to recognize the limitations of our attempts at semantic distinction. Although these efforts are worthwhile, it is clear that we (the scientific–mathematical–engineering community) are *not* going to achieve uniform, non-overlapping terminology. For example, in the classification framework of Sindir et al (1996), their "Validation" procedure includes what is herein and elsewhere defined as aspects of Verification, so that their "Validation" is more like a total code Quality Assurance program (see Chapter 11). Likewise, Bogdonoff (1993), Grasso et al (1994), Marvin (1995), Jameson and Martinelli (1996) refer to "Validation" of both numerical simulations and experiments, i.e., what we here define as Verification and Validation. (Also,

Validation	–	Supported by an accepted authority, such as an archival benchmark experiment that provides specified requirements
Verification	–	Implies that accuracy has been ensured, for instance with regard to coding
Calibration	–	Implies that deviations from some standard, such as an archival benchmark experiment, have been determined
Certification	–	Implies official recognition that a standard has been met

Figure 2.20.1. Connotations Associated with Key CFD Terms. (From Figure 2 of Porter, 1996).

these authors often use the word "verification" in a non-specific sense when discussing Validation, i.e., they do not make the distinction in terms as used herein.) Beyond the "mere" semantics, Oberkampf (1994) includes comparisons of calculations with carefully performed experiments as part of "Verification" of a code. In an extreme case of semantic confusion, the otherwise informative general-audience article titled "Verifying Analytical Methods" (Puttre, 1994) should have been titled "Validating Numerical Codes"!

Nevertheless, following the advice of Chuang Tzu (see quote at the beginning of this chapter), it is worthwhile to keep in mind the ideas *behind* Verification and Validation. Porter (1996) makes the limitations of languages clear by using the word "connotations" instead of "definitions" for the key terms, as shown in Figure 2.20.1.

Porter (1996) also quotes Alice from *Through the Looking Glass* (an English fantasy written, appropriately enough, by a mathematician, Lewis Carroll): "Don't listen to what I say, listen to what I mean." This wise advice agrees with the Taoist spirit of Chuang Tzu. In reading journal papers and reports, it is not a good idea to try to enforce terminology or taxonomies. It is better to try to detect the authors' terminology and/or taxonomies (sometimes defined only implicitly, and perhaps used with some forgivable inconsistency) and then go on to learn from the authors' experience and perspective.

Finally, techniques are already available to convincingly Verify the numerical accuracy of Codes and Calculations without undue stress on computer resources, as will be demonstrated in the succeeding chapters. These techniques are applicable to commercial codes as well as specialized scientific codes.

REFERENCES FOR CHAPTER 2

Aeschliman, D. P., Oberkampf, W. L., and Blottner, F. G. (1995), "A Proposed Methodology for Computational Fluid Dynamics Code Verification, Calibration, and Validation," *Proceedings of the International Congress on Instrumentation in Aerospace Simulation Facilities*, WPAFB, Ohio, 18–21 July 1995.

AGARD (1988), *Validation of Computational Fluid Dynamics*, Lisbon, Portugal, 2–5 May 1988, NATO Advisory Group for Aeronautical Research and Development, AGARD CP 437, Dec. 1988. Volume 1: Symposium Papers and Round Table Discussion. Volume 2: Poster Papers.

AIAA (1994), "Editorial Policy Statement on Numerical Accuracy and Experimental Uncertainty," *AIAA Journal*, Vol. 32, No. 1, January 1994, p. 3.

Allen, J. S. and Cheng, S. I. (1970), "Numerical Solution s of the Compressible Navier-Stokes Equations for the Laminar Near Wake," *Physics of Fluids*, Vol. 13, No. 1, January 1970, pp. 37–52.

American Nuclear Society (1987), *Guidelines for the Verification and Validation of Scientific and Engineering Computer Programs for the Nuclear Industry*, ANSI/ANS–10.4–1987, American Nuclear Society, 555 North Kensington Avenue, La Grange Park, Illinois 60525, USA.

Barber, T. J. (1996), "The Role of Code Validation and Certification in the Design Environment," AIAA Paper 96–2033, 27th AIAA Fluid Dynamics Conference, New Orleans, Louisiana, 17–20 June 1996.

Blottner, F. G. (1990), "Accurate Navier–Stokes Results for the Hypersonic Flow over a Spherical Nosetip," *AIAA Journal of Spacecraft and Rockets*, Vol. 27, No. 2, March–April 1990, pp. 113–122.

Bobbitt, P. J. (1988), "The Pros and Cons of Code Validation," AIAA Paper 88–2535. Also, NASA TM 100657, July 1988.

Boehm, B. W. (1981), *Software Engineering Economics*, Prentice-Hall, 1981.

Bogdonoff, S. M. (1993), "Computational Validation for Hypersonics: An Experimentalist's Perspective," *Workshop on Hypersonic Flows for Reentry Problems*, Antibes, France, January 1993.

Bradley, R. G. (1988), "CFD Validation Philosophy," in AGARD (1988).

Bradshaw, P. (1992), "Final Report on AFOSR 90–0154, Collaborative Testing of Turbulence Models," Mechanical Engineering Department, Stanford University, Stanford, California.

Celik, I. and Freitas, C. J. (1990), Eds., *Benchmark Test Cases for Computational Fluid Dynamics*, ASME FED Vol. 93, Book No. H00598, June 1990.

Celik., I, Chen, C. J., Roache, P. J., and Scheurer, G. (1993), Eds., *Quantification of Uncertainty in Computational Fluid Dynamics*, ASME Fluids Engineering Division Summer Meeting, Washington, DC, ASME FED–Vol. 158, 20–24 June 1993.

Celik, I. and Zhang, W-M. (1995), "Calculation of Numerical Uncertainty Using Richardson Extrapolation: Application to Some Simple Turbulent Flow Calculations," *ASME Journal of Fluids Engineering*, Vol. 117, September 1995, pp. 439–445.

Cheng, S. I. (1970), "Numerical Integration of the Navier–Stokes Equations, *AIAA Journal*, Vol. 8, No. 12, December 1970, pp. 2115–2122.

Coakley, T. J. and Huang, P. G. (1992), "Turbulence Modeling for High Speed Flows," AIAA Paper 92–0436, January 1992.

de Marsily, G., Combes, P., and Goblet, P. (1992), "Comment on 'Ground-Water Models Cannot be Validated," *Advances in Water Resources*, Vol. 15, 1992, pp. 367–369.

de Vahl Davis, G. (1983), "Natural Convection of Air in a Square Cavity: A Bench Mark Numerical Solution," *International Journal for Numerical Methods in Fluids*, Vol. 3, No. 3, 1983, pp. 249–264.

de Vahl Davis, G. (1994), "Going for Quality in CFD Codes," *Proceedings 2nd World Conference in Applied Computational Fluid Dynamics*, World User Association–CFD, Basel, Switzerland, 1–5 May 1994, pp. 18.1–18.6.

Dudebout, R. and Fahs, S. (1996), "Evaluation of Workstation Performance by Running a CFD Benchmark Code," CFD Society of Canada Bulletin, No. 7, Spring 1996, pp. 1–5.

Editorial Board (1994), "Journal of Heat Transfer Editorial Policy Statement on Numerical Accuracy," *ASME Journal of Heat Transfer*, Vol. 116, Nov. 1994, pp. 797–798.

Ferziger, J. H. (1993), "Estimation and Reduction of Numerical Error," ASME FED–Vol. 158, *Quantification of Uncertainty in Computational Fluid Dynamics*, ASME Fluids Engineering Division Summer Meeting, Washington, DC, 20–24 June 1993. Celik., I, Chen, C. J., Roache, P. J., and Scheurer, G., Eds., pp. 18.

Freitas, C. J. (1993a), "Editorial Policy Statement on the Control of Numerical Accuracy," *ASME Journal of Fluids Engineering*, Vol. 115, No. 2, June 1993, p. 339.

Freitas, C. J. (1993b), Ed., *The CFD Triathlon: Three Laminar Flow Simulations by Commercial CFD Codes*, FED Vol. 160, ASME Fluids Engineering Conference, Washington, DC, June 20–24, 1993.

Freitas, C. J. (1995), "Perspective: Selected Benchmarks for Commercial CFD Codes," *ASME Journal of Fluids Engineering*, Vol. 117, June 1995, pp. 208–218.

Ghia, U., Ghia, K. N., and Shin, C. T. (1982), "High–Re Solutions for Incompressible Flows Using the Navier–Stokes Equations and a Multigrid Method," *Journal of Computational Physics*, Vol. 48, 1982, pp. 387–411.

Golub, G. H. and Ortega, J. M. (1992), *Scientific Computing and Differential Equations*, Academic Press, San Diego.

Grasso, F., Leone, G., and Delery, J. M. (1994), "Validation Procedure for the Analysis of Shock–Wave/Boundary–Layer Interaction Problems," *AIAA Journal*, Vol. 32, No. 9, September 1994, pp. 1820–1827.

Gray, W. (1993), Ed., Special Issue devoted to "Research Perspectives in Hydrology," *Advances in Water Resources*, Vol. 17, 1994.

Gresho, P. M. and Taylor, C. (1994), "Editorial," *International Journal for Numerical Methods in Fluids*, Vol. 19, 1994, p. iii.

Haynes, T. S., Reed, H. L., and Saric, W. S. (1996), "CFD Validation Issues in Transition Modeling," AIAA Paper 96–2051, 27 AIAA Fluid Dynamics Conference, 17–20 June 1996, New Orleans, Louisiana.

Helton, J. C., Anderson, D. R., Roache, P. J., et al (1995), "Effect of Alternative Conceptual Models in a Preliminary Performance Assessment for the Waste Isolation Pilot Plant," *Nuclear Engineering and Design*, Vol. 154, 1995, pp. 251–344.

Helton, J. C., Anderson, D. R., Roache, P. J., et al (1996), "Uncertainty and Sensitivity Analysis Results Obtained in the 1992 Performance Assessment for the Waste Isolation Pilot Plant," *Reliability Engineering and System Safety*, Vol. 51, 1996, pp. 251–53–100.

Horgan, J. (1995), "From Complexity to Perplexity," *Scientific American*, Vol. 272, No. 6, June 1995, pp. 104–109.

Hull, M. (1997), The Creationism Debate: Behind the Ignorance that Confuses Religion with Science," *Crosswinds*, Albuquerque, New Mexico, U.S.A, March 1997, pp. 6–7.

Hutchings, B. and Iannuzzelli, R. (1987), "Benchmark Problems for Fluid Dynamics Codes," *Mechanical Engineering*, Vol. 109, No. 6, June 1987, pp. 54–58.

IAHR (1982), "Proceedings IAHR Workshop on Expanding Channel Problem," Rome, Italy, 24 June 1982.

ISO (1991), *Quality Management and Quality Assurance Standards–Part 3: Guidelines for the Application of ISO 9001 to the Development, Supply, and Maintenance of Software*, First

Edition, International Organization for Standardization, ISO 9000–3, Geneva, Switzerland, 1991, reprinted 1993.

Jameson, A. and Martinelli, L. (1996), "Mesh Refinement and Modeling Errors in Flow Simulation," AIAA Paper 96–2050, 27th AIAA Fluid Dynamics Conference, June 17–20, 1996, New Orleans, Louisiana.

Jay, F. (1984), Ed., *IEEE Standard Dictionary of Electrical and Electronic Terms*, Institute of Electrical and Electronic Engineers, ANSI/IEEE Std. 100–1984.

Johnson, C., Rannacher, R., and Boman, M. (1995), "Numerics and Hydrodynamic Stability: Toward Error Control in Computational Fluid Dynamics," *SIAM Journal on Numerical Analysis*, Vol. 32, 1995, pp. 1058–1079.

Johnson, R. W. and Hughes, E. D. (1995), eds., *Quantification of Uncertainty in Computational Fluids Dynamics–1995,* Joint JSME–ASME Fluid Mechanics Meeting, ASME FED Vol. 213, Hilton Head, South Carolina, August 14–18, 1995.

Karniadakis, G. E. (1995), "Toward a Numerical Error Bar in CFD," *ASME Journal of Fluids Engineering*, Vol. 117, No. 1, March 1995, pp. 7–9.

Kline, S. J., Cantwell, B. J., and Lilley, G. M. Eds. (1981), *Objectives, Evaluation of Data, Specifications of Test Cases, Discussion, and Position Papers. The 1980–81 AFOSR/HTTM–Stanford Conference on Complex Turbulent Flows*, Vol. 1, ThermoSciences Division, Mechanical Engineering Department, Stanford University.

Knupp, P. and Luczak, R. (1995), "Truncation Error in Grid Generation: A Case Study," *Numerical Methods for Partial Differential Equations*, Vol. 11, pp. 561–571, 1995.

Knupp, P. (1996), "A Moving Grid Algorithm for Three-Dimensional Regional Groundwater Flow Calculations with Water Table and Seepage Face," *Advances in Water Resources*, Vol. 19, No. 2, 1996, pp. 83–95.

Knupp, P., Rucker, D. and Corbet, T. F. (1996), "Groundwater Modeling with the Phreatic Boundary Condition: Is It Necessary?," Ecodynamics Research Associates, Albuquerque, New Mexico, December 1996. Also, to appear.

Konikow, L. F. and Bredehoeft, J. D. (1992), "Groundwater Models Cannot be Validated," *Advances in Water Resources*, Vol. 15, 1992, pp. 75–83.

Larsson, A. (1992), "The International Projects INTRACOIN, HYDROCOIN and INTRAVAL," *Advances in Water Resources*, Vol. 15, 1995, pp. 85–87.

Lax, P. (1957), "Hyperbolic Systems of Conservation Laws II," *Communications on Pure and Applied Mathematics*, Vol. 10, pp. 537–566.

Leijnse, A. and Hassanizadeh, S. M. (1994), "Model Definition and Model Validation," *Advances in Water Resources*, Vol. 17, 1994, pp. 197–200.

Levy, Y., Hesselink, L. and Degani, D. (1995), "Anomalous Asymmetries in Flows Generated by Algorithms that Fail to Conserve Symmetry," *AIAA Journal*, Vol. 33, 1995, pp. 999–1007.

Marvin, J. G. (1988), "Accuracy Requirements and Benchmark Experiments for CFD Validation," in AGARD (1988), Paper No. 2.

Marvin, J. G. (1995), "Perspective on Computational Fluid Dynamics Validation," *AIAA Journal*, Vol. 33, No. 10, October 1995, pp. 1778–1787.

Mehta, U. (1988), "Flight Performance Estimation Utilizing Computational Fluid Dynamics," *Proceedings of the 5th National Aerospace Plane Technology Symposium*, Volume 1, NASP CP–5028, October 1988.

Mehta, U. (1989), "Computational Requirements for Hypersonic Flight Performance Estimates," AIAA Paper 89–1670, June 1989.

Mehta, U. B. (1991), "Some Aspects of Uncertainty in Computational Fluid Dynamics Results," *ASME Journal of Fluids Engineering*, Vol. 113, 1991, pp. 538–543. (See also NASA TM 102804, March 1990.)

Mehta, U. B. (1992a), "NASP and SDI Spearhead CFD Developments," *Aerospace America*, February 1992.

Mehta, U. B. (1992b), "Aerospace Plane Design Challenge: Credible Computations," *Journal of Aircraft*, Vol. 30, No. 4, July–August 1993, pp. 519–525.

Mehta, U. (1995), "Guide to Credible Computational Fluid Dynamics Simulations," AIAA Paper 95–2225, June 1995.

Mehta, U. (1996), "Guide to Credible Computational Fluid Dynamics Simulations," *AIAA Journal of Propulsion and Power*, Vol. 12, No. 5, October 1996, pp. 940–948.

Melnik, R. E., Siclari, M. J., Marconi, F., Barber, T. and Verhoff, A. (1995), "An Overview of a Recent Industry Effort at CFD Code Validation," AIAA Paper 95–2229, 26th AIAA Fluid Dynamics Conference, San Diego, California, June 19–23, 1995.

Nies, A., Laurens, J.-M., Galson, D. A., and Webster, S. (1990), *PSACOIN Level 1A Intercomparison*, Nuclear Energy Agency, Organisation for Economic Co-Operation and Development, Paris, November 1990.

Oberkampf, W. L. (1994), "A Proposed Framework for Computational Fluid Dynamics Code Calibration/Validation," AIAA Paper 94–2540, 18th AIAA Aerospace Ground Testing Conference, Colorado Springs, Colorado, 20–23 June 1994.

Oberkampf, W. L., Blottner, F. G., and Aeschliman, D. P. (1995), "Methodology for Computational Fluid Dynamics Code Verification/Validation," AIAA Paper 95–2226, 26th AIAA Fluid Dynamics Conference, San Diego, California, 19–22 June 1995.

Oberkampf, W. L. (1998), personal communication and draft of AIAA Standard Guide.

OECD (1988), The International HYDROCOIN Project: Groundwater Hydrology Modelling Strategies for Performance Assessment of Nuclear Waste Disposal. Level 1: Code Verification, Nuclear Energy Agency, Organisation for Economic Co-Operation and Development (OECD), Paris.

Oreskes, N., Shrader-Frechette, K., and Belitz, K. (1994), "Verification, Validation, and Confirmation of Numerical Models in the Earth Sciences," *Science*, Vol. 263, 4 February 1994, pp. 641–646.

Patankar, S. V. (1980), *Numerical Heat Transfer and Fluid Flow*, McGraw-Hill, New York, 1980.

Porter, J. L. (1996), "A Summary/Overview of Selected Computational Fluid Dynamics (CFD) Code Validation/Calibration Activities," AIAA Paper 96–2053, 27th AIAA Fluid Dynamics Conference, New Orleans, Louisiana, June 17–20, 1996.

Puttre, M. (1994), "Verifying Analytical Models," *Mechanical Engineering*, Vol. 116, No. 11, November 1994, pp. 70–74.

Richtmyer, R. D. and Morton, K. W. (1967), *Difference Methods for Initial-Value Problems*, Interscience Publishers, New York.

Rizzi, A. and Vos, J. (1996), "Towards Establishing Credibility in CFD Simulations," AIAA Paper 96–2029, 27th AIAA Fluid Dynamics Conference, New Orleans, Louisiana, June 17–20, 1996.

Rizzo, A. R. (1991), "Estimating Errors in FE Analysis," *Mechanical Engineering*, Vol. 113, No. 5, May 1991, pp. 61–63.

Roache, P. J. and Mueller, T. J. (1970), "Numerical Solutions of Laminar Separated Flows," *AIAA Journal*, Vol. 8, No. 3, March 1970, pp. 530–538. See also AIAA Paper 68–741, June 1968.

Roache, Patrick J. (1972), *Computational Fluid Dynamics*, First Edition, Hermosa Publishers, P. O. Box 9110, Albuquerque, New Mexico 87119-9110, USA. http://www.hermosa-pub.com/hermosa. Revised Printing 1976. See also Roache (1998).

Roache, P. J. and Steinberg, S. (1984), "Symbolic Manipulation and Computational Fluid Dynamics," *AIAA Journal*, Vol. 22, October 1984, pp. 1390–1394. See also AIAA Paper No. 83–1952.

Roache, P. J., Ghia, K. and White, F. (1986), "Editorial Policy Statement on the Control of Numerical Accuracy," *ASME Journal of Fluids Engineering*, Vol. 108, No. 1, March 1986, p. 2.

Roache, P. J., Knupp, P. M., Steinberg, S. and Blaine, R. L. (1990), "Experience with Benchmark Test Cases for Groundwater Flow," ASME FED–Vol. 93, *Benchmark Test Cases for Computational Fluid Dynamics*, I. Celik and C. J. Freitas, eds., Book No. H00598–1990, pp. 49–56.

Roache, P. J. (1990), "Need for Control of Numerical Accuracy," *Journal of Spacecraft and Rockets*, Vol. 27, No. 2, March–April 1990, pp. 98–102. See also, AIAA Paper No. 89–1669.

Roache, P. J. (1994), "Response: To the Comments by Drs. W. Shyy and M. Sindir," *ASME Journal of Fluids Engineering*, Vol. 116, No. 2, June 1994, pp. 198–199.

Roache, P. J. (1995), "Verification of Codes and Calculations," AIAA Paper No. 95–2224, 26th AIAA Fluid Dynamics Conference, San Diego, California, June 19–23, 1995. See also AIAA Journal, May 1998, to appear.

Roache, P. J., Blaine, R. L., and Baker, B. (1996), "SECOFL2D User's Manual," Appendix I of "WIPP PA User's Manual for SECOFL2D, Version 3.03," WPO # 37271, Sandia National Laboratories, Albuquerque, New Mexico, May 7, 1996.

Roache, P. J. (1997), "Quantification of Uncertainty in Computational Fluid Dynamics," *Annual Review of Fluid Mechanics*, Vol. 29, 1997, pp. 123–160.

Roache, P. J. (1998), *Fundamentals of Computational Fluid Dynamics*, Hermosa Publishers, P. O. Box 9110, Albuquerque, New Mexico 87119-9110, USA. http://www.hermosa-pub.com/hermosa. To appear.

Muller, A. (1994), *Proceedings of the Basel World CFD User Days 1994*, Second World Conference in Applied Computational Fluid Dynamics, Basel, Switzerland, 1–5 May 1994.

Roberts, L. (1988), in Round Table Discussion in AGARD (1988).

Rumsey, C. L. and Anderson, W. K. (1988), "Parametric Study of Grid Size, Time Step, and Turbulence Modeling on Navier-Stokes Computations Over Airfoils," Paper 5 in AGARD (1988).

Rykiel, E. J. Jr. (1994), in "The Meaning of Models," Letters, *Science*, Vol. 264, 15 April 1994, pp. 330–331.

Sani, R. L. and Gresho, P. M. (1994), "Resume' and Remarks on the Open Boundary Condition Minisymposium," *International Journal for Numerical Methods in Fluids*, Vol. 18, 1994, pp. 983–1008.

Schlesinger, S. (1979), Terminology for Model Credibility," *Simulation*, Vol. 32, No. 3, pp. 103–104.

Sengupta, R., Orkwis, P. D., and Davis, S. M. (1995), "Unsteady Solutions, Solution Transients and Local Grid Resolution Requirements for Asymmetric Vortex Formation," AIAA Paper 95–0364, AIAA 33rd Aerospace Sciences Meeting, Reno, Nevada, January 9–12, 1995.

Serrano, S. E. (1995), "Analytical Solutions of the Nonlinear Groundwater Flow Equation in Unconfined Aquifers and the Effect of Heterogeneity," *Water Resources Research*, Vol. 31, No. 11, November 1995, pp. 2733–2742.

Sindir, M. M., Barson, S. L., Chan, D. C., and Lin, W. H. (1996), "On the Development and Demonstration of a Code Validation Process for Industrial Applications," AIAA Paper 96–2032, 27 AIAA Fluid Dynamics Conference, New Orleans, Louisiana, June 17–20, 1996.

Steinberg, S. L. and Roache, P. J. (1985), "Symbolic Manipulation and Computational Fluid Dynamics," *Journal of Computational Physics*, Vol. 57, pp. 251–284.

Stephens, A. B and Shubin, G. R. (1981), "Multiple Solutions and Bifurcation of Finite Difference Approximations to Some Steady Problems in Fluid Dynamics," *SIAM Journal of Scientific and Statistical Computing*, Vol. 2, No. 4, 1981, pp. 404–415.

Sudicky, E. A. and Frind, E. O. (1982), "Contaminant Transport in Fractured Porous Media: Analytical Solutions for a System of Parallel Fractures," *Water Resources Research*, Vol. 18, No. 6, December 1982, pp. 1634–1642.

Thomas, J. and Salas, M. (1985), "Far-Field Boundary Conditions for Transonic Lifting Solutions to the Euler Equations," AIAA Paper 85–0020.

Tsang, C.-F. (1991), "The Modeling Process and Model Validation," *Ground Water*, November–December 1991, pp. 825–831.

Van Dyke, M. (1975), *Perturbation Methods in Fluid Dynamics*, Parabolic Press, Stanford, California.

Vanka, P. (1995), "Comment on 'Toward a Numerical Error Bar in CFD'," *ASME Journal of Fluids Engineering*, Vol. 117, No. 1, 1995, pp. 9–10.

Wilcox, D. C. (1995), *Perturbation Methods in the Computer Age*, DCW Industries, La Cañada, California.

Williams, P. T. and Baker, A. J. (1996), "Incompressible Computational Fluid Dynamics and the Continuity Constraint Method for the Three-Dimensional Navier-Stokes Equations," *Numerical Heat Transfer, Part B*, Vol. 29, 1996, pp. 137–273.

WIPP PA Dept. (1992), *Annual Performance Assessment for the Waste Isolation Pilot Plant*, Vol. 1, Preliminary Comparison with 40 CFR 191, Subpart B, Vol. 2: Technical Basis, Performance Assessment Department, Sandia National Laboratories, Albuquerque, New Mexico, December 1992.

Zha, G-C. and Knight, D. (1996), "Three-Dimensional Shock/Boundary-Layer Interaction Using Reynolds Stress Equation Turbulence Model," *AIAA Journal*, Vol. 34, No. 7, July 1996, pp. 1313–1321.

Zingg, D. W. (1992), Grid Studies for Thin-Layer Navier-Stokes Computations of Airfoil Flowfield," *AIAA Journal*, Vol. 30, No. 10, October 1992, pp. 2561–2564.

PART II
VERIFICATION

Part II will cover the principle emphasis of this book, procedures for the rigorous accuracy Verification of Codes and for Verification of individual calculations. These two activities are distinct. The code must be convincingly Verified before Verification of an individual calculation can proceed with any confidence. Verification of a calculation involves error *estimation* or banding, whereas Verification of a code involves error *evaluation* from a known solution. Included in the Verification of calculations will be a taxonomy for sources of additional information needed for the error estimation.

Since this Part II concerns accuracy Verification only, we will not be concerned herein with the accuracy of physical laws but only with mathematics. Part III will cover Validation, i.e., the agreement of the mathematics with science.

3

A METHODOLOGY FOR ACCURACY VERIFICATION OF CODES: THE METHOD OF MANUFACTURED SOLUTIONS

3.1 INTRODUCTION

This chapter is based primarily on the methodology I presented in Steinberg and Roache (1985), "Symbolic Manipulation and Computational Fluid Dynamics" and expanded in Roache et al (1990). The methodology provides for convincing, rigorous Verification of the numerical accuracy of a code via systematic grid convergence testing. This procedure is straightforward though somewhat tedious to apply, and verifies all accuracy aspects of the code: formulation of the discrete equations (interior and boundary conditions) and their order of accuracy, the accuracy of the solution procedure, and the user instructions.

3.2 WARNINGS: THE DIVISION OF LABOR IN CODE DEVELOPMENT AND USE

As complete as this may sound, a rigorous Verification of the numerical accuracy of a code does *not* by any means imply that all numerical questions about a particular *simulation* are finished. Far from it. To keep the distinction in mind, it is helpful to think of the *division of labor* or division of responsibility, with a code builder Verifying the code with no knowledge of the application. If this sounds strange, realize that there are a multitude of applications possible for a code. I want to Verify my code, without waiting to see how many applications will develop.

I build a code: what can I claim about it, and communicate to you the possible user, without knowing anything about your intended application? I can tell you what equations my code solves. I cannot tell you what equations you need to solve for your problem. I can tell you a theoretical order of convergence for my code, and the observed order of convergence of my code for one or more well-behaved problems, and what grid refinement level was sufficient to attain asymptotic performance on those well-behaved problems. I cannot tell you what grid will be required for your problem, and (unfortunately) I cannot guarantee that the observed order of convergence for your problem will be the same as that observed on my well-behaved problems. You as modeler and code user will have to decide if the equations that I solve are appropriate for your problem. You will have to decide whether my Verified order of convergence on well-behaved problems can be trusted for your problem, or whether you will Verify the rate of convergence on your problem. In either case, you cannot simply perform a single grid calculation, but rather you must Verify the calculation for your problem, which involves error estimation and/or banding, as discussed in Chapters 4 through 8.

We are concerned in this Chapter 3 with accuracy Verification of the *code* only. Strictly speaking, this will involve error *evaluation* rather than error *estimation*. However, the Verification of the *order of convergence* (or the *rate of convergence*) of the code involves the same definition of "order" that is at the basis of error estimation. Thus, there are many connections, but as usual, it will be useful to maintain the distinctions, this time between error *evaluation* and error *estimation*.

3.3 ORDER OF CONVERGENCE

The principle definition of "order of convergence" that we use here is based on behavior of the error of the discrete solution. There are various measures of error, but in some sense we are always referring to the difference between the discrete solution $f(\Delta)$ (or a functional of the solution) and the exact (continuum) solution,

$$E = f(\Delta) - f^{exact} \tag{3.3.1}$$

For an order p method, and for a well-behaved problem (exceptions to be discussed in Chapters 6 and 8), the error in the solution E asymptotically will be proportional to Δ^p, where Δ is the grid spacing or other measure of the discretization. This terminology applies to every "consistent" methodology: finite difference methods (FDM), finite volume methods (FVM), finite element methods (FEM), spectral, pseudo-spectral, vortex-in-cell, etc., regardless of solution smoothness. By "consistent" we mean, of course, that the continuum equations are recovered in the limit of infinite discretization. ("Lattice gas" methods may or may not qualify as consistent, and therefore are highly suspect.) Thus,

$$E = f(\Delta) - f^{exact} = C \cdot \Delta^p + \text{H.O.T.} \tag{3.3.2}$$

where H.O.T. are higher order terms. For smooth problems, it may be possible in principle to evaluate the coefficient *C* and the H.O.T. from the continuum solution, but as a practical matter, we do not do this in the accuracy Verification procedure.

3.4 A GENERAL PROCEDURE FOR GENERATING AN ANALYTICAL SOLUTION FOR CODE ACCURACY VERIFICATION: THE METHOD OF MANUFACTURED SOLUTIONS

The well-known procedure for grid convergence testing described earlier in Chapter 1 is applicable only when an exact solution is known. In the anecdote cited there, a known exact solution for electrode potential fields was very useful in uncovering a previously unsuspected code bug. However, the procedure Verified the code only for those features exercised for the analytical solution. Thus, the code accuracy Verification was limited to the vacuum electrode case; the solution was for linear steady state only and constant properties, whereas the code also treated the more general problem.

It has often been stated in research journal articles that general accuracy Verification of codes for difficult problems, e.g. the full Navier-Stokes equations of fluid dynamics, is not possible because exact solutions exist only for relatively simple problems that do not fully exercise the code. Many papers and reports approach accuracy Verification of codes in a haphazard way, comparing single-grid results for a few exact solutions on problems of reduced complexity. In fact, a very general procedure exists (Steinberg and Roache, 1985) for generating analytical solutions for code accuracy Verification.

The basic idea of the procedure is to simply *manufacture* an exact solution, without being concerned about its physical realism. The "realism" or lack thereof has nothing to do with the mathematics, and Verification is a *purely mathematical* exercise. In the original, most straightforward and most universally applicable version of the method , one simply includes in the code a general source term, $Q(x, y, z, t)$ and uses it to generate a non-trivial but known solution structure. We follow the counsel of G. Polya (1957):

Only a fool starts at the beginning; the wise one starts at the end.

We first pick a continuum *solution*. Interestingly enough, we can pick a solution virtually independent of the code or the hosted equations. That is, we can pick a solution, then use it to verify an incompressible Navier-Stokes code, a Darcy flow in porous media code, a heat conduction code, an electrode design code, a materials code,...("Pick a card–any card!")

We want a solution that is non-trivial but analytic, and that exercises all ordered derivatives in the error expansion and all terms, e.g., cross-derivative terms. For example, chose a solution involving the hyperbolic tangent function *tanh*. This solution also defines boundary conditions, to be applied in any (all) forms, i.e., Dirichlet, Neuman, Robin, etc. *Then the solution is passed through the governing PDE's to give the production term $Q(x, y, z, t)$ that produces this solution.* In Steinberg and Roache (1985) we used Symbolic Manipulation to generate Q; the detailed results will be given shortly.

This procedure is much easier and more general than looking for solutions to real problems (see additional references in Roache et al, 1990.) We then monitor the numerical error as the grid is systematically refined. Successive grid *halving* is not required, just *refinement*. Thorough iterative convergence is required; inadequate iterative convergence will corrupt the results. Theoretically [from Equation (3.3.2)], values of $C = error/\Delta^p$ should become constant as the grid is refined for a uniformly *p-th* order method ("uniformly" implying at all points for all derivatives).

When this systematic grid convergence test is Verified (for all point-by-point values), we have verified

1. any equation transformations used (e.g., nonorthogonal boundary fitted coordinates),
2. the order of the discretization,
3. the encoding of the discretization, and
4. the matrix solution procedure.

This technique was originally applied in Steinberg and Roache (1985) to long Fortran code produced by Artificial Intelligence (Symbol Manipulation) methods. (See also Roache and Steinberg, 1984; Steinberg and Roache, 1986.) The first versions of the code produced extremely long subroutines because the Symbol Manipulation code Macsyma at that time did not know the rules for intermediate expressions in the chain rule expansions and for the derivative of the inverse of a matrix function. (S. Steinberg later installed these rules in Macsyma; see Steinberg and Roache, 1985.) The original three-dimensional nonorthogonal coordinate code contained about 1800 lines of dense Fortran. It would be impossible to check this by reading the source code, yet the procedure described Verified the code convincingly. (Surprisingly, roundoff error was not a problem.)

The arbitrary solution, produced inversely by the specification of the source term Q, is aptly described by Oberkampf et al (1995) and Haynes et al (1996) as a "Manufactured Solution." The fact that the Manufactured Solution may bear no relation to any physical problem does not affect the rigor of the accuracy Verification. The only important point is that the solution (manufactured or otherwise) be non-trivial, i.e., that it exercise all the terms in the error expansion. (Ethier and Steinman, 1994, follow a similar philosophy and procedure.) The algebraic complexity may be something of a difficulty, but is not insurmountable, and in practice has been easily handled using Symbolic Manipulation packages like Macsyma, Mathematica, Maple, etc. Using the source-code (Fortran) writing capability of Macsyma, it is not even necessary for the analyst to look at the form of Q; rather, the specification of the solution (e.g., *tanh* function) to the Symbolic Manipulation code results in some possibly complicated analytical expression that can be directly converted by the Symbolic Manipulation code to a

Fortran (or Pascal, C, etc.) source code segment, which is then readily emplaced in a source code module (subroutine, function, etc.) that then is called in the accuracy Verification procedure. (This "emplacement" can be performed by hand by the analyst, without actually reading the complicated source code expressions, or can itself be automated in the Symbolic Manipulation code.)

The procedure may also be applied to systems of equations, with separate Q's generated for each equation. Both unsteady and steady solutions are possible. (It may be useful to avoid exponential solution growth in time so as to avoid confusion with instabilities; e.g., see the fully three-dimensional incompressible Navier-Stokes analytical solutions of Ethier and Steinman, 1994.) Nonlinearity is an issue only because of uniqueness questions; otherwise, the source term complexity may be worse, but that is the job of the Symbolic Manipulation code. Non-uniqueness could be an issue because the code could converge to another legitimate solution other than the Manufactured Solution, producing a false-negative accuracy Verification test for a correct code. However, it would be difficult to contrive a situation in which a false positive accuracy Verification was obtained. In Steinberg and Roache (1985), we applied the procedure to the nonlinear (quasi-linear) PDE's of the elliptic grid generation method for non-orthogonal coordinates (Thompson et al, 1974). The "Manufactured Solution" here was an analytical three-dimensional coordinate transformation; see examples below.

The only disadvantage of the procedure is the requirement that the code being Verified must include accurate treatment of a source term and that its boundary condition values not be hard-wired. Many codes are built with source terms included, and many algorithms allow trivial extension to include Q's. However, in directionally-split algorithms such as Approximate Factorization (Briley and McDonald, 1977; Beam and Warming, 1976) the time-accurate treatment of $Q(x, y, z, t)$ involves subtleties and complexities at boundaries, especially for non-orthogonal coordinates (e.g. see Salari and Roache, 1990). Thus, significant code extensions may be required in order to apply this procedure involving Manufactured Solutions for code accuracy Verification. Likewise, some old groundwater flow codes were built with hard-wired homogeneous Neumann boundary conditions, $\partial f/\partial n = 0$. In order to use an arbitrary solution function, non-homogeneous boundary values like, $\partial f/\partial n = g$ would be required. (The Q could be modified to set g.)Alternately, one could restrict the choice of Manufactured Solution functions to fit the hard-wired values. Likewise, to test periodic boundary conditions, one must chose a periodic function for the Manufactured Solution.

Note that it is not necessary to develop different Manufactured Solutions to test different boundary condition *types*. The same Manufactured Solution can be evaluated at boundaries to produce function values to test Dirichlet boundary condition options, normal gradient values to test Neumann boundary condition options, and both for mixed (Robin) boundary condition options, with an arbitrary non-zero weighting function β in

$$f + \beta \cdot \frac{\partial f}{\partial n} = g \tag{3.4.1}$$

Similarly for angle conditions encountered in linearized potential flow and magnetohydrodynamics,

$$\frac{\partial f}{\partial x} = g\frac{\partial f}{\partial y} \tag{3.4.2}$$

Note that the method will Verify convergence rate of the solution, but will not Verify that a code is algebraically conservative (i.e., "strongly conservative" so that mass is identically conserved at all grid resolutions).

3.5 EXAMPLE: VERIFICATION OF A 3-D POISSON EQUATION CODE IN NONORTHOGONAL GRID AND A 3-D GRID GENERATION CODE

The following examples are taken almost verbatim from my original work in Steinberg and Roache (1985), pp. 277–283, with some repetition of material presented earlier in the chapter. (One departure from the "verbatim" use is that I used "Validation" instead of "Verification" in 1985.) The codes being tested had been generated by Prof. Steinberg using an early VAX computer version of the Symbolic Manipulation code MACSYMA.

3.5.1 Verification of Code Generated by Symbolic Manipulation

The potential for errors in either the problem formulation or the encoding procedure always exists in complex codes such as three dimensional boundary fitted coordinate codes. The need for Code Verification was emphasized in the present work because Symbolic Manipulation was used to generate the codes (e.g., see Roache and Steinberg, 1984, Steinberg and Roache, 1985, 1986a,b). The resulting "psychological distance" from the work made it less likely to be satisfied with superficial plausibility exercises based on intuitive ideas of acceptable levels of absolute error.

We Verify the codes by performing numerical tests. The approach is to choose a (manufactured) continuum solution for the class of problems treated by the code, and to Verify the convergence of the results to this continuum solution by systematic discretization error convergence testing over a sequence of grid sizes.

3.5.2 Hosted Equation Convergence Testing Method

To test the equation $L(F) = Q$, we tested a modified equation whose manufactured solution was chosen to be

$$F_m = \left(x_1^{m+1} x_2^{m+2} x_3^{m+3}\right) / F_{mn} \quad \text{on} \quad R \tag{3.5.2.1}$$

where

$$m = nl + na - 1 \tag{3.5.2.2}$$

and nl = degree of L (e.g., $nl = 2$ for L = Laplacian) and na = order of accuracy of the finite-difference expression (e.g., $na = 2$ for centered differences). F_{mn} is the normalizing value for the manufactured solution F_m. The motivation in the selected form of the solution is of course to insure that all the derivative terms in L are exercised, and that there is non-zero truncation error for finite h_i, even without the transformation of coordinates. (For example, a parabolic solution will show no truncation error using 2nd-order accurate solutions with the identity transformation. Several published "Verifications" of the accuracy of upwind differencing are inadequate because the chosen solution structure does not exercise the meaningful terms in the truncation error, giving a false indication of accuracy.)

The modified problem to be tested was then

$$L(F) = Qs \quad \text{on} \quad R \tag{3.5.2.3}$$

where

$$Qs = Q + L(F_m). \tag{3.5.2.4}$$

The boundary equations for the modified problem are

$$F = F_m \quad \text{on boundary of} \quad R. \tag{3.5.2.5}$$

Qs can be obtained in x by elementary operations as

$$Qs = Q + F_m \left[\frac{m(m+1)}{x_1^2} + \frac{(m+1)(m+2)}{x_2^2} + \frac{(m+2)(m+3)}{x_3^2} \right]. \tag{3.5.2.6}$$

The solution was obtained in a stretched coordinate system $\xi = (\xi_1, \xi_2, \xi_3)$ where the ξ_i values are linear from 0 to 1, $\xi_1 = h1 * (i - 1)$, etc. The chosen coordinate transformation was given by

$$x_i = \xi_s + \xi_i + \tanh(d_i \xi_1 \xi_2 \xi_3) \tag{3.5.2.7}$$

ξ_s is a shift, necessary to avoid a singularity in the chosen solution at the origin. The parameters d_i control the severity of the coordinate stretching; for $d_i = 0$, there is no stretch in x_i direction. The form of the coordinate transformation is chosen to exercise all terms in the transformed equations, notably the cross-derivative terms, which would be zero if the coordinate transformations were independent in each direction. For non-zero d_i the *tanh* term in the transformation assures non-zero values for all derivatives.

The above solution and transformation generate non-trivial (non-zero and non-repeating or canceling) derivatives of all orders, including cross-derivatives.

The discrete solution F_d is then obtained on a family of grids, using any standard method. We used the spatial marching methods in the GEM codes (Roache, 1995) for 2-D, and hopscotch SOR in 3-D. It is important that iterative convergence criteria be stringent, so as not to confuse the incomplete iteration (residual) error with the discretization error $te = F_m - F_d$. We required iterative convergence to essentially the single precision accuracy of the 32-bit VAX computer used, about 7–8 decimal figures. We also forced a minimum number of iterations to be performed, somewhat greater than the maximum grid index, to be sure that a false indicator of iterative convergence was not obtained due to good initial conditions available at the fine grid spacing.

The discretization error convergence is then monitored as the grid is systematically refined. Theoretically, the local values of $p = te/h^2$ should become constant as the grid size is refined. For a uniformly na-th order finite difference discretization ("uniformly" implying at all points for all derivatives) the truncation error can be written as

$$te = C_1 h_1^{na} + C_2 h_2^{na} + C_3 h_3^{na} + \text{H.O.T.} \tag{3.5.2.8}$$

We define the index I of predicted order of discretization error

$$I = N^{na}. \tag{3.5.2.9}$$

With $h_1 = \xi_{1\max} / N = 1/N$, etc. we have

$$I = C_1 \xi_{1\max}^{na} + C_2 \xi_{1\max}^{na} + C_3 \xi_{1\max}^{na} + \text{H.O.T.} \tag{3.5.2.10}$$

which is constant to order na approximation.

We then monitor I as N increases, using I at specified points (e.g., Icr at the center of the grid) and $Imx = \max_{i,j,k} I$. If all $I = constant$ as N increases, we have Verified (1) the equation transformation, (2) the order of finite difference expressions, and (incidentally) (3) the solution procedure.

This procedure does not require a grid halving sequence, $N = 2^m$, although this sequence does make the judgment easier. Also, we note from experience, even on trivial 1-D problems, that I = "constant" as the grid is halved can be significantly relaxed; a 5% variation is very safe. (The results presented herein are more convincing, as will be seen.)

For a coarse-mesh solution, it would be advantageous to eliminate the iterative residual from the calculation of te by using a direct method, i.e., Gaussian elimination. However, in 3-D this will fail for large N by round-off error and/or CPU time demands; likewise for the marching methods. Iterative methods are dangerous in that their convergence depends on L, and on the coordinate transformation. Note, however, that it is not necessary, nor even advisable, to test the small-parameter (high Reynolds number) problem. Errors will show up more readily when all coefficients in L are of the same order. Once the above procedure has Verified the algebra and coding accuracy, one can proceed with alternate solutions techniques for the small-parameter problem.

Note that this procedure can be followed for mixed-order finite difference equations (e.g., upwind differencing on first derivative terms and 2nd-order differencing on other terms) with the modification that we look only for convergence in the sense of zero error as N increases. This modified procedure will *not* Verify the mixed orders of accuracy separately. Ultimately the convergence rate will be limited by the lowest order discretization used, i.e., 1st-order, which will be Verified by the procedure. However the procedure will not Verify that the non-advection terms were correctly discretized to 2nd-order. Also, mixed-order codes will probably require finer grids since the convergence of these is difficult to judge.

3.5.3 Hosted Equation Convergence Results in 3-D

Table 3.5.3.1 presents the results of testing for the constant-coefficient Laplacian equation in 3-D, using moderate stretching parameters $d_i = 0.1$. The mesh size was successively halved three times, from 5^3 to 3^{33}. The numerically calculated Jacobian for the 5^3 grid ranged from 1.0187 to 1.1684, and for the 3^{33} grid ranged from 1.0003 to 1.2792. The table presents the maximum discretization error over the grid, *temax*, and the grid location at which it occurred; the discretization error at the center of the grid, *tectr*; and the corresponding values $Imx = temax/h^2$ and $Ier = tectr/h^2$.

The near-constancy of the coefficient $Imx = temax/h^2$ indicates that the entire solution, including the coordinate transformation and the finite difference expressions, is 2nd-order accurate. The erratic behavior of *Icr* at the finest grid indicates round-off error problems with the short word-length computer used.

Table 3.5.3.2 presents results of testing for the variable-coefficient Laplacian equation in 3-D, using both moderate stretching $d_i = 0.1$ and strong stretching $d_i = 10$. The hosted equation solved was

$$\nabla\sigma \cdot \nabla F = 0 \tag{3.5.3.1}$$

$$\sigma = \sigma_0 + \sigma_m \sin(b) \tag{3.5.3.2}$$

grid	*temax*	*at*	*tectr*	*Imx*	*Ier*
5^3	0.402E–03	4,4,4	0.832E–05	0.646E–02	0.133E–03
9^3	0.172E–03	8,8,8	0.650E–06	0.111E–01	0.416E–04
17^3	0.492E–04	15,15,15	0.340E–07	0.126E–01	0.870E–05
33^3	0.134E–04	29,29,29	–0.169E–06	0.137E–01	–0.173E–03

Table 3.5.3.1 3-D Hosted Equation Convergence Testing for $\nabla^2 F = 0$. The Stretching Parameters $d_i = 0.1$. (From Table I, Steinberg and Roache, 1985.)

	grid =	5^3	9^3	17^3	33^3
di = 0.1	*Imx* =	0.390E–02	0.877E–02	1.00E–02	–
di= 10.0	*Imx* =	.433	.691	1.21	1.53

Table 3.5.3.2 3-D Hosted Equation Convergence Testing for $\nabla \cdot \sigma \nabla F = 0$. (From Table II, Steinberg and Roache, 1985.)

$$b = x_1^n x_2^n x_3^n \tag{3.5.3.3}$$

$$x_{1n} = \frac{x_1 - x_{1\min}}{x_{1\max} - x_{1\min}}, \text{ etc.} \tag{3.5.3.4}$$

The code, of necessity, used the expanded (non-conservation form),

$$\sigma \nabla^2 F + \nabla \sigma \cdot \nabla F = 0. \tag{3.5.3.5}$$

(In 1985, Prof. Steinberg had not yet developed symbolic manipulation codes with the ability to retain the unexpanded or conservation form of the equation.)

The manufactured solution chosen was

$$F_m = \frac{1}{F_{mn}} \left(x_1^{m+1} x_2^{m+2} x_3^{m+3} \right) \tag{3.5.3.6}$$

giving the forcing term

$$\begin{aligned} Qs = \frac{1}{F_{mn}} \Big[& \sigma m(m+1) x_1^{m-1} x_2^{m+2} x_3^{m+3} + \sigma(m+1)(m+2) x_1^{m+1} x_2^{m} x_3^{m+3} \\ & + \sigma(m+2)(m+3) x_1^{m+1} x_2^{m+2} x_3^{m+1} + \sigma_x (m+1) x_1^{m} x_2^{m+2} x_3^{m+3} \\ & + \sigma_y (m+2) x_1^{m+1} x_2^{m+1} x_3^{m+3} + \sigma_z (m+3) x_1^{m+1} x_2^{m+2} x_3^{m+2} \Big] \end{aligned} \tag{3.5.3.7}$$

with

$$\sigma_x = \frac{\sigma_m \cos(b) x_2^n x_3^n}{bd} \tag{3.5.3.8}$$

$$\sigma_y = \frac{\sigma_m \cos(b) x_1^n x_3^n}{bd} \tag{3.5.3.9}$$

$$\sigma_z = \frac{\sigma_m \cos(b) x_1^n x_2^n}{bd} \tag{3.5.3.10}$$

$$bd = \left(x_{1\max} - x_{1\min}\right)\left(x_{2\max} - x_{2\min}\right)\left(x_{3\max} - x_{3\min}\right) \qquad (3.5.3.11)$$

The results for a Verification test using $\sigma_m / \sigma_0 = 0.1$ are shown in Table 3.5.3.2. The results for *Imx* show that the Code Verification occurs even at the grid 17^3 for moderate stretching with $d_i = 0.1$, but that there is some variation in *Imx* even at the grid 33^3 for strong stretching with $d_i = 10$. The convergence is clear, however.

3.5.4 Hosted Equation Convergence Results for Strong Stretching

The behavior of the discretization error for the strong stretching is of interest. The values of *temax* for just the coarse grid are plotted in Table 3.5.4.1 for a range of stretching parameters d_i.

For this coarse grid, $d_i = 0.1$ minimizes the maximum discretization error. The table shows that large and inappropriate stretching values do indeed increase the discretization error by almost two orders of magnitude compared to the best transformation used. However, the table shows that the results are still $O(h^2)$ accurate. By definition, we mean that the method is 2nd-order accurate if a reduction in h by a factor of 1/2 will (asymptotically) reduce the discretization error by a factor of 1/4.

We also Verified this retention of 2nd-order accuracy for *very* strong 1-D coordinate stretching. Using double precision calculations, we experimented with a range of stretching parameters for reasonable coordinate transformations based on hyperbolic tangent and exponential functions, and even for *un*reasonable transformations based on exponentials of exponentials. In the extreme case, we used a transformation of

$$x = \xi + \frac{\exp\left(d1 * \left(f(xi) - 1\right)\right) - \exp(-d1)}{1 - \exp(d1)} \qquad (3.5.4.1)$$

where

$$f(x_i) = \frac{\exp\left(d1 * (xi - 1)\right) - \exp(-d1)}{1 - \exp(d1)} \qquad (3.5.4.1)$$

with $d1$ ranging from 0 to 100. For $d1 = 10$, *Imx* = *temax*/*n***2 reaches its asymptotic value of 225 to two significant figures at $n = 257$. For $d1 = 100$, the truncation error is 6 orders of

dj =	0.	0.1	0.5	1.0	10.0
temax =	0.830E–03	0.244E–03	0.489E–02	0.737E–02	0.271E–01

Table 3.5.4.1 3-D Hosted Equation Convergence Testing for $\nabla \cdot \sigma \nabla F = 0$ in Coarse Grid = 5^3.
(From Table III, Steinberg and Roache, 1985.)

magnitude higher than the no transformation case; nevertheless, *Imx* is constant to at least the first figure at $n = 8193$ and 16385.

These results are at variance with analyses which focus on the coefficients of the Taylor series terms in the original ("physical") independent variables. The present results also Verify the claim (in Roache, 1972) that an analytical transformation of the equations does not change the order of the accuracy. If the coefficients of the transformation are also evaluated to $O(h^2)$ accuracy, as well as the hosted equation, then the overall results will be $O(h^2)$ accurate.

3.5.5 Grid Generation Results in 3-D

A similar procedure was followed to Verify the grid generation codes. The grid generation method (Thompson et al, 1974) involves solving a set of three elliptic equations for the new coordinates $\xi(\mathbf{x})$, written in the original ("physical") coordinates as

$$L(\xi_i) = P_i \tag{3.5.5.1}$$

where L is the Laplacian operator in $\mathbf{x}$, and the P_i may be chosen to give some control of grid position in the interior. Since the system is transformed to ξ space, resulting in a set of three coupled nonlinear (quasilinear) equations, the Code Verification procedure is more complicated. We chose the manufactured "solution" grid to be the transformation

$$\xi_i = x_i + \tanh(d_i x_1 x_2 x_3) \tag{3.5.5.2}$$

which is the obverse of the transformation used for the hosted equation testing above. (Note: when discrete grid generation equations are taken to the limit of $\Delta \to 0$, the result is a "continuum grid," i.e. a transformation or parameterization.) Since the ξ_i are known at each grid point, being selected as just linear variables in the grid indexes, the above equation must be solved for the inverse values of x_1, x_2, x_3 at each grid point. This nonlinear 3×3 system was solved by coupled 3×3 Newton-Raphson iteration at each grid point to establish the continuum grid solution. Again, tight convergence criteria were used, assuring accuracy to essentially the single precision of the machine.

The P_i necessary to produce this transformation are then obtained by operating on the solution ξ with the Laplacian L, giving

$$P_i = -2d_i^2\left[x_2^2\left(x_3^2 + x_1^2\right) + x_1^2 x_3^2\right]\operatorname{sech}^2(d_i x_1 x_2 x_3)\tanh(d_i x_1 x_2 x_3). \tag{3.5.5.3}$$

The numerical solution was obtained using hopscotch SOR to solve linearized equations sequentially for x_1, x_2, and x_3, followed by Picard outer iterations to update the linearized coefficients. A relaxation factor of 1/2 was used in the outer (nonlinear) iterations, and loose iterative convergence criteria (actually, a limit on maximum number of inner iterations) were used in the early nonlinear stages. Again, overall iterative convergence was tight, essentially to

grid =	5^3	9^3
temax =	0.394E–02	0.205E–02

Table 3.5.5.1 Grid Generation Tests, d_i = 3, with an Error in Coding in the Test Driver. (From Table IV, Steinberg and Roache, 1985.)

	grid =	5^3	9^3	17^3	33^3
original	*temax* ∗ 100=	0.394	0.205	0.169	0.160
corrected	*temax* ∗ 100 =	0.304	0.0821	0.0208	0.00533
original	*Imx* =	0.0630	0.131	0.431	1.646
corrected	*Imx* =	0.0486	0.0525	0.0533	0.0545

Table 3.5.5.2 Complete Grid Generation Tests, d_i = 3. (From Table V, Steinberg and Roache, 1985.)

the single precision of the machine, to clarify the discretization error behavior. The cost of solving the nonlinear 3×3 system for the grid test was more expensive than solving the scalar linear equation for the hosted equation test, but was still reasonably obtained because the exact continuum solution was available for use as initial conditions.

The result for an early test in two grids is given in Table 3.5.5.1 for strong stretching values $d_i = 3$. The values of *temax* seem small, and in fact decreased by almost a factor of 2 with the mesh size halving. This test could likely pass for a Code Verification. The test was actually inconclusive and, in fact, the test driver code was in error.

The false result is included here to emphasize that systematic convergence testing over a *sequence* of grids is required for Code Verification. The error was in fact discovered with such systematic testing, the results of which are shown in Table 3.5.5.2, for both the original and corrected test drivers.

The complete, systematic convergence test in Table 3.5.5.2 clearly indicate a persistent error in the original test, and clearly indicate 2nd-order convergence in the corrected test as *Imx* becomes virtually constant.

3.5.6 Discussion of the Code Verification Procedure Using Manufactured Solutions

We note that the Code Verification procedure can be complicated by the use of 1st-order differencing, since this slows convergence and makes it difficult to judge. Also, the use of mixed 1st- and 2nd-order differencing (boundary conditions, advection terms at high Reynolds numbers, etc.) and/or any sort of conditional differencing (e.g., upstream differencing) will complicate the Code Verification procedure. The technique is applicable to Verifying the hosted equation codes as long as the method of grid generation can systematically refine the grid, which is the case for elliptic grid generating systems themselves. Particularly, the technique is applicable to Verifying code for other grid generation techniques that provide control over the grid properties of smoothness and orthogonality (Knupp and Steinberg, 1993).

The cost of computer time to do the systematic convergence testing can be significant, especially in 3-D problems. However, it should be noted that the kind of Code Verification described here does not address the small parameter (high Reynolds number) problem, so the costs and difficulties associated with it do not contribute to the expense. In fact, in designing the test problems, one specifically avoids large or small parameters, since these could mask errors in the treatment of the negligible terms. Also, if the code and the Verification code are indeed correct, the use of the correct continuum solution as the initial condition for the iterative discrete solution will speed the testing process. (The test driver must require some minimal number of passes through an iterative solver, to avoid a possible false indication of iterative convergence.) Thus, the computer CPU time to perform these Code Verification computations in the finest grid will be much less than that required to obtain a realistic (unknown) problem solution in that grid, when iterative solution methods are used.

The results obtained strongly confirm that strong and inappropriate coordinate stretching can indeed increase the size of the discretization error, as is well known, but that the asymptotic order of the accuracy, as indicated by the reduction in discretization error resulting from systematic reduction in grid size, remains $O(\Delta^2)$ if centered differences are used in the transformed equations. Thus, the Method of Manufactured Solutions is not disabled by strong coordinate stretching.

3.5.7 Debugging with Manufactured Solutions

As described above, the Code Verification procedure simply gives a "yes" or "no" answer to the accuracy of the entire code: the transformation of the hosted equation(s), the substitution of the finite difference forms, the ensuing algebraic grouping, the Fortran encoding, the solution procedure for the discretized equations, and the correct formulation of the test problem. If the Code Verification procedure produces a negative result, it is conceivable that the only information obtained is that there is an error, *somewhere*. For example, an error in the formulation could destroy the diagonal dominance of the matrix equation, resulting in an unstable or extremely slowly converging numerical procedure.

In actual experience, the procedure has been readily modified, on an *ad hoc* basis, and has successfully helped in isolating and identifying coding mistakes. By judiciously building up or modifying the test problem, one can selectively turn off the cross-derivative terms, perform the trivial identity transformation, etc. Direct solvers for the hosted equation testing may also be used if iterative convergence is not attainable. We have found that printing and inspection of all values of the stencil at one internal grid point aided debugging and Code Verification. Likewise, Ethier and Steinman (1994) also found the use of their manufactured incompressible solutions to be very helpful for debugging purposes by analytically computing individual terms in the Navier-Stokes equations and comparing them with their numerical counterparts.

3.6 ANOTHER PATH TO MANUFACTURED SOLUTIONS

Another path to generating Manufactured Solutions is to chose a functional form of variable coefficients in the PDE's instead of a source term $Q(x, y, z, t)$ to generate a non-trivial solution. This approach may not be as general as that utilizing $Q(x, y, z, t)$, but it may also produce more realistic-looking solutions. Although this is not a mathematical advantage, it may provide a political advantage and inspire more confidence, depending on the mathematical sophistication of the audience (e.g., regulatory agencies, code QA committees, etc.).

Further examples of Accuracy Verification of Codes will be given in Chapter 6.

3.7 CODE VERIFICATION INCLUDING SHOCK WAVES

Without special considerations, the Method of Manufactured Solutions can be used for supersonic flow codes, or other PDE's in the smooth solution regime, but treatments for shocks (e.g., TVD, FCT, explicit dissipation methods) and other shock-like near-discontinuities may not be fully exercised. It appears to be possible to extend the method to include arbitrary shocks by including jumps in Q, but to my knowledge this has not been accomplished at the time of this writing. A straightforward application is simply to pick a solution with sufficiently steep gradients/curvatures to activate the shock algorithm at coarse grids, but the behavior of grid convergence may be difficult to judge.

However, a very similar approach, independently developed by J. M. Powers and colleagues (Powers and Stewart, 1992; Powers and Gonthier, 1992; Grismer and Powers, 1992, 1996), has been applied to detonation shocked flows thoroughly and successfully, although special considerations apply.

The simplest multidimensional shocked solution for accuracy Verification of codes is obtained by simply translating a one-dimensional normal shock solution tangentially to obtain the oblique shock solutions. For perfect gas dynamics, this solution exercises the shock-capturing algorithm, but little else, since the flow before and after the shock is uniform. One could then use a non-shocked flow to exercise and Verify treatment of other terms, and could perhaps claim code accuracy Verification. But it would be more convincing to have a single exact solution that exercised shocks and non-uniform flow in the same problem. Grismer and Powers (1996) obtain more structure in their solution to a detonation shock problem, since the solution is not simply uniform before and after the shock. In that problem, the one-dimensional solution is obtained numerically by integration of ordinary differential equations (ODE's). Since ODE solvers are so highly developed and reliable, this is considered to be an "exact solution," in the same way that the Blasius boundary layer solution, although numerical, has been considered to be a primitive or closed-form solution in fluid dynamics. (For example, if a solution to a compressible flow problem can be expressed in terms of the incompressible Blasius solution, it is considered to be a closed-form analytical solution. After all, solutions in terms of exponential and trigonometric functions also require numerical evaluations; these are closed form solutions, though not algebraic solutions.) This solution, translated parallel to the shock, gives a two-dimensional solution with shock structure before and after the shock. However, cross-derivative terms are identically zero.

A two-dimensional detonation solution with more structure in the shock-free regions, but still containing a straight shock wave, was obtained by Powers and Gonthier (1992). Beginning with a straight shock, they solved backwards for the final shape of a wall that would generate the straight shock. This solution structure is more general and does contain cross derivatives in the non-shock regions. However, the shock itself is straight; curved shocks generate entropy-vorticity gradients downstream, and are a more demanding exercise of shock-capturing and shock-fitting algorithms.

The additional generality of weakly curved shocks is included in the asymptotic solutions of premixed reactive ideal-gas oblique detonations developed by Powers and Stewart (1992) and used as a numerical benchmark by Grismer and Powers (1992). Although strictly applicable only in the asymptotic limits of weak shock curvature and high free-stream Mach number M [the perturbation parameter is $\varepsilon = 1 / M^2$ and the asymptotic error is $O(\varepsilon^2)$], they provide a practical, reliable benchmark solution for attainable M. The solution for curved shocks and straight walls is *not* restricted to small wedge angles. (They also present an elegant solution for the somewhat simpler problem of straight shocks and curved walls.) The solution method follows the spirit of the pioneering work of Van Dyke (1958) on the supersonic blunt body problem. The solutions for the pressure and velocity fields are first parameterized by the shock position, and then boundary conditions at the surface of a wedge are chosen to fix the shock position. The solution procedure is non-trivial and the answer involves infinite series, but the work already has been done, and the series converges absolutely and uniformly. The solution provides a good Benchmark problem for curved shock flows with reactions. (The flow may properly be described as both a shock-induced combustion and an oblique detonation.)

Any remaining doubts about the applicability of an asymptotic solution at a finite M could also be addressed systematically, by obtaining code numerical solutions for a sequence of increasing M and extrapolating to the high M limit. Note again the important distinction between Verification and Validation in regard to the high M limit. The facts that the user may be interested in lower M, and that ideal gas assumptions are not *physically* valid at $M = 20$, or that the solutions may not be physically stable, are irrelevant to the issue of Verification of the code, just as we noted previously that the requirement for non-physically-realizable source terms in the Method of Manufactured Solutions is irrelevant. Verification is a *purely mathematical* exercise, so questions like applicability of perfect gas assumptions should be postponed until Validation exercises, which will presumably be performed at more reasonable M and will demarcate the limits of applicability.

If complete generality is deemed necessary, with curved shocks but without the asymptotic limitation on the solution, one could pursue the generation of a Manufactured Solution with general source terms, as noted earlier. But the work of Powers and colleagues provides reliable Benchmark solutions that allow convincing accuracy Verification of codes including the phenomena of shock waves.

3.8 NEED FOR A THEOREM

There seems to be no chance for a sweeping theorem proving correctness of PDE codes by this Method of Manufactured Solutions in any general sense. However, like any analysis, it seems that there could be a useful theorem for a properly defined and limited scope. For codes like the one described above, treating only the well behaved Poisson equation in general nonorthogonal coordinates, the exercise is compelling. I claim that this technique applied to such PDE codes (which class of codes I unfortunately cannot define with sufficient mathematical theorem-like precision) is correct, i.e., the numerical accuracy of the code is Verified, beyond a reasonable doubt.

3.9 SPECIFIC ANALYTICAL SOLUTIONS

Manufactured solutions often have an advantage over other analytical solutions because the boundary conditions are often more general. It is not necessary, from the purely mathematical considerations, that the flows be realistic or physically realizable. However, as noted, realistic looking analytical solutions have a psychological and political advantage. Fortunately, there are many manufactured or otherwise contrived analytical solutions in the literature. Also, as noted above in Section 3.7, approximate but highly accurate solutions (often obtained by perturbation methods) can also be utilized in Code Verification. Each discipline will have its own literature, which I certainly cannot cover, but the following list, though far from complete, may be useful to some workers. (See also Chapter 7 of Roache, 1998.)

- Rogowski electrode, as used in Roache et al (1984); see Lorrain and Corson (1962).
- Darcy flow in 2-D, variable properties, as used in Roache et al (1990); see Chapter 6, Section 6.2.2, Darcy Flow with Tensor Conductivity in Non-Orthogonal Coordinates.
- Particle tracking in 2-D and 3-D, as used in Roache et al (1990); see Chapter 8, Section 8.6.
- Colloid Transport in saturated fractures; see Abdel-Salam and Chrysikopoulos (1994).
- Driven Cavity Flow. Huang and Li (1997) present two simple analytic solutions for streamline flows that are suggestive of their problem of (classical) incompressible flow in a driven cavity, including a secondary (recirculation) vortex. They note that one of their methods with apparently 1st-order boundary conditions produced 2nd-order performance on one of the test solutions, but correctly explain it as a "coincidence" [or artifact] of the accident that the solution produced homogeneous boundary conditions. As noted earlier, it is important for the manufactured (or other analytic) solution to exercise all the terms in the equations, in order to avoid misleading and over-optimistic results.
- Nonlinear Free-Surface Boundary Condition. Serrano (1995) obtained new analytical solutions of the nonlinear Boussinesq flow equation and of the exact two-dimensional groundwater flow equation subject to a nonlinear free-surface boundary condition. The generality of the solution allowed testing (or "Justification" as in Chapter 2) of the simpler linearized equation with the Dupuit assumptions, showing that these commonly used approximations produce discrepancies in the presence of high regional hydraulic gradients, unusually high recharge rates, or regions of low conductivity.

- Analytic Elements. The "analytic element modeling" of Haitjema (1995) is a method of producing simulations that involve analytical solutions applied element-wise, and could also be used for exact solutions in accuracy Verification exercises.
- For fluid dynamics solutions, see also the compilation of exact solutions of the steady-state Navier-Stokes equations by Wang (1991) and the use of analytical solutions by Grenda et al (1996) for Verifying (in the present terminology) unsteady CFD codes.

3.10 MANUFACTURED SOLUTIONS VS. INFINITE SERIES SOLUTIONS

Sudicky and Frind (1982) present analytical solutions for contaminant transport in a system of parallel fractures in porous media. Their solution is very well known and widely used for Benchmarking groundwater transport codes. What is not so well known is that the general solution as originally presented is incorrect, in that it appears to be given for more general boundary conditions but in fact it only satisfies homogeneous boundary conditions, as pointed out by Davies and Johnston (1984). They also noted that the infinite series "solution" involves a summation that is only conditionally convergent, and present a more strongly convergent (robust) summation. Van Gulick's (1994) experience is that several other numerical problems exist in the Sudicky–Frind solution, e.g. the series solutions for high diffusivity or early time converge slowly or not at all.

The situation with the Sudicky–Frind solution is not unusual. Even if "correct" mathematically, any such analytical solution that involves series needs to be evaluated numerically, and subtle problems often arise. Also, it is difficult and not confidence-inspiring to confirm the mathematical correctness of a non-trivial analytical solution. In my fairly extensive experience with colleagues in compressible and incompressible aerodynamics, heat transfer, groundwater flow, groundwater contaminant transport, and laser electrode design, I can say categorically that the numerical evaluation of these "analytical solutions" *invariably* has involved more frustration than the computational codes being Verified. In this regard, as well as in generality of the initial and boundary conditions, Manufactured Solutions (which do not involve infinite series) have an advantage over these analytical solutions.

3.11 THE SENSITIVITY OF GRID CONVERGENCE TESTING

We have had considerable experience in using such systematic grid convergence tests, both with "realistic" and with more general "manufactured" solutions, for accuracy Verification of Codes as well as Verification of individual Calculations (involving error estimation). Many of these cases have utilized the Generalized Richardson Extrapolation and the Grid Convergence Index (GCI) to be described in Chapter 5. In our experience, this method of code accuracy Verification via systematic grid convergence testing (whether or not the GCI is used) is *remarkably sensitive* in revealing code problems, as indicated by the following examples. (Some of these examples will be covered in more detail in later chapters.)

1. In accuracy Verification tests of a groundwater flow commercial code, a 1st-order error in a *single corner cell* in a strongly elliptic problem caused the observed *global* convergence to be 1st-order accurate. (Roache et al, 1990.)
2. In accuracy Verification tests of our SECO_FLOW_2-D variable density (dissolved salt) groundwater flow code, 1st-order extrapolation for *ghost cell values of only one quantity* (aquifer thickness) along one boundary caused the observed convergence to be 1st-order accurate. (Roache, 1993.)
3. In groundwater contaminant transport calculations (advection-diffusion + decay, retardation, and matrix diffusion), use of a plausible single-grid-block representation for a point source as the grid is refined introduces error in a finite volume (or block-centered finite difference) formulation. In this cell configuration, the cell faces align with the boundaries of the computational domain, and doubling the number of cells requires the *location* of the single cell representing the source to *shift by* $\Delta/2$. It is to be expected that the solution accuracy in the neighborhood of the source would be affected. But surprisingly, the accuracy of time-integrated *discharge* across boundaries far from the source was also degraded to 1st-order accuracy. See Chapter 6, Section 6.12, and Salari et al (1995).
4. In tests of the two-dimensional SECO_FLOW groundwater flow code using the option for node (cell center) placement on the boundary (rather than cell edges on the boundary), the domain corner nodes enter into the calculation only when cross derivative terms are present. (These can arise from either non-orthogonal coordinate transformations or tensor properties.) When this value (say, at the lower left corner *ic, jc*) was set by plausible averaging, as in

$$f(ic,jc) = 1/2\,[f(ic+1,jc) + f(ic,jc+1)] \tag{3.11.1}$$

   the result was global 1st-order convergence rate. When the following form (recommended by S. Steinberg)

$$f(ic,jc) = f(ic+1,jc) + f(ic,jc+1) - f(ic+1,jc+1) \tag{3.11.2}$$

   was used, the result was global 2nd-order convergence rate. (Roache et al, 1990.)
5. The observed convergence rate of ostensibly 2nd-order accurate turbulent boundary layer codes (Wilcox, 1993) can be degraded, apparently by *conditional statements* limiting eddy viscosity and defining the boundary layer edge. (Wilcox, 1995.)
6. Airfoil codes can exhibit the expected 2nd-order convergence rates for lift and drag, but less for moment, possibly because of approximations involved in applying *quasi-periodicity* across cut-planes of a C-grid. (Salari, 1995.)

3.12 EXAMPLES OF UNANTICIPATED CONVERGENCE RATES DETERMINED BY SYSTEMATIC GRID CONVERGENCE TESTS

Note that these systematic grid convergence tests can Verify a code, not only in the sense of identifying presence of coding errors, but also in the sense of Verifying the order of convergence. Contrary to common belief, the theoretical order of convergence may not be so obvious for complicated methods. Jameson and Martinelli (1996) note that "the nonlinearity of conservation laws, and the geometric complexity of the flow domain of interest in most applications, make it difficult to devise elegant analytical tools for the accuracy Verification of a numerical scheme."

Furthermore, "*the* formal analysis" may be a misnomer, since different levels of analysis may provide different approximate solutions. (Not all the world is governed by the Laplace equation in a square domain.) Here I give two examples.

3.12.1 Reduction to Periodicity Method: Unequal Orders of Accuracy for Derivatives

In Roache (1978), the "reduction to periodicity" (RTP) method was devised to use pseudo-spectral FFT (Fast Fourier Transform) methods for non-periodic problems by pre- and post-processing the solution with polynomials of various degrees. The accuracy was not confidently predictable *a priori*. The testing involved parametric combinations (4 methods of reduction-to-periodicity, 5 values of the reducing polynomial function N), 3 error indices of two derivatives, and comparison with 2nd-, 4th- and 6th-order finite difference methods (FDM). The results were given in rankings compared to the FDM, for two static functions and the transient Burgers equation. An example of the accuracy ranking obtained for the case of using exact values of boundary derivatives in the RTP method is shown in Table 3.12.1.1. The details of the technique are not required here; the point is the unusual and unanticipated grid convergence behavior.

Note the unusual convergence rates exhibited by this family of methods. The convergence rate for the second derivative $f''(x)$ is two orders lower than that of the first derivative $f'(x)$ for RTP with $N = 3$ or 5, but is equal for $N = 2$ or 4. For $f'(x)$, the accuracy jumps two orders when N is increased from 2 to 3, and from 4 to 5, but does not improve when N is increased from 3 to

$N = 2$	O(2) < RTP < O(4)
$N = 3$	O(4) < RTP < O(6)
$N = 4$	O(4) < RTP < O(6)
$N = 5$	O(6) < RTP

a. Accuracy ranking for $f'(x)$ for the RTP method.

$N = 2$	O(2) < RTP < O(4)
$N = 3$	O(2) < RTP < O(4)
$N = 4$	O(4) < RTP < O(6)
$N = 5$	O(4) < RTP < O(6)

b. Accuracy ranking for $f''(x)$ for the RTP method.

Table 3.12.1.1. Order of accuracy ranking for the Reduction to Periodicity method (RTP) applied to a static function. The mesh spacing was varied from $\Delta = 1/8$ to 1/128. The notation O(2) < RTP < O(4) indicates that the RTP technique had an accuracy between 2nd-order and 4th-order FDM. From Table I of Roache (1978).

4. This *modulo* 2 behavior is shifted for $f''(x)$, which jumps two orders when N increases from 3 to 4, but does not improve when N increases from 2 to 3, nor from 4 to 5.

Successful analysis of this technique is possible, but far from conventional. Lyness (1974) had analyzed the similar Lanczos representation of a function, and showed theoretically and in general what the above tests showed experimentally for only specific functions: (1) that the degree of the reducing polynomial is important to the accuracy, (2) that the accuracy of $f'(x)$ would improve as N was increased from $N = 1$ by increments of 2. His theory also showed other results (not shown in the Table) found only experimentally in Roache (1978), that (3) when boundary derivatives for the RTP technique are evaluated by FDM (rather than using exact values as in the above example), the order of the FDM was not so important, and that (4) the use of a 5-th degree reducing polynomial with 2nd-order FDM at boundaries would give $f'(x)$ to overall 4th-order accuracy. At my request, Lyness also extended his previous work to prove another aspect found only experimentally, that (5) $f''(x)$ is two orders less accurate than $f'(x)$. (This is not a characteristic of pseudo-spectral or spectral methods themselves, but of the RTP method.) This imbalance in the order of accuracy of $f'(x)$ and $f''(x)$ is in fact acceptable, giving the chance of a "balanced truncation error method" (Roache, 1978) in which the *size* rather than the *order* of truncation errors from advection and diffusion terms are roughly balanced for high Reynolds number (Peclet number) problems.

This complicated and unanticipated convergence behavior would not likely be determined without the kind of systematic grid convergence tests recommended herein. Note, however, that with the numerical grid convergence tests providing motivation, theoretical justification of this behavior was found using the independent results of Lyness (1974).

3.12.2 Completed Richardson Extrapolation Method: Higher Order Truncation Error Interaction

Richardson Extrapolation will be covered extensively (and generalized) in Chapter 5. For now, we note that the original method (Richardson, 1908) combines two 2nd-order solutions on two grids, the fine grid having twice the resolution of the coarse grid, to produce a 4th-order accurate solution on the coarse grid only. I devised a method (Roache and Knupp, 1993) for easily obtaining a 4th-order solution on the entire fine grid, evaluating the solution at the "skipped" fine grid points by linear interpolation, not of the solution, but of the coarse grid *correction* between the 2nd- and 4th-order solutions. Recently, Richards (1997) has extended the procedure systematically to mixed time and space PDE's with more general integer refinement.

Without going into the details here, the original formal analysis of the "completed" Richardson extrapolation indicated that all fine grid nodes should converge at 4th-order rate, i.e., $p = 4$. Systematic grid convergence studies did show this experimental rate $p = 4$, but originally only for the Laplace and Poisson equations (with non-trivial source terms). For the advection-diffusion equation, the "skipped" fine grid points mysteriously showed roughly 3rd-order convergence, $p = 3$. It was not known if the source of discrepancy was an error in the formal analysis of the convergence rate, or a coding error.

N	Maximum Error	E_3	E_4
4	0.63299798320	40.512	162.05
8	0.09835113825	50.356	402.85
16	0.01602753711	65.649	1050.38
32	0.00208620873	68.361	2187.55
64	0.00019995963	52.429	3354.77
128	0.00001580555	33.135	4242.77
256	0.00000111805	18.758	4802.00
512	0.00000007447	9.995	5117.81
1024	0.00000000481	5.165	5285.92

Table 3.12.2.1. Grid Convergence Study for Completed Richardson Extrapolation. Linear advection-diffusion equation with *Re* = 16. E_3 = max error/Δ^3 and E_4 = max error/Δ^4. (From Table IVb, Roache and Knupp, 1993.)

The problem lay unexplained for a few years, until P. Knupp extended the formal analysis to show (Roache and Knupp, 1993) that the asymptotic rate $p = 4$ indicated by the original formal analysis was indeed correct, but would not emerge until the grid refinement had proceeded enough to reduce the cell Reynolds number $Rc << 3$. The cause is higher order truncation error interaction determined by the exp(x) form of the exact solution. This is demonstrated in Table 3.12.2.1, for $Re = 16$. For the coarse grid, from $\Delta = 1/N = 1/4$ to roughly 1/64, E_3 = max error/Δ^3 is *roughly* constant, which might suggest $p = 3$. Only for finer grids is the true asymptotic behavior of $p = 4$ approached, with E_4 increasing only 3.2% from $N = 512$ to 1024. Note also the difficulty of judging the rate of convergence just from a presentation of max error, especially for $p > 2$.

The truncation error interaction is indicated by Knupp's analysis. The one-dimensional steady state equation solved is the following ODE for $f(x)$.

$$f'' - Re\, f' = 0, f(0) = 1, f(1) = 0 \tag{3.12.2.1}$$

The exact solution is

$$f(x) = (e^{-Rex} - e^{-Re}) / (1 - e^{-Re}) \tag{3.12.2.2}$$

From the exact solution, the form for any k-th derivative is

$$f^{(k)}(x) = \frac{(-1)^k Re^k e^{-Rex}}{1 - e^{-Re}} \tag{3.12.2.3}$$

In particular, the relation between any k-th derivative and the 2nd derivative is

$$\frac{f^{(k)}}{f^{(2)}} = (-1)^k Re^{k-2} \tag{3.12.2.4}$$

This is the aspect of the solution that causes the truncation error interaction. The Taylor Series for (any) f is

$$f(x+\Delta) = \sum_{k=0}^{\infty} \frac{\Delta^k}{k!} f^{(k)}(x) \tag{3.12.2.5}$$

Using the relation between any kth derivative and the 2nd derivative, which is a feature only of this solution, we obtain

$$f(x+\Delta) = [f(x) + \Delta f'(x)] + f^{(2)}(x) \left\{ \frac{\Delta^2}{2!} - \frac{\Delta^3}{3!} Re + \frac{\Delta^4}{4!} Re^2 + \cdots \right\} \tag{3.12.2.6}$$

The significance of this expression is that the higher-order error terms contain not only the grid spacing Δ, but also the continuum parameter Re. Now,

$$F(x) = f(x) + \Delta f'(x) \tag{3.12.2.7}$$

acts like a 2nd-order approximation to $f(x+\Delta)$ only if

$$\Delta^2 / 2 >> \Delta^3 Re / 6, \cdots \tag{3.12.2.8}$$

This translates into the requirement that $1/\Delta = N >> Re/3$. Therefore, if Re is large, a large N (small Δ) is needed to make F behave in a 2nd-order accurate manner. Otherwise, the assumption necessary for the completed Richardson extrapolation is violated, and the fine-grid results are not 4th-order accurate.

As in the previous example of the Reduction To Periodicity method, this unanticipated convergence behavior would not have been determined without the kind of systematic grid convergence tests recommended herein. And again, with the numerical grid convergence tests providing motivation, theoretical justification of the behavior was determined using more detailed formal analysis. Note that the original formal analysis was not *wrong*, i.e. it applied asymptotically, but neither was it complete enough to adequately describe the experimental results.

These two examples illustrate an important and usually overlooked aspect of numerical analysis. Authors typically speak of "*the* formal convergence rate" as though only one analysis were possible. In fact, formal analyses of convergence rates can be at various levels, with various results. The second example above shows unanticipated structure in the analysis, even for a linear, one-dimensional, steady state problem with uniform grid increments Δx.

3.12.3 Two Other Examples

For two other examples showing the difficulty of "the" formal truncation error analysis, consider first that the *formal* accuracy of the QUICK schemes, first published in 1979, was still being debated in 1993, and "considerable confusion remains in the literature" (Leonard, 1993). In another paper (not related to this question), Leonard and MacVean (1995) provide a formal analysis that illuminates grid convergence effects due to nonlinearities.

3.13 SPECIAL CONSIDERATIONS FOR PROBLEMS WITH MULTIPLE SCALES: TURBULENCE MODELING

There are special considerations required for turbulence modeling, and for other fields with *multiple scales*. Here, the code theoretical performance can be Verified (within a tolerance) for a range of parameters, but could fail in another range.

It is necessary to get the grid resolution into the asymptotic range in order to do grid convergence testing. Virtually any grid is in the asymptotic range for a simple Laplace equation. For any boundary layer calculation, it is clear that the initial (coarse) grid must get some points into the boundary layer. For turbulence modeling without wall functions (Shirazi and Truman, 1989; Wilcox, 1993) the grid must get some points into the wall layer. For turbulence modeling with wall functions, it would seem that the grid preferably should not get into the wall layer (Celik and Zhang, 1995). In my interpretation, effectively the wall functions should be viewed as an elaborate nonlinear boundary condition, and the grid convergence exercise should be done from the edge of the wall layer out. (However, see Celik and Karatekin, 1995 for a counter-example.) Similarly, for large eddy simulations (LES) as used in aerodynamic turbulence research and in atmosphere and ocean modeling with sub-grid turbulence modeling, the grid convergence must not go to zero, or else the Reynolds stresses will be counted twice, once from the full Navier-Stokes terms and again modeled from the LES terms. Also, the presence of any switching functions, such as length determinations for the Baldwin-Lomax turbulence model (Wilcox, 1993) can easily corrupt 2nd-order convergence rates.

Finally, the grid resolution requirements are much more demanding for turbulent boundary layers, just as laminar boundary layers are much more demanding than inviscid flows. For example, Claus and Vanka (1992) found that 2.4 million nodes ($256 \times 96 \times 96$) did not demonstrate grid independence of the computed velocity and turbulence fields of crossflow jets.

3.14 WARNINGS: WHAT THE METHOD DOES NOT "VERIFY"

The Method of Manufactured Solutions convincingly Verifies the *accuracy* of the solution produced by a code, and this is the technical definition of "Verify" adopted herein. As already noted, the method cannot Verify the coding accuracy of individual terms in mixed-order methods, nor can it Verify algebraic conservation.

Also, the method does not "verify" (in the more general sense) that everything about the coding is correct, since not all coding mistakes affect accuracy. In particular, the method does not verify that the *solution procedure* encoded actually follows the algorithm intended. It verifies that the procedure produces the correct numerical solution, but there can be coding mistakes that decrease the iterative convergence rate yet still produce the correct solution; the mistake causes a loss of *efficiency*, rather than a loss of *accuracy*. (Note that such coding mistakes should not be a concern to regulatory agencies or stakeholders in public policy projects like geologic waste disposal, since the numerical accuracy of the answers is not affected.)

Examples of such coding mistakes are plentiful. An SOR iteration scheme can have an mistake in the determination of the optimum relaxation factor without (significantly) changing the answer produced. If the mistake is gross, the slow iterative convergence will probably bring attention to the mistake and lead to successful debugging, but smaller mistakes may go unnoticed permanently. Multigrid methods (Brandt, 1977; McCormick, 1989) are known to be prone to subtle coding mistakes that diminish the theoretical iterative convergence rate. Verification of a theoretical *iterative* convergence rate is a good debugging tool, and builds high confidence for linear problems, but for nonlinear problems, the conclusions about correctness of the coding are not as firm as they are for Verification testing of numerical accuracy. Other esoteric mistakes that are difficult to detect will be described in Chapter 8.

3.15 ROBUSTNESS AND CONFIDENCE

Likewise, code/algorithm robustness is not "Verified" by the procedure. Robustness is necessarily a qualitative concept, so it does not make sense to ask for a definitive statement about it, especially for general purpose commercial codes.

One technique for building confidence (rather than a rigorous demonstration) in robustness and coding accuracy is given in Roache (1995). The questions were (1) robustness of direct elliptic marching methods (which are sensitive to computer round-off error) to various boundary conditions, and (2) correctness of the coding for mixed boundary conditions. A general boundary condition had been formulated and coded as

$$\alpha f + \beta \frac{\partial f}{\partial n} = \gamma \tag{3.15.1}$$

where f is the dependent variable and n is the direction normal to the boundary. For various combinations of α, β, and γ one obtains the usual homogeneous or non-homogeneous Dirichlet, Neuman, or Robin (mixed) boundary conditions. (Because of algorithmic quirks, the formulation is not so general as to include tangential derivatives at the boundary.) Robustness and coding were demonstrated by running many calculations with α, β, and γ arrays determined by a pseudo-random number generator. The values were not constant along a boundary, but changed randomly along the boundary. The range was limited so as to avoid indeterminacies in the discrete formulation. Since the elliptic marching methods are direct, i.e. non-iterative, the stencil evaluation for the error should be algebraically zero, except for round-off error, and the tests

Verified this convincingly. (Note this interesting generality: solution methods that are sensitive to computer word length, i.e., to round-off error, are easy to debug or to demonstrate coding correctness, whereas solution methods that are not sensitive to word length are difficult.)

The approach of testing the general Robin boundary condition by pseudo-random number generation of coefficients can be combined with the Method of Manufactured Solutions. With the analytical solution evaluated for f and $\partial f / \partial n$ at every point, two of the three coefficients α, β, and γ can be determined by pseudo-random number generation, and the third solved algebraically. This test will provide a convincing demonstration of both accuracy Verification and robustness to boundary condition type.

REFERENCES FOR CHAPTER 3

Abdel-Salam, A. and Chrysikopoulos, C. V. (1994), "Analytical Solutions for One-Dimensional Colloid Transport in Saturated Fractures," *Advances in Water Resources*, Vol. 17, 1994, pp. 283–396.

Beam, R. and Warming, R. F. (1976), "An Implicit Finite Difference Algorithm for Hyperbolic Systems in Conservation Law Form," *Journal of Computational Physics*, Volume 22, 1977, pp. 87–110.

Brandt, A. (1977), "Multi-Level Adaptive Solutions to Boundary-Value Problems," *Mathematics of Computation*, Vol. 31, 1977, pp. 333–390.

Briley, W. R. and McDonald, H. (1977), "An Implicit Numerical Method for Multi-Dimensional Compressible Navier-Stokes Equations," *Journal of Computational Physics*, Vol. 24, No. 4, 1977, pp. 372–397.

Celik, I. and Karatekin, O. (1995), "Numerical Experiments on Application of Richardson Extrapolation to a Turbulent Flow with Separation," in R. W. Johnson and E. D. Hughes (1995), eds., *Quantification of Uncertainty in Computational Fluids Dynamics–1995,* Joint JSME–ASME Fluid Mechanics Meeting, August 14–18, 1995, ASME FED Vol. 213, Hilton Head, South Carolina, pp. 37–43. See also *ASME Journal of Fluids Engineering*, Vol. 119, No. 3, September 1997, pp. 584–590.

Claus, R. W. and Vanka, S. P. (1992), "Multigrid Calculations of a Jet in Crossflow," *Journal of Propulsion Power*, Vol. 8, pp. 185–193, 1992.

Davies, G. B. and Johnston, C. D. (1984), "Comment on 'Contaminant Transport in Fractured Porous Media: Analytical Solutions for a System of Parallel Fractures'," *Water Resources Research*, Vol. 20, No. 9, September 1984, pp. 1321–1322.

Ethier, C. R. and Steinman, D. A. (1994), "Exact Fully 3-D Navier-Stokes Solutions for Benchmarking," *International Journal for Numerical Methods in Fluids*, Vol. 19, 1994, pp. 369–375.

Grenda, J. M., Schwer, D. A. and Merkle, C. L. (1996), "Use of Analytical Solutions in Validating Unsteady CFD Codes," AIAA Paper No. 96–2164, June 1996.

Grismer, M. J. and Powers, J. M. (1992), "Comparison of Numerical Oblique Detonation Solutions with an Asymptotic Benchmark," *AIAA Journal*, Vol. 30, No. 12, December 1992, pp. 2985–2987.

Grismer, M. J. and Powers, J. M. (1996), "Numerical Predictions of Oblique Detonation Stability Boundaries," *Shock Waves*, Vol. 6, pp. 147–156.

Haitjema, H. M. (1995), *Analytic Element Modeling of Groundwater Flow*, Academic Press, New York.

Haynes, T. S., Reed, H. L., and Saric, W. S. (1996), "CFD Validation Issues in Transition Modeling," AIAA Paper 96–2051, 27 AIAA Fluid Dynamics Conference, New Orleans, Louisiana, 17–20 June 1996.

Huang, H. and Li, M. (1997), "Finite-Difference Approximation for the Velocity-Vorticity Formulation on Staggered and Non-Staggered Grids," *Computers and Fluids*, Vol. 26, No. 1, 1997, pp. 59–82.

Jameson, A. and Martinelli, L. (1996), "Mesh Refinement and Modeling Errors in Flow Simulation," AIAA Paper 96–2050, 27th AIAA Fluid Dynamics Conference, New Orleans, Louisiana, June 17–20, 1996.

Knupp, P.M. and Steinberg, S. L. (1993), *Fundamentals of Grid Generation*, CRC Press, Boca Raton, Florida.

Leonard, B. P. (1993), "Order of Accuracy of QUICK and Related Convection-Diffusion Schemes," NASA TM 106402, ICOMP–93–47, November 1993.

Leonard, B. P. and MacVean, M. K. (1995), "Grid Refinement Study of a Benchmark Problem Using the Flux–Integral Method," in R. W. Johnson and E. D. Hughes (1995), Eds., *Quantification of Uncertainty in Computational Fluids Dynamics–1995,* Joint JSME–ASME Fluid Mechanics Meeting, 1995, ASME FED Vol. 213, Hilton Head, South Carolina, August 14–18, pp. 45–50.

Lorrain, P. and Corson, D. P. (1962), *Introduction to Electromagnetic Fields and Waves*, W. H. Freeman Company, San Francisco.

Lyness, J. N. (1974), *Mathematics of Computation*, Vol. 28, p. 81.

McCormick, S. F. (1989), *Multilevel Adaptive Methods for Partial Differential Equations*, SIAM Frontiers in Applied Mathematics, Vol. 6, SIAM, Philadelphia, Pennsylvania.

Oberkampf, W. L., Blottner, F. G., and Aeschliman, D. P. (1995), "Methodology for Computational Fluid Dynamics Code Verification/Validation," AIAA Paper 95–2226, 26th AIAA Fluid Dynamics Conference, San Diego, California, 19–22 June 1995.

Polya, G. (1957), *How to Solve It; A New Aspect of Mathematical Method*, Princeton University Press, Princeton, New Jersey.

Powers, J. M. and Stewart, D. S. (1992), "Approximate Solutions for Oblique Detonations in the Hypersonic Limit," *AIAA Journal*, Vol. 30, No. 3, March 1992, pp. 726–736.

Powers, J. M. and Gonthier, K. A. (1992), "Reaction Zone Structure for Strong, Weak Overdriven and Weak Underdriven Oblique Detonations," *Physics of Fluids A*, Vol. 4, No. 9, September 1992, pp. 2082–2089.

Richards, S. A. (1997), "Completed Richardson Extrapolation in Space and Time, *Communications in Numerical Methods in Engineering*, Vol. 13, No. 7, 573–582.

Richardson, L. F. (1908), The Approximate Arithmetical Solution by Finite Differences of Physical Problems Involving Differential Equations, with an Application to the Stresses in a Masonry Dam," *Transactions of the Royal Society of London*, Series A, Vol. 210, pp. 307–357.

Roache, Patrick J. (1972), *Computational Fluid Dynamics*, First Edition, Hermosa Publishers, P. O. Box 9110, Albuquerque, New Mexico 87119-9110, USA. http://www.hermosa-pub.com/hermosa. Revised Printing 1976. See also Roache (1998).

Roache, P. J. (1978), "A Pseudo-Spectral FFT Technique for Non-Periodic Problems," *Journal of Computational Physics*, Vol. 27, No. 2, May 1978.

Roache, P. J., Steinberg, S., and Moeny, W. M. (1984), "Interactive Electric Field Calculations for Lasers," AIAA Paper 84–1655, AIAA 17th Fluid Dynamics, Plasma Dynamics, and Lasers Conference, Snowmass, Colorado, June 25–27, 1984.

Roache, P. J. and Steinberg, S. (1984), "Symbolic Manipulation and Computational Fluid Dynamics," AIAA Paper No. 83–1952. See also *AIAA Journal*, Vol. 22, October 1984, pp. 1390–1394.

Roache, P. J. (1990), "Need for Control of Numerical Accuracy," *Journal of Spacecraft and Rockets*, Vol. 27, No. 2, March–April 1990, pp. 98–102. See also AIAA Paper No. 89–1669.

Roache, P. J., Knupp, P. M., Steinberg, S. and Blaine, R. L. (1990), "Experience with Benchmark Test Cases for Groundwater Flow," ASME FED–Vol. 93, *Benchmark Test Cases for Computational Fluid Dynamics*, I. Celik and C. J. Freitas, Eds., Book No. H00598–1990, pp. 49–56.

Roache, P. J. and Knupp, P. M. (1993), "Completed Richardson Extrapolation," *Communications in Applied Numerical Methods*, Volume 9, pp. 365–374.

Roache, P. J. (1993), "The SECO Code Algorithms for Groundwater Flow and Transport," *Finite Elements in Fluids: New Trends and Applications*, Part II, Proceedings of the VIII International Conference on Finite Elements in Fluids, Univ. Politécnica de Cataluña, Barcelona, Spain, K. Morgan, E. Onate, J. Periaux, J. Peraire, and O. C. Zienkiewicz, Eds., 20–23 Sept. 1993, pp. 939–948.

Roache, P. J. (1995), *Elliptic Marching Methods for Domain Decomposition*, CRC Press, Boca Raton, Florida.

Roache, P. J. (1998), *Fundamentals of Computational Fluid Dynamics*, Hermosa Publishers, P. O. Box 9110, Albuquerque, New Mexico 87119-9110, USA. http://www.hermosa-pub.com/hermosa. To appear.

Salari, K. and Roache, P. J. (1990), "The Influence of Sweep on Dynamic Stall Produced by a Rapidly Pitching Wing," AIAA Paper 90–0581.

Salari, K., Blaine, R. L., Economy, K., and Roache, P. J. (1995), "Grid Resolution Studies of Radionuclide Transport in Fractured Porous Media," in R. W. Johnson and E. D. Hughes (1995), eds., *Quantification of Uncertainty in Computational Fluids Dynamics–1995,* Joint JSME–ASME Fluid Mechanics Meeting, ASME FED Vol. 213, Hilton Head, South Carolina, August 14–18, 1995, pp. 25–30.

K. Salari (1995), personal communication.

Serrano, S. E. (1995), "Analytical Solutions of the Nonlinear Groundwater Flow Equation in Unconfined Aquifers and the Effect of Heterogeneity," *Water Resources Research*, Vol. 31, No. 11, November 1995, pp. 2733–2742.

Steinberg, S. and Roache, P. J. (1985), "Symbolic Manipulation and Computational Fluid Dynamics," *Journal of Computational Physics*, Vol. 57, No. 2, January 1985, pp. 251–284.

Steinberg, S. and Roache, P. J. (1986a), "A Tool Kit of Symbolic Manipulation Programs for Variational Grid Generation," AIAA Paper No. 86–0241, AIAA 24th Aerospace Sciences Meeting, Reno, Nevada, 6–9 January 1986.

Steinberg, S. and Roache, P., 1986b, "Using MACSYMA to Write FORTRAN Subroutines," *Journal of Symbolic Computation*, Vol. 2, pp. 213–216.

Sudicky, E. A. and Frind, E. O. (1982), "Contaminant Transport in Fractured Porous Media: Analytical Solutions for a System of Parallel Fractures," *Water Resources Research*, Vol. 18, No. 6, Dec. 1982, pp. 1634–1642.

Thompson, J. F., Thames, F. C., and Mastin, C. W. (1974), "Automatic Numerical Generation of Body-Fitted Curvilinear Coordinate System for Field Containing Any Number of Arbitrary Two-Dimensional Bodies," *Journal of Computational Physics*, Vol. 15, No. 3, July 1974, pp. 299–319.

Van Dyke, M. (1958), "The Supersonic Blunt Body Problem–Review and Extension," *Journal of the Aeronautical Sciences*, Vol. 25, p. 485.

Van Gulick, P. (1994), personal communication.

Wang, C. Y. (1991), "Exact Solutions of the Steady-State Navier-Stokes Equations," *Annual Review of Fluid Mechanics*, Vol. 23, 1991, pp. 159–177.

Wilcox, D. C. (1993), *Turbulence Modeling for CFD*, DCW Industries, La Cañada, California, 1993.

D. C. Wilcox (1995), personal communication.

4

ERROR ESTIMATION FOR QUANTIFICATION OF UNCERTAINTY; VERIFICATION OF CALCULATIONS

4.1 INTRODUCTION

This chapter is taken primarily from Roache (1997), "Quantification of Uncertainty in CFD."

As noted in the previous chapter, the task of Code Verification as presented does not, strictly speaking, involve error *estimation* but simple error evaluation, because the Verification is done with respect to a known analytical solution. The solution may be unrealistic, i.e. a "manufactured" solution, but is nonetheless useful for Code Verification. Verification of an individual Calculation, on the other hand, strictly involves error estimation. Yet the latter term needs some distinction because of its use in different contexts.

4.2 ERROR ESTIMATION FOR GRID ADAPTATION VS. QUANTIFICATION OF UNCERTAINTY

The focus of this book is the "quantification of uncertainty," the estimation or "banding" of the numerical error of a "final" calculation in Computational Fluid Dynamics (CFD) and similar fields in Computational Science, Engineering, and Mathematics. By "final calculation" I mean one which is considered to be used "as is." This provides a different motivation than that of an error estimate to be used for the process of solution adaptive grid generation. Although the present methods may be applicable to grid adaptation, and the developments of some of the methods described herein were motivated by that problem, the grid adaptation problem has vastly different (though usually unacknowledged) requirements.

The key word is "quantification" of uncertainty, as opposed to vague and all too common qualitative assessments. "Quantification of Uncertainty" may also involve more than just obtaining a good error estimate; in fact, the more conservative procedure based on the Grid Convergence Index (described in the following Chapter 5) reports an error band equal to three times an error estimate. Further, I consider herein only *a posteriori* error estimation, being of the opinion that useful *a priori* estimation is not possible for non-trivial problems.

Local "error" estimators are used for solution adaptive grid generation algorithms, and they are usually successful for this purpose (e.g., see Oden et al, 1993). However, almost anything intuitive is successful for adaptation purposes, e.g. minimizing solution curvature, or adapting to solution gradients, even though solution gradients *per se* cause no error in most discretization schemes. In one-dimensional problems (e.g., Salari and Steinberg, 1994) or quasi-one-dimensional problems (Dwyer et al, 1980), the gains in computational efficiency from solution adaptivity of the *r*-type (redistribution) in structured grids are very impressive. In strongly multidimensional problems, gains are usually modest (Roache et al, 1984; Hall and Zingg, 1995) but are somewhat more significant for unstructured grid adaptation (e.g., see Morgan et al, 1991; Lohner, 1989; Hetu and Pelletier, 1992; Pelletier and Ignat, 1995; Coorevits et al, 1995; Ilinca et al, 1997).

Unfortunately, this task of grid adaptation as typically practiced has little connection to the quantification of uncertainty for a final calculation with any useful error measure. Thus, it is important to recognize that the successes of these local "error" estimators in guiding grid adaptation must not be taken as demonstrations of their efficacy for the quantification of uncertainty. Even for grid adaptation, the success of local "error" estimates is only a partial success. Hagen et al (1997) have given examples of strong non-localness in tide-driven simulations of ocean model problems. Increased element resolution in the deep ocean, where local error estimators would not suggest increased resolution, in fact improves accuracy far away from the deep ocean, on the continental slope and shelf. Such behavior may be expected for any wave-dominated problem exhibiting amphidromes, "hanging" shocks, or other local features of strong solution structure with no obvious geometrical cause in the immediate neighborhood.

Lee and Yeh (1994a,b) have shown good correlations between local (normalized) weight functions based on solution-gradient estimators and (normalized) true solution errors as obtained with multiple grid solutions and Richardson Extrapolation, for both laminar and turbulent steady flows (although somewhat less for swirling flows). Their hybrid solution adaptive grid generation (combining a global, *r*-type adaptation followed by a local *h*-type adaptation) obtained significant efficiencies. Nevertheless, Lee and Yeh (1994b) correctly note the distinction between an error "indicator" and an error "estimator," and note that "the weight function is employed as an error 'indicator' rather than an 'estimator', in that the accuracy of the predicted values is not a primary concern." Clearly, *normalization* is the key concept. The distribution of errors, provided by the correlation between normalized values, is definitely of interest but is not sufficient for the task of the quantification of uncertainty. Rather, absolute values are required, and correlation of these with local weight functions (and their use as Surrogate Estimators for measures of interest) must be established anew for each new problem and each error metric of interest.

We need *global* error estimates for Verification of Calculations. By "global" we do not here mean just a global summing up of local values (as sometimes used in the literature of Finite Element Methods) but an evaluation which includes non-local effects, i.e., which takes account of the fact that errors are advected, diffused, "beat" in wave resonance problems, etc.

4.3 TAXONOMY FOR ADDITIONAL INFORMATION FOR ERROR ESTIMATES

Once we have produced a discrete solution of the governing partial differential equations, it is clear that we require some *additional* information in order to quantitatively estimate the uncertainty or numerical accuracy. The following taxonomy of sources in Table 4.3.1 of this additional information will provide a framework for the discussion. By the word "grid," I refer to any measure of discretization, i.e., Cartesian grid, non-orthogonal grid, number of Fourier modes in a spectral solution, number of discrete vortices, etc.

The following are brief remarks on this taxonomy, some of which will be justified and amplified in later discussion.

Categories A and B, Additional Solution(s) of the Governing Equations (on Other Grids or the Same Grid) involve the direct, unambiguous evaluation of any error measure of engineering or scientific interest. Note that *only* these procedures can be used, as in the previous chapter, for Verification of Codes, and as a practical matter, only A.1, Grid Refinement, provides for rigorous Verification, because only it can be continued to any desired level of accuracy (like the classical ε-δ limit proofs of calculus).

For Category A, Additional Solution(s) of the Governing Equations on Other Grids, no additional code development or modifications are required.

For Category B, Additional Solution(s) of the Governing Equations on the Same Grid, no additional grid generation is required.

A. Additional Solution(s) of the Governing Equations on Other Grids
   A.1 Grid Refinement
   A.2 Grid Coarsening
   A.3 Other Unrelated Grid(s)
B. Additional Solution(s) of the Governing Equations on the Same Grid
   B.1 Higher Order Accuracy Solution(s)
   B.2 Lower Order Accuracy Solution(s)
C. Auxiliary PDE Solutions on the Same Grid
D. Auxiliary Algebraic Evaluations on the Same Grid; Surrogate Estimators
   D.1 Non-Conservation of Conservation Variables
   D.2 Non-Conservation of Higher Moments
   D.3 Zhu-Zienkiewicz Type Estimators
   D.4 Convergence of Higher Order Quadratures

TABLE 4.3.1. Sources of Additional Information for Error Estimation, Given a Discrete Solution of the Governing Partial Differential Equations on a Grid. (From Roache, 1997.)

Category C, Auxiliary PDE Solutions on the Same Grid (e.g., Van Straalen et al, 1995) does not simply involve a local evaluation of something. The key aspect here is that errors are transported, advected, diffused, etc. (However, it is also true that a simple local evaluation of something, without advection, is just what you want to guide solution adaptation; hence the different needs of error estimation for solution adaptation vs. quantification of uncertainty for a final calculation, as noted above.)

Category D methods, Auxiliary Algebraic Evaluations on the Same Grid, are relatively cheap, need no additional grid generation, and (sometimes) use no significant dynamic memory. However, the "error" evaluated usually has no direct relation to any error measure of engineering or scientific interest; hence my use of the term "Surrogate Estimators."

Category B, C and D methods, all of which do not require multiple grids, will be discussed in Chapter 7.

4.4 GRID REFINING AND COARSENING

Systematic grid convergence studies are the most common and straightforward methods, and arguably constitute the most reliable technique for the quantification of numerical uncertainty. Unlike the other methods available, this approach can be used to dependably consider the convergence of any quantity of interest, as well as the usual L_2 and L_∞ norms.

By "grid convergence studies" people usually mean Category A.1, Grid Refinement, but Grid Coarsening, Category A.2, usually would make more sense. If completely solved solutions are obtained on two grids, presumably the finer grid solution would be used, so the coarse grid solution could be used to estimate the error of the fine grid solution. Whether one "refines" or "coarsens" just depends on which grid was calculated first! So for completely solved solutions, grid refining and coarsening are identical.

A disadvantage of Category A methods is that multiple grid generations are required. Cartesian grids obviously pose no problem. For boundary-fitted structured grids, the simplest method for grid doubling (halving) is to generate the finest grid first, using whatever method is preferred (e.g., see Thompson et al, 1985 or Knupp and Steinberg, 1993) and then obtain the coarser grids by removing every other point (e.g., see Zingg, 1991,1992). For non-integer grid refinement (coarsening), the same generating equations and parameters should be used. (See Chapters 5 and 6 for further discussion including non-structured grid refinement and structured refinement of non-structured grids.)

Error estimation using other unrelated grids, Category A.3, pose an interesting challenge. By "unrelated grids" I mean two or more grids that are overlapping but not simply obtained one from the other; say grid A is finer than grid B in some regions but coarser in others, such as might be obtained in two steps of an *r*-type (redistribution) solution-adaptive grid. It would seem that the two solutions on unrelated grids would provide the additional information necessary to estimate the uncertainty in either, but to my knowledge a method for doing so has not been invented.

In order to quantify the uncertainty with systematically refined (coarsened) grids, we need the convergence rate p to estimate the error. Initially, we will assume that p is known, i.e., that we are using a rigorously Verified code on a well behaved problem, and are now concerned with quantification of the uncertainty of a particular calculation using two grid solutions. The same methods can be converted to Verify a code, i.e., to Verify (or determine) p.

It will be convenient to present the techniques of using systematic grid convergence studies (i.e., grid refinement and/or coarsening) to estimate errors in the next chapter on the Grid Convergence Index, or GCI. The GCI includes not only the error estimation *per se* but also a recommended level of conservatism, i.e. a "factor of safety" or "error banding," which provides some uniformity in the reporting of the "Quantification of Uncertainty." However, the techniques for systematic grid convergence studies covered in the next Chapter do not depend on the use of the GCI and can be used independently.

4.5 LEVELS OF SIMULATION USE

Simulations can be used at different levels, with corresponding appropriate error estimates. Benek et al (1996) note three stages of the evolution of aerodynamic CFD applications, with increasing accuracy requirements:

1. provide diagnostic information,
2. supply incremental data,
3. generate baseline data for the performance model data base.

At Level 1, Benek et al note that "CFD is an ideal diagnostic tool" because of its high information content and variety of data manipulation and presentation, useful for determining the origin of disturbances and flow instabilities, and location and strength of flow structures (e.g., vortices). They note that the absolute accuracy requirement is not high for diagnostics, because the results will be confirmed by additional analyses and/or experimental simulation. The basic requirement is that the predominant physics be reasonably represented; "Validation concerns are minimal." [The accuracy requirement at Level 1 is basically qualitative.]

At Level 2, quantitative accuracy is required, but only for increments. For example, transonic flow about a wing both in a wind tunnel and in a free stream might be simulated with the expectation not of predicting lift and drag, but only of predicting the correction to experimental data from a wind tunnel to free stream conditions. Often [not always], "highly accurate increments may be obtained from simulations that are less accurate than those required for base line data." Benek et al cite examples of "excellent agreement between measured and computed corrections" for transonic wall interference, model sting (support) interference, and mismatch between flight and wind tunnel Reynolds numbers. They note the potential for "evaluation of increments that cannot be readily obtained experimentally," such as geometric compromises between wind tunnel models and flight models, and Reynolds number corrections for inlet swirl. Also, CFD results applied incrementally can be used [essentially as an elaborate interpolating function] for providing higher resolution to low spatial resolution experimental data, e.g. engine inlet pressure data. (They also note that, in the world of aerodynamics, "the major stumbling block to Validation of wind tunnel correction methodology is that the majority of data is either classified or proprietary, and therefore, not generally available.")

At Level 3, quantitative accuracy is required for the absolute quantities, which is generally most demanding. However, the methodology for error estimation, the sources of additional information (above), etc., are the same, whether the focus is qualitative data, incremental data, or absolute base-line data. Even for absolute data, different requirements will occur, as various quantities will have various accuracy requirements and various sensitivities to grid resolution, etc. Thus the methods described herein are applicable to all three levels.

4.6 VERIFICATION OF COMPUTER ROUND-OFF ERRORS

Round-off error is sometimes confused with discretization error, and to emphasize the distinction people often speak of "computer round-off error" or "machine error" or "floating-point error." Round-off errors arise from the fact that computers do not work in the real number system, but only with a finite word-length subset of the real number system. This finite subset of floating-point numbers becomes an increasingly sparse (and therefore poorer) approximation to the complete real number line as the size of numbers increases. Different computers can have different representations of numbers, although this situation has vastly improved with the widespread adoption of IEEE standards. Even with identical floating-point representations, different computers can have different hardware implementations of arithmetic operations; again, this situation has improved with the adoption of IEEE standards. Although arguably distinct from round-off errors, for practical user purposes we can also include in this category errors in computer system evaluation of primitive functions such as trigonometric functions, exponentials and logs, as well as higher level functions which are not the domain of the computer manufacturer but of local system libraries, such as Bessel functions, error functions, Gaussian distribution functions, etc.

The basic concept can be demonstrated by the following example. For real numbers, the result of A = 3 × (1/3) gives A = 1, but we cannot *generally* trust the computer implementation of these two operations (first, dividing 1 by 3, then multiplying the result by 3) to give A = 1; it may give something like A = 0.9999997. It may work for an isolated calculation for modern computers, but we cannot depend on it. Some compilers even give a warning if the code uses an IF test for equality of floating-point numbers, since these may not be meaningful; in the above example, the test "IF (A .EQ. 1.0) ..." fails. Likewise, especially when the operations are widely separated with intermediate calculations and storage (actualization of intermediate results) in various subprogram modules, we cannot depend on obtaining the same answer from different computers, nor even on the same computer with different languages, nor even on the same computer and same language but different compilers or simply different options on the same compiler. What we *can* insist is that the various representations be good approximations of the correct, real-number answer and therefore close to one another, "close" being a function of the computer word-lengths. However, elaborate sequences of calculations can produce significant deviations; it is not unusual for calculations performed in double precision to lose enough to floating-point errors to be reduced to single-precision accuracy in the end.

It is amazing that this simple, virtually kindergarten fact of computer life is not recognized by some managers and regulators, who want all calculations to be identical across computer systems. Other sources of discrepancy include variations in user-adjustable numerical parameters such as iterative relaxation factors, iterative convergence criteria, etc.

On the other hand, it is perhaps surprising that round-off errors *most* often are either not a problem at all, or are easily controlled to a acceptable level, in physical simulations like CFD. The reason is the historical fact that computer hardware and system software have not evolved in isolation, but have co-evolved with simulation methods and applications.

Round-off errors in discretization *per se* (as opposed to calculations such as an equation of state evaluation that does not involve discretization) can be identified by grid convergence studies or other, often *ad hoc* approaches, but often they are simply demonstrated to not be significant. When they are significant, they manifest themselves as repeatable but *erratic* behavior as the grid is refined. (The behavior is not *chaotic* in the technical sense of *chaos*, *fractals*, etc., nor *random* in the technical sense of probability distributions, but just *erratic*.) It is not unusual to achieve iterative convergence to "machine zero," e.g. an identically steady state discrete solution, and to demonstrate grid convergence (of discretization error or truncation error) to 5 or 6 figures, but to then encounter divergence of results as the grid is further refined due to round-off error. This is easy to demonstrate in 1D, wherein extreme grid refinement is inexpensive. The example in Table 4.6.1 is from Roache and Knupp (1993).

N	Maximum Error	E_4
4	0.00022824765	0.058431
8	0.00001781814	0.072983
16	0.00000125631	0.082333
32	0.00000008359	0.087655
64	0.00000000539	0.090497
128	0.00000000034	0.091967
256	0.00000000002	0.092717
512	0.00000000000	0.093193
1024	0.00000000000	0.537369

Table 4.6.1. Demonstration of Round-Off Error in a Grid Convergence Study for Completed Richardson Extrapolation. Linear advection-diffusion equation with $Re = 1$. Grid spacing $\Delta = 1 / N$ and E_4 = max error / Δ^4. The approximate constancy (convergence) of E_4 up to N = 512 indicates the method is 4th-order accurate; the erratic deviation at N = 1024 is symptomatic of round-off error accumulation. (From Table IVa, Roache and Knupp, 1993)

Round-off error will become evident at more practical grid sizes when highly stretched grids are used, e.g. in turbulent boundary layer grids. Some algorithms are notoriously sensitive to round-off error, e.g. elliptic marching methods (Roache, 1995). Some physical problems are sensitive to and limited by round-off error, e.g. boundary layer transition calculations beginning from small initial perturbations cannot be carried through to full transition (Haynes et al, 1996). Often round-off errors can be isolated, or demonstrated to be not a problem, by changing word-length. Most modern engineering and scientific codes are designed to be compiled and run in double precision. In fact, most workstations actually run slightly *faster* in double precision than in single precision, being designed for the intended normal case of double precision calculations. This change from double to single precision, or vice versa, can sometimes be accomplished by a simple compiler option on modern computers–otherwise, it can be a surprisingly difficult and clumsy test to implement. Another approach is to introduce *imitation* round-off error by additive or multiplicative pseudo-random number noise generation on selected quantities.

Round-off errors can in principle manifest themselves during Verification of a Code, but are more likely to rear their heads in application calculations because of (typically) more extreme parameter variations and solution scales. Therefore, Verification for round-off error effects is best considered as part of Verification of a Calculation (here in Chapter 4) rather than Verification of a Code (Chapter 2).

A impressive test of an iterative solver is to demonstrate that it can achieve iterative convergence to "machine zero," i.e. to reach a point on the iterative path at which there are identically zero changes in all variables. Also, it can be a valuable debugging exercise on a coarse grid, and Oberkampf et al (1995) state that "it is highly recommended that this iterative convergence [to machine zero] be demonstrated on coarse grid solutions." However, in my opinion, it is not a necessary requirement for a code or an algorithm, as some authors suggest. Reaching machine zero in an iteration is *itself* an artifact of round-off error; a real number implementation would approach real zero only asymptotically. The fact that a particular code/algorithm actually reaches machine zero is somewhat accidental, since the noise of finite word-length arithmetic means that iterative changes near the limit of precision do not follow the real number arithmetic rules on which the iterative algorithm is based. Experience suggests that the simplest iterative algorithms (e.g., direct substitution iteration) are more likely to converge to machine zero, while more complicated algorithms (with more non-real-number arithmetic operations and intermediate storage) will often produce persistent, erratic fluctuations in the last one or two decimal significant figures of the variables. This should not disqualify these algorithms and codes, especially in view of the fact that the more complicated algorithms are often much more efficient than the simple algorithms.

4.7 EFFECT OF DIFFERING FORMULATIONS

One sometimes reads in the literature and in advertisements for commercial codes that different simulation results are to be expected from different formulations, e.g. different discretization methods (e.g., FEM vs. FDM vs. FVM), or different but mathematically equivalent forms of the continuum equations (e.g., incompressible flow formulated in terms of primitive velocity-pressure variables, vs. vorticity-stream function variables, vs. vorticity-velocity variables), or different equation solvers (direct Gaussian elimination vs. SOR vs. PCG vs. multigrid). Clearly, every different formulation will affect numerical accuracy, because of differences in discretization error, incomplete iteration error, and round-off error, so the answers will not be exactly identical. However, once numerical accuracy is quantified, the error bands from different simulations *must* overlap. If they do not, then a counter-example to the Verification of Calculations has been produced. *Either* Calculation A and its error estimate (say, a second-order FEM simulation using velocity - pressure variables with 100×100 quadrilateral elements and a direct solver), *or* Calculation B (say, a 4th-order vorticity-velocity simulation using 200×200 cells and a multigrid solver), *or* the comparison exercise itself is wrong. It should not be necessary to state this, but it apparently is.

Numerical methods are not part of the continuum parameters!

REFERENCES FOR CHAPTER 4

Benek, J. A., Kraft, E. M., and Lauer, R. F. (1996), "Validation Issues for Engine/Airframe Integration," AIAA Paper 96-2031, 27th AIAA Fluid Dynamics Conference, New Orleans, Louisiana, June 17–20, 1996.

Coorevits, P., Ladeveze, P., and Pelle, J.-P. (1995), "An Automatic Procedure with a Control of Accuracy for Finite Element Analysis in 2D Elasticity," *Computer Methods in Applied Mechanics and Engineering*, Vol. 121, 1995, pp. 91–120.

Dwyer, H. A., Kee, R., and Sanders, B. (1980), "Adaptive Grid Method for Problems in Fluid Mechanics and Heat Transfer," *AIAA Journal*, Vol. 18, pp. 1205–1212.

Hagen, S. C., Westerink, J. J. and Kolar, R. L. (1997), "Finite Element Grids based on a Localized Truncation Error Analysis," *International Journal for Numerical Methods in Fluids*, in revision.

Hall, D. J. and Zingg, D. W. (1995), "Viscous Airfoil Computations Using Adaptive Grid Redistribution," *AIAA Journal*, Vol. 33, No. 7, July 1995, pp. 1205–1210.

Haynes, T. S., Reed, H. L., and Saric, W. S. (1996), "CFD Validation Issues in Transition Modeling," AIAA Paper 96–2051, 27 AIAA Fluid Dynamics Conference, New Orleans, Louisiana, 17–20 June 1996.

Hetu, J-F and Pelletier, D. H. (1992), "Adaptive Remeshing for Viscous Incompressible Flows," *AIAA Journal*, Vol. 30, 1992, pp. 1986–1992.

Ilinca, F., Pelletier, D. and Garon, A. (1997), "An Adaptive Finite Element Method for a Two-Equation Turbulence Model in Wall-Bounded Flows," *International Journal for Numerical Methods in Fluids*, Vol. 24, 1997, pp. 101–120.

Knupp, P.M. and Steinberg, S. L. (1993), *Fundamentals of Grid Generation*, CRC Press, Boca Raton, Florida.

Lee, D. and Yeh, C. L. (1994a), "A Hybrid Adaptive Gridding Procedure for Three-Dimensional Flow Problems," *Computers and Fluids*, Vol. 23, 1994, pp. 39–53.

Lee, D. and Yeh, C. L. (1994b), "Computation of Turbulent Recirculating Flows Using a Hybrid Adaptive Grid," *Numerical Heat Transfer, Part A*, Vol. 26, 1994, pp. 415–430.

Lohner, R. (1989), "Adaptive Remeshing for Transient Problems," *Computer Methods in Applied Mechanics and Engineering*, Vol. 61, 1989, pp. 323–338.

Morgan, K., Peraire, J., Peiro, J., Hassan, O., and Probert, E. J. (1991), "Unstructured Grid Methods for Compressible Flows," *Computational Mechanics*, Y. K. Cheung, J. H. W. Lee and A. Y. Y. Leung, Eds., Vol. 2, Balkema, Rotterdam, pp. 1489–1497.

Oberkampf, W. L., Blottner, F. G., and Aeschliman, D. P. (1995), "Methodology for Computational Fluid Dynamics Code Verification/Validation," AIAA Paper 95–2226, 26th AIAA Fluid Dynamics Conference, San Diego, California, 19–22 June 1995.

Oden, J. T., Kennon, S. R., Tworzydlo, W. W., Bass, J. M., and Berry, C. (1993), "Progress on Adaptive h-p Finite Element Methods for the Incompressible Navier-Stokes Equations," *Computational Mechanics*, Vol. 11, 1993, pp. 421–432.

Pelletier, D. and Ignat, L. (1995), "On the Accuracy of the Grid Convergence Index and the Zhu-Zienkiewicz Error Estimator," Joint JSME–ASME Fluid Mechanics Meeting, August 14–18, 1995, *Quantification of Uncertainty in Computational Fluids Dynamics–1995*, ASME FED Vol. 213, Hilton Head, South Carolina, R. W. Johnson and E. D. Hughes, Eds., pp. 31–36.

Roache, P. J., Steinberg, S., and Moeny, W. M. (1984), "Interactive Electric Field Calculations for Lasers," AIAA Paper 84–1655, AIAA 17th Fluid Dynamics, Plasma Physics, and Lasers Conference, Snowmass, Colorado, 25–27 June 1984.

Roache, P. J. and Knupp, P. M. (1993), "Completed Richardson Extrapolation," *Communications in Applied Numerical Methods*, Vol. 9, pp. 365–374.

Roache, P. J. (1995), *Elliptic Marching Methods for Domain Decomposition*, CRC Press, Boca Raton, Florida.

Roache, P. J. (1997), "Quantification of Uncertainty in Computational Fluid Dynamics," *Annual Review of Fluid Mechanics*, Vol. 29, 1997, pp. 123–160.

Salari, K. and Steinberg, S. (1994), "Flux-Corrected Transport in a Moving Grid," *Journal of Computational Physics*, Vol. 111, No. 1, March 1994, pp. 24–32.

Thompson, J. F., Warsi, Z. U. A., and Mastin, C. W. (1985), *Numerical Grid Generation Foundations and Applications,* North-Holland, New York.

Van Straalen, B. P., Simpson, R. B., and Stubley, G. D. (1995), "A Posteriori Error Estimation for Finite Volume Simulations of Fluid Flow Transport," *Proceedings of the Third Annual Conference of the CFD Society of Canada*, Banff, Alberta, P. A. Thibault and D. M. Bergeron, Eds., Vol. I, 25–27 June 1995.

Zingg, D. W. (1991), "Viscous Airfoil Computations Using Richardson Extrapolation," AIAA Paper 91–1559–CP, AIAA 30th Aerospace Sciences Meeting, Reno, Nevada, January 6–9, 1992.

Zingg, D. W. (1992), "Grid Studies for Thin-Layer Navier-Stokes Computations of Airfoil Flowfields," *AIAA Journal*, Vol. 30, No. 10, October 1993, pp. 2561–2564. See also AIAA Paper 92–0184.

5

SYSTEMATIC GRID CONVERGENCE STUDIES AND THE GRID CONVERGENCE INDEX (GCI)

"If you want a new idea, read an old book."

Anonymous

5.1 INTRODUCTION

This chapter is taken primarily from Roache (1994), " Perspective: A Method For Uniform Reporting Of Grid Refinement Studies." Additional material is added from Roache (1997), "Quantification of Uncertainty in CFD," Roache (1995a), "Verification of Codes and Calculations," and Westerink and Roache (1995), "Issues in Convergence Studies in Geophysical Flow Computations." The chapter organization repeats some of the material in previous chapters but is used here to make the chapter more nearly self contained, and because the material deserves repetition, especially since it is applicable to commercial codes.

In (Roache, 1994) I proposed the use of a Grid Convergence Index (GCI) for the uniform reporting of grid convergence studies in Computational Fluid Dynamics and related disciplines.

The method provides an objective asymptotic approach to quantification of uncertainty of grid convergence. The basic idea is to approximately relate the results from any grid convergence test to the expected results from a grid doubling using a 2nd-order method. The GCI is based upon a grid convergence error estimator derived from the theory of generalized Richardson Extrapolation. It is recommended for use whether or not Richardson Extrapolation is actually used to improve the accuracy. A different form of the GCI applies to reporting coarse grid solutions when the GCI is evaluated from a "nearby" problem. The simple formulas may be applied *a posteriori* by editors and reviewers, even if authors are reluctant to do so.

The generalized Richardson Extrapolation described can be used independently of the GCI, for either Verification of Calculations or Verification of Codes (see Chapter 4). Likewise, as will become clear, the GCI is recommended for use in uniform reporting of grid convergence tests even if the conditions for the theory do not hold strictly.

5.2 BACKGROUND ON GRID CONVERGENCE REPORTING

There are other possible techniques for the quantification of numerical uncertainty, but systematic grid convergence studies are the most common, most straightforward and arguably the most reliable. The motivation for development of the uniform Grid Convergence Index was the inconsistent and confusing reporting of grid convergence studies in the engineering and scientific literature. The following hypothetical examples will suffice to illustrate the confusing reporting.

One paper states that the grid density was increased by 50%, resulting in a difference in some solution norm of 4% (of the fine grid solution) using a 1st-order accurate method. In another paper, grid density was doubled, resulting in a difference of 6%, using a 2nd-order method.

Which fine grid solution is more reliable?, i.e. better converged? More importantly, can the reader have any reasonable expectation that these numbers represent a "% accuracy error band," i.e. that the fine-grid calculations are probably accurate to within 4% or 6% of the true solution of the continuum equations?

Note that we are concerned herein with Verification of a particular Calculation, i.e. estimating and banding the grid convergence accuracy of a particular discretized solution. We assume that the code itself has already been Verified for the same class of problems, so that coding errors are not an issue and, if done properly, the order of accuracy has been Verified (see Chapter 3). Nor are we concerned with code Validation, e.g. that a turbulence model is adequate. (See Chapter 2 for terminology.) Also, it is worth repeating that this chapter (and similar grid convergence studies) address only "ordered" discretization errors, which by definition vanish as grid spacing Δ or $h \rightarrow 0$. Specifically, the errors introduced by the use of far-field computational boundaries must be assessed separately (see Chapters 2, 6, or Roache, 1972). Further, I consider herein only *a posteriori* error estimation, being of the opinion that useful *a priori* estimation is not possible for non-trivial problems.

5.3 RICHARDSON EXTRAPOLATION

Richardson Extrapolation, also known as "h^2 extrapolation" and "the deferred approach to the limit" and "iterated extrapolation," was first used by Richardson in 1910 (in a structures problem of determining the stresses in a masonry dam) and later embellished in 1927. The discrete solutions f are assumed to have a series representation in the grid spacing h (or Δ) of

$$f = f_{exact} + g_1 h + g_2 h^2 + g_3 h^3 + \cdots \tag{5.3.1}$$

The functions g_1, g_2, etc. are defined in the continuum and do not depend on any discretization. For infinitely differentiable solutions, they are related to all orders to the solution derivatives through the elementary Taylor series expansions, but this is not a necessary assumption for Richardson Extrapolation, nor is the infinite series indicated in Equation 5.3.1. It is only necessary that Equation 5.3.1 be a valid definition for the order of the discretization. Thus, the extrapolation may be valid for finite volume solutions, finite element solutions, etc.

For a 2nd-order method, $g_1 = 0$. Then the idea is to combine two separate discrete solutions f_1 and f_2, on two different grids with (uniform) discrete spacings of h_1 fine grid) and h_2 (coarse grid), so as to eliminate the leading order error terms in the assumed error expansion, i.e. to solve for g_2 at the grid points in Equation 5.3.1, substitute this into Equation 5.3.1 and obtain a more accurate estimate of f_{exact}. The result is the original statement (Richardson, 1927) for h^2 extrapolation.

$$f_{exact} = \frac{h_2^2 f_1 - h_1^2 f_2}{h_2^2 - h_1^2} + \text{H.O.T.} \tag{5.3.2}$$

where H.O.T. are higher order terms. Using the grid refinement ratio $r = h_2 / h_1$, this can be conveniently expressed in terms of a correction to the fine grid solution f_1, dropping H.O.T.

$$f_{exact} \cong f_1 + \frac{f_1 - f_2}{r^2 - 1} \tag{5.3.3}$$

The most common use of this method is with a grid doubling, or halving. (These are *identical*. Both use two grids, one twice as fine as the other, i.e. we have a coarse grid and a fine grid. Whether we "doubled" or "halved" just depends on which calculation came first.) With $r = 2$, Equation 5.3.3 becomes

$$f_{exact} \cong 4/3\, f_1 - 1/3\, f_2 \tag{5.3.4}$$

It is often stated that Equation 5.3.4 is 4th-order accurate if f_1 and f_2 are 2nd-order accurate. Actually, as known by Richardson, it is true only if odd powers are absent in the expansion (5.3.1), which he achieved by assuming the exclusive use of 2nd-order centered differences. If uncentered differences are used, e.g. upstream weighting of advection terms, even if these are 2nd-order accurate (3-point upstream), the h^2 extrapolation is 3rd-order accurate, not 4th. As a practical limitation, even extrapolations based on centered differences do not display the anticipated 4th-order accuracy until the cell Reynolds number Rc is reduced; for the one-dimensional advection-diffusion equation with Dirichlet boundary conditions, $Rc < 3$ is required (Roache and Knupp, 1993; see also Chapter 3).

Although Richardson Extrapolation is most commonly applied to grid doubling, and is often stated to be *only* applicable to integer grid refinement (e.g., Conte and DeBoor, 1965), grid doubling is not required. In order to use Equation 5.3.3 it is necessary to have values of f_1 and f_2 at the same points, which would seem to require commonality of the discrete solutions, and therefore integer grid refinement ratios r (grid doubling/halving, tripling, etc.). However, even in his 1910 paper, Richardson looked forward to defining a continuum f_2 by higher order interpolation, and in the 1927 paper he had a specific approach worked out. Ferziger (1993) alludes to this approach with less detail but more generality. Similarly, Richardson Extrapolation is commonly applied only to obtaining a higher-order estimate on the coarse grid with $h_2 = 2h_1$, but Roache and Knupp (1993) showed how to obtain 4th-order accuracy on *all* fine-grid points by simple 2nd-order interpolation, not of the solution values f_2, but of the extrapolated *correction* from Equation 5.3.4, i.e. by 2nd-order interpolation of 1/3 $(f_1 - f_2)$. The use of simple 2nd-order interpolation avoids complexities with non-uniform grids and near-boundary points. S. A. Richards (1997) extended the method for "Completed Richardson Extrapolation" to space and time, including possible mixed space-time order of accuracy and more generality than grid doubling.

Richardson (1910,1927) also considered 6th-order extrapolation (using 3 grid solutions to eliminate g_2 and g_4), parabolic and elliptic equations, staggered grids (then called "interpenetrating lattices"), rapid oscillations and the $2h$ wavelength limit (the "cell Reynolds number" limit for "wiggles" in CFD; see Roache, 1972), *a priori* error estimates, singularities, integral equations, statistical problems, Fourier coefficients, and other non-calculus problems. For example, Richardson (1927) showed the power of the method in an elegant example of extrapolating two very crude approximations to a circle, namely an inscribed square and an inscribed hexagon, to get an estimate of π with 3-figure accuracy, without using any trigonometry. (This calculation will be given later in this chapter.)

The usual assumptions of smoothness apply, as well as the assumption (often verified) that the local error order is indicative of the global error order. The extrapolation must be used with considerable caution, since it involves the additional assumption of monotone truncation error convergence in the mesh spacing h. This assumption may not be valid for coarse grids, or possibly other conditions. The discretization method must be 2nd order everywhere, including boundaries; for example, Gourlay and Morris (1968) showed how to successfully apply Richardson extrapolation in nonlinear hyperbolic systems with over-determined boundary information, and gave references to earlier work on Richardson Extrapolation involving starting values in multi-step ODE solvers. Also, the extrapolation magnifies machine round-off errors and incomplete iteration errors (Roache, 1972). In spite of these caveats, the method is extremely convenient to use compared to forming and solving direct 4th-order discretizations, which involve more complicated stencils, wider bandwidth matrices, special considerations for near-boundary points and non-Dirichlet boundary conditions, additional stability analyses, etc., especially in non-orthogonal coordinates which generate cross-derivative terms and generally complicated equations. Such an application was given by the present author in Roache (1982); see also Chapter 6.

The method is in fact oblivious to the equations being discretized and to the dimensionality of the problem, and can easily be applied as a *postprocessor* (Roache, 1982) to solutions on two grids with no reference to the codes, algorithms or governing equations which produced the solutions, as long as the original solutions are indeed "2nd-order accurate" or more generally, "p-th order accurate." I use the common but somewhat abusive terminology of "2nd-order accurate solution" to mean a solution obtained by a Verified 2nd-order accurate method applied in the asymptotic range of grid spacing. Even the expression "2nd-order accurate method" itself can be misleading by unrealistically raising expectations that the method will be 2nd-order accurate applied to any problem. A clumsy but more precise description is "a method whose leading order term in the Taylor series or other convergence analysis is 2nd order."

The difference between the 2nd-order solution and the extrapolated 4th-order solution is itself a useful diagnostic tool, obviously being an error estimator (although it does not provide a true *bound* on the error except possibly for certain trivial problems). It was used very carefully, with an experimental determination rather than an assumption of the *local* order of convergence, by de Vahl Davis (1983) in his classic Benchmark study of a model free convection problem. See Nguyen and Maclaine-Cross (1988) for application to heat exchanger pressure drop coefficients. Zingg (1991,1992) applied the Richardson error estimator to airfoil lift and drag calculations in body-fitted grids. Also, his data indicate that Richardson Extrapolation can be applied to the estimation of far-field boundary errors, with the error being 1st-order in the inverse of distance to the boundary. (See Chapter 6, Section 6.10.) Blottner (1990) used Richardson Extrapolation to estimate effects of artificial dissipation terms in hypersonic flow calculations. (See Chapter 6, Section 6.11.) See other applications by Anderson and Batina (1988), Caruso et al (1985), Kessler et al (1988), Williams (1989).

An important aspect of Richardson Extrapolation is that it applies not only to point-by-point solution values, but also to solution functionals, e.g., lift coefficient C_L for an aerodynamics problem or integrated discharge for a groundwater flow problem, provided that consistent or higher-order methods are used in the evaluations (e.g., 2nd- or higher order quadratures for lift) as well as the basic assumption that the "order" of the method applies globally as well as locally. If Richardson Extrapolation is applied to produce (say) 4th- order accurate grid values, one could in principle calculate a 4th-order accurate functional like C_L from the grid values, but it would require careful implementation of 4th-order accurate quadratures. It is much simpler to apply the extrapolation directly to the C_L's obtained in each grid, requiring only 2nd-order quadratures. Indeed, this is a major attraction of Richardson Extrapolation compared to using 4th-order accurate stencils solved either directly or by deferred corrections. (Note, however, that the two approaches yield different answers, although both are 4th-order accurate if done properly.)

A very significant yet overlooked *disadvantage* of Richardson Extrapolation is that the extrapolated solution generally is not "conservative" in the sense of maintaining conservation properties (e.g., Roache, 1972). This could well dictate that Richardson Extrapolation not be used. For example, if it were used on the ground-water flow simulations for the Waste Isolation Pilot Plant (WIPP PA Dept., 1992; Helton et al, 1995,1996), it would be "more accurate" in some metric, but would introduce additional non-conservative (i.e., lack of conservation property) source terms into the radionuclide transport equation. It is also noteworthy that Richardson (1927) pointed out that the accuracy of the extrapolation does not apply to arbitrarily high derivatives of the solution. The extrapolation can introduce noise to the solution which, although low level, may decrease the accuracy of the solution higher derivatives.

Thus, it is not advocated here that Richardson Extrapolation necessarily be used to improve the reported solution, since that decision involves these considerations and possibly others. What is advocated is that, regardless of whether Richardson Extrapolation is used to improve the solution, it can be used to estimate the error, and the Grid Convergence Index (defined herein and based on the generalized theory of Richardson Extrapolation) can be used to uniformly *report* grid convergence studies.

5.4 A GENERALIZATION OF RICHARDSON EXTRAPOLATION

Without assuming the absence of odd powers in the expansion of Equation 5.3.1, we can generalize the usual Richardson Extrapolation (5.3.4) to p-th order methods and r-value of grid ratio, again eliminating the leading term in the error expansion, as follows.

$$f_{exact} \cong f_1 + \frac{f_1 - f_2}{r^p - 1} \tag{5.4.1}$$

If the next term in the series (Equation 5.3.1) is zero, e.g. if centered differences were used, then the extrapolation is $(p + 2)$ order accurate. But generally, and notably if upstream-weighted methods for advection have been used, the extrapolation is $(p + 1)$-order accurate.

It may easily be verified that Equation 5.4.1 is valid for multidimensions in any coordinates, including space and time, provided that the same grid refinement ratio r is applied, and the order p is uniform, in all space and time directions.

In Equation 5.4.1, the correction to the fine grid solution f_1 is obviously an error estimator of the fine grid solution. Expressing this as an **E**stimated fractional error E_1 for the fine grid solution f_1, we have

$$E_1[\text{fine grid}] = \frac{\varepsilon}{r^p - 1} \tag{5.4.2}$$

$$\varepsilon = \frac{f_2 - f_1}{f_1} \tag{5.4.3}$$

Defining the **A**ctual fractional error A_1 of the fine-grid solution as usual,

$$A_1[\text{fine grid}] = \frac{f_1 - f_{exact}}{f_{exact}} \tag{5.4.4}$$

and using Equations 5.4.1–4 and the binomial expansion gives

$$A_1[\text{fine grid}] = E_1 + \mathrm{O}\left(h^{p+m}, E_1^2\right) \tag{5.4.5}$$

where $m = 1$ generally or $m = 2$ if centered differences have been used. Thus, E_1 is an *ordered error estimator*, i.e. an ordered approximation to the actual fractional error of the fine grid solution. E_1 is a good approximation when the solution is of reasonable accuracy, i.e. when $E_1 << 1$.

This is generally *not* true of ε in Equation 5.4.3, which is the quantity commonly reported in past grid convergence studies. That is, ε is not always an error estimator since it does not take into account r or p. For $r < 2$ and $p = 1$, ε alone is optimistic, under-estimating the grid convergence error compared to E_1 (by a factor of 2 for $r = 1.5$). For $r = 2$ and $p = 2$, ε alone is conservative, over-estimating the grid convergence error compared to E_1 (by a factor of 3). Note that ε can be made (almost) arbitrarily small, just by choosing $r \cong 1$. [The only restriction is that r is limited by the integer character of the number of grid points, so the smallest possible $r = N/(N-1)$ where N is the number of grid points in each direction of the fine grid.] This is analogous to the situation wherein an arbitrarily small tolerance on *iterative* convergence can always be met by using an arbitrarily small relaxation factor, belying the adequacy of such an iterative convergence criterion (Roache, 1972; Ferziger, 1993, Ferziger and Peric, 1996).

E_1 may of course be expressed as a relative error, either a fraction or a % of some normalizing value. If the normalizing value is taken as the local discrete value itself (f_1) or (if known) the local (or global) exact value (f_{exact}), then, like any *relative* error indicator, it will become meaningless when f_1 or f_{exact} is zero or small relative to ($f_2 - f_1$). In such a case, the denominator of Equation 5.4.3 should be replaced with some suitable normalizing value for the problem at hand, as would the usual definition of actual relative error A_1 in Equation 5.4.4.

As described earlier, one may chose to not use Richardson Extrapolation for good reasons, e.g., due to concern over the actual order of the method, or accumulation of round-off error, or incomplete iterative convergence error, or uncertainty that the asymptotic range has been reached, or lack of the conservation property in the extrapolated results, etc. But whether or not one chooses to use or report the results of the extrapolation, one can still use the theoretical basis to *consistently* report the results of the grid convergence study.

5.5 RICHARDSON'S EXTRAPOLATION FOR π

Richardson (1927) showed the power of his extrapolation method in an entertaining example of extrapolating two very crude approximations to a circle, namely an inscribed square and an inscribed hexagon, to get an estimate of π with 3-figure accuracy, without using any trigonometry. This exemplary calculation, repeated here, shows the generality of the concept of "grid spacing" and "discretization measure" h or Δ.

We use the fundamental definition of π, the ratio of the circumference of the circle to its diameter, and consider the circle of unit radius to be the limit as $N \to \infty$ of the same ratio defined for a regular inscribed N-sided polygon. Call that ratio Π_I where I indicates "inscribed" polygon. For $N = 4$, $\Pi_{I,4} = 2\sqrt{2}$. For $N = 6$, each of the 6 triangles is equilateral, and (without using trigonometry) we get $\Pi_{I,6} = 3$ exactly. At this point, lacking trigonometry, we have to just assume that convergence is quadratic in the discretization measure $\Delta = 1/N$, i.e. $p = 2$. [Trigonometry would show that this is indeed true; the exact formula for $\Pi_{I,N} = N \sin \pi / N$ has the series expansion $= \pi - 2/3\, \pi^3 \Delta^2 + O(\Delta^4)$. See also Ferziger, 1981.] In terms of the grid discretization ratio r, for Equation 5.4.1, with coarse "grid" $N = 4$ and fine "grid" $N = 6$, we have $r = (1/4)/(1/6) = 3/2$, so $r^2 = 9/4$, and $1/(r^2 - 1) = 4/5$. Applying this to Equation 5.4.1, we obtain the extrapolated estimate $\Pi_I = 3 + (4/5)(3 - 2\sqrt{2}) = 3.1372583$, which is only 0.138% low, or correct to better than 3 figures significant figures.

Richardson notes that this error is only 1/33 of the error of the $N = 6$ estimate. To obtain as good a result without extrapolation would require $N = 35$, and you would have to know some trigonometry and do *much* more calculating.

[The result for the *circumscribed* polygons is $\Pi_C = 2\sqrt{3} + (4/5)(2\sqrt{3} - 4) = 3.0353829$, which is 3.381% low, or correct to only 1 significant figure. The problem is that the leading error in the series expansion for the *tan* required for circumscribed polygons is twice as large as that for *sin* used for the inscribed polygons.]

5.6 GRID CONVERGENCE INDEX FOR THE FINE GRID SOLUTION

Although the error estimator E_1 of Equation 5.4.2 is based on a rational and consistent theory, it is certainly not a *bound* on the error. (Nor is a reliable and practically tight bound on solution error for nontrivial and nonlinear problems likely to be forthcoming, in my opinion.) What is generally sought in engineering calculations is not a true "error bound" but just an "error band," i.e. a tolerance on the accuracy of the solution which may in fact be exceeded, but in which the reader/user can have some practical level of confidence. The error estimator E_1 itself does not provide a very good confidence interval. One might expect that it is equally probable that E_1 be optimistic as conservative, i.e. it is just as likely that the actual error A_1 be greater than E_1 as less than E_1. This would correspond roughly to a 50% confidence band. A well-founded probability statement on the error estimate, such as a statistician would prefer (e.g., a 2σ limit) is not likely forthcoming for practical PDE problems. However, based on cumulative experience in the CFD community, at least a practical confidence level exists for the ε of Equation 5.4.3 obtained using a grid doubling and a verified 2nd-order accuracy code.

That is, for a grid doubling with a 2nd-order method, and some indication that the calculations are within the asymptotic range of convergence, most practitioners would accept the ε of Equation 5.4.3 as a reasonable error band, in the flavor of a statistician's 2σ range or an experimentalist's 20:1 odds (Kline and McClinstock, 1953). An ε of (say) 6% would be taken to indicate (not absolutely, but with reasonable confidence) that the fine grid solution was within 6% of the asymptotic answer. This confidence is well justified by the theory of Richardson Extrapolation, which shows, from Equations 5.4.2–3 with $r = 2$ and $p = 2$, that the error estimate E_1 is only 1/3 of this error band, or 2%.

The idea behind the Grid Convergence Index is to approximately relate the ε of Equation 5.4.3 obtained by whatever grid convergence study is performed (whatever p and r) to the ε that would be expected from a grid convergence study of the same problem with the same fine grid using $p = 2$ and $r = 2$, i.e. a grid doubling with a 2nd-order method. The relation is based on equality of the error estimates. Given an ε from an actual grid convergence test, the GCI is derived by calculating the error estimate E_1 from Equations 5.4.2–3, then calculating an equivalent ε that would produce approximately the same E_1 with $p = 2$ and $r = 2$. The absolute value of that equivalent ε is the (relative error) Grid Convergence Index for the fine grid solution, which is conveniently expressed as

$$\mathrm{GCI}\left[\text{fine grid}\right] = F_s \frac{|\varepsilon|}{r^p - 1} \quad , \quad F_s = 3 \qquad (5.6.1)$$

where ε is defined in Equation 5.4.3. Obviously, if the denominator of Equation 5.4.3 is small, ε should be normalized by some other characteristic quantity for the calculation (as discussed earlier) or alternately it can be evaluated as an absolute (rather than relative) quantity (i.e., without the division by f_1 or any normalizing value), in which case Equation 10 produces an "absolute error" GCI.

We note immediately that for a grid doubling ($r = 2$) with a 2nd-order method ($p = 2$), the denominator = 3, and we obtain GCI = $|\varepsilon|$, as intended. For any r and p, if the coefficient F_s is chosen as $F_s = 1$, then the GCI = E_1. Thus F_s may be interpreted as a "factor of safety" over the Richardson Error Estimator E_1.

The purpose of the GCI is not to preclude more convincing grid convergence tests (such as using Richardson Extrapolation over several grid refinements, e.g., Shirazi and Truman, 1989; Blottner, 1990; Zingg, 1991,1992; see Chapters 4 and 6). The modest purpose herein is just to get minimal two-calculation grid convergence exercises onto a uniform reporting basis.

The GCI, like the theory of Richardson Extrapolation on which it is based, is equally applicable not only to grid values, but also to solution functionals (e.g., C_L) and to plotted curves, wherein ε may be read visually or calculated from interpolated tabular values. Thus it may be used to produce plots of the estimated error band about a fine grid solution by post-processing the results of any two grid solutions. Non-physical oscillations in the solutions ("wiggles," e.g. see Roache, 1972) are of course a cue that the solutions are not in the asymptotic range, Richardson Extrapolation is not accurate, E_1 of Equation 5.4.2 is not a valid error estimator, and confidence in the GCI as an error band is not justifiable.

Applying Equation 5.6.1 to the hypothetical cases in the second paragraph of Section 5.2, we see that a 4% difference from a grid refined by 50% using a 1st-order method gives a fine-grid GCI = 24%, whereas a 6% difference from a doubled grid using a 2nd-order method gives a fine-grid GCI = 6%. Even though the first paper's reported raw deviation ε from coarse to fine grid calculations might appear at first glance to be better than that reported in the second paper (4% compared to 6%), it is in fact not nearly as well converged (24% compared to 6%), as indicated by the Grid Convergence Index. (The ratio does not depend on the value of the "factor of safety" F_s used in Equation 5.6.1, but would also apply if the Richardson Error Estimator E_1 were used, i.e. with $F_s = 1$.)

For a less hypothetical example, consider the grid convergence results reported by the present author (Roache, 1982; Chapter 16 of Roache, 1998) for Benchmark calculations of weakly separated flows obtained using Richardson Extrapolation applied with grid doubling. The reported quantification of convergence was the maximum fractional deviation ε_4 between the fine-grid 2nd-order and the extrapolated 4th-order solution f_4 from Equation 5.3.4,

$$\varepsilon_4 = \frac{f_1 - f_4}{f_4} \tag{5.6.2}$$

The values reported were $\varepsilon_4 = 0.17\%$ for wall vorticity and 0.13% for a velocity profile at a longitudinal station traversing the separation bubble. This ε_4 is easily related to the ε of Equation 5.4.3; combining Equations 5.3.4, 5.4.3 and 5.6.2 shows

$$\varepsilon = 3 - \frac{3}{1+\varepsilon_4} = 3\varepsilon_4 + O\left(\varepsilon_4^2\right) \tag{5.6.3a}$$

$$\text{GCI}\left[\text{fine grid}\right] = 9\varepsilon_4 \tag{5.6.3b}$$

The reported grid convergence criteria (Roache, 1982) of 0.17% for wall vorticity and 0.13% for velocity *for the fine grid 2nd-order solution* would now be replaced by the much more conservative GCI [fine grid] = 1.53% and 1.17%.

In such cases wherein Richardson Extrapolation is actually used to produce a higher order accurate solution, rather than just to estimate the error of the 2nd-order fine grid solution, the GCI of Equation 5.6.1 (or 5.6.3b) appears to be unfairly conservative. The solution used is the (say) 4th-order accurate solution, but the reported GCI would be the same even if only the 2nd-order accurate fine grid solution were used. That is, E_1 and GCI are respectively the Richardson Error Estimator and Grid Convergence Index for the fine grid 2nd-order solution, not for the 4th-order solution. Although we expect the extrapolated solution to be more accurate than the 2nd-order fine grid solution, we would need additional information (a solution on a third grid) to estimate the error of the extrapolated solution itself. Such a third grid solution could be used in principle (possibly not in practice, for difficult nonlinear problems) to extrapolate a 6th-order accurate solution. The error estimate (and therefore the GCI) will always lag the best solution estimate. This is quite conservative when the conditions for validity of Richardson Extrapolation have been convincingly demonstrated by numerical experiments (e.g. Roache, 1982; Shirazi and Truman, 1989; Blottner, 1990; Roache and Knupp, 1993). A heuristic extension for such situations (Roache, 1994) is to report the GCI for the extrapolated solution based on Equation 5.6.1 with ε replaced by ε_4 from Equation 5.6.2, giving

$$\begin{aligned} \text{GCI}\left[\text{extrapolated solution}\right] &\cong \left|E_1\left[\text{fine grid}\right]\right| \\ &= \frac{|\varepsilon|}{r^p - 1} \end{aligned} \tag{5.6.4}$$

This agrees with the reported grid convergence criteria (Roache, 1982) of 0.17% for wall vorticity and 0.13% for velocity.

On the other extreme, it is recognized that ostensibly 2nd-order algorithms may fail to attain 2nd-order performance in a particular calculation, due to coding quirks or errors, subtleties in nonlinear problems, overly strong grid stretching, failure to attain the asymptotic range, etc. (See, e.g., de Vahl Davis, 1983; Steinberg and Roache, 1985; Shirazi and Truman, 1989; Roache et al, 1990; Westerink and Roache, 1995; see also Chapter 6.) Unless the author has convincingly Verified that the code actually attains the theoretical order, at least on a "nearby" problem, the more conservative value of $F_s = 3$ should be used in reporting the GCI in Equation 5.6.1.

Two calculations of the same problem with the same value for GCI, say a 1st-order calculation on a finer grid and a 2nd-order calculation on a coarser grid, are not quite indifferent as to the uncertainty of the calculations. The GCI of the 1st-order calculations is based on an only 2nd-order accurate error estimator, whereas the GCI of the 2nd-order calculations is based on a 3rd- or 4th-order accurate error estimator. Thus, even with the same GCI, the 2nd-order calculations have less uncertainty (in their uncertainty estimates) than the 1st-order calculations. Generally, it is more difficult to judge grid convergence for 1st-order methods; e.g., see Leonard and Drummond (1995).

5.7 GRID CONVERGENCE INDEX FOR THE COARSE GRID SOLUTION

Seemingly, if we have a fine grid and a coarse grid solution, we would be expected to use the fine grid solution, so reporting of the above fine-grid GCI of Equation 5.6.1 would apply.

However, a practical scenario occurs for which the contrary situation applies, i.e. we use the coarse grid solution.

Consider a parametric study in which hundreds or thousands of variations are to be run. For example, consider a three-dimensional time-dependent study of dynamic stall, with perhaps 3 Mach numbers, 6 Reynolds numbers, 6 airfoil thickness ratios, 3 rotor tip designs, and 2 turbulence models: a total of 648 combinations. For another, consider the Monte Carlo type study of groundwater transport with thousands of runs (WIPP PA Dept., 1992; Helton et al, 1995,1996). A scrupulous approach would require a grid convergence study for each case, but most engineers would be satisfied with one or a few good grid convergence tests, expecting, e.g., that a grid adequate for a NACA 0012 airfoil could be assumed to be adequate for a NACA 0015 airfoil. (In fact, this is often not justified by experience, e.g. stall characteristics can be quite sensitive to thickness ratio.) So for the bulk of the stack of calculations, we would be using the coarse grid solution, and we want a Grid Convergence Index for it. That is, we derive the (relative error) GCI from Equation 5.4.1, not as the correction to the fine grid solution f_1, but as the correction to the coarse grid solution f_2. In this case, the error estimate changes and *must be less optimistic.*

$$f_{exact} \cong f_2 + (f_1 - f_2)\, r^p / (r^p - 1) \tag{5.7.1}$$

The coarse-grid GCI is then

$$GCI\,[coarse\ grid\,] = F_s\,|\varepsilon|\, r^p / (r^p - 1), \;\; F_s = 3 \tag{5.7.2}$$

$$\text{GCI}\left[\text{coarse grid}\right] = r^p\text{GCI}\left[\text{fine grid}\right] \tag{5.7.3a}$$

$$\text{GCI}\left[\text{coarse grid}\right] = r^p\text{GCI}\left[\text{fine grid}\right] + F_s|\varepsilon| \tag{5.7.3b}$$

The last of these is easy to interpret; the error estimate for the coarse grid solution is just the error estimate for the fine grid solution, plus the difference between the solutions, which is ε. The GCI difference is then $F_s\,\varepsilon$.

Applying this equation to the hypothetical cases in the second paragraph of Section 5.2, we see that a 4% difference from a grid refined by 50% using a 1st-order method gives a coarse-grid GCI = 36%, whereas a 6% difference from a doubled grid using a 2nd-order method gives a coarse-grid GCI = 24%. Note that the higher-order method appears to be working against us here, because we are coarsening, rather than refining, the grid. In actuality, the ε for the higher order method will be smaller for the same grid refinement close to convergence.

GCI values for some common combinations of r and p, normalized to $\varepsilon = 1\%$, are given in Table 5.7.1.

	Fine Grid GCI				Coarse Grid GCI		
p	$r = 2$	1.5	1.1	p	$r = 2$	1.5	1.1
1	3.00%	6.00%	30.00%	1	6.00%	9.00%	33.00%
2	1.00%	2.40%	14.29%	2	4.00%	5.40%	17.29%
3	0.43%	1.26%	9.06%	3	3.43%	4.26%	12.06%
4	0.20%	0.74%	6.46%	4	3.20%	3.74%	9.46%

Table 5.7.1. Grid Convergence Index (GCI) with F_s = 3 calculated from Equations 5.6.1 and 5.7.2 for common values of grid ratios (r) and orders of the basic numerical method (p), for both coarse grid solutions and fine grid solutions, normalized to ε = 1%.

5.8 EXAMPLE GCI CALCULATION

A simple example of the calculation of a Grid Convergence Index follows. I choose the easily reproduced case of a steady-state Burgers equation

$$-U\,U_x + U_{xx}\,/\,Re = 0,\;\; U(0) = 1,\;\; U(1) = 0 \tag{5.8.1}$$

for Re = 1000 solved with 2nd-order centered differences on a uniform grid, and evaluate the one-dimensional "shear" $f = dU/dx$ at x = 1. Using a fine grid calculation with 2000 cells, we obtain $f_1 = -529.41$. Then we coarsen the grid to 1600 cells (r = 1.25) and obtain $f_2 = -544.48$. The quantity typically reported (from Equation 5.4.3) would be $|\varepsilon| = 100\% \times (f_2 - f_1)/f_1 =$ 2.85%. The factor $(r^p - 1)$ is $(1.25^2 - 1) = 0.5625$. The magnitude of the Richardson Extrapolation error estimator for the fine grid solution from Equation 5.4.2 is $|E_1| = |\varepsilon| / (r^p - 1)$ = 2.85% / 0.5625 = 5.07%. The fine grid value of the Grid Convergence Index from Equation 5.6.1 with F_s = 3 is GCI [fine grid] = $3\,|\varepsilon|/(r^p - 1) = 3 \times 2.85\% / 0.5625 = 15.20\%$. Comparison with the exact solution $f_{exact} = -500.00$ indicates that the exact magnitude of the fine-grid error A_1 is $100\% \times |(f_2 - f_{exact})/f_{exact}| = 100\% \times |-529.41 + 500.00| / 500.00 = 5.88\%$. As is typical, the Richardson Extrapolation error estimator E_1 is not conservative (5.07% < 5.88%), whereas the GCI is conservative and quite so with F_s = 3 (15.20% > 5.88%), in the spirit of a 2σ error band. With F_s = 1.25 (see the following Section 5.9), the GCI = 6.34% > 5.88%, still conservative but more palatable.

If the coarse grid solution (or the coarse grid solution to a nearby problem) were to be used, the Richardson Extrapolation error estimator would be increased to $|E_1| + |\varepsilon|$ = 5.07% + 2.85% = 7.92%. The GCI with F_s = 3 would be increased by $3|\varepsilon|$ as in Equation 5.7.3b to GCI [coarse grid] = 15.20% + 3 × 2.85% = 23.75%. The actual magnitude of the coarse grid error is 100% × $|(f_2 - f_{exact})/f_{exact}| = 100\% \times |-544.48 + 500.00| / 500.00 = 8.90\%$. Again, the Richardson Extrapolation error estimator E_1 is not conservative for the coarse grid (7.92% < 8.90%), whereas the GCI is conservative and quite so with F_s =3 (23.75% > 8.90%). With F_s = 1.25, the GCI = 9.90% > 8.90%, still adequately conservative.

5.9 SHOULD THE COEFFICIENT BE "1" OR "3" OR "1.25"?

The functional form of the definition of the GCI (Equations 5.6.1. 5.7.2, 5.7.3) is rational and objective, but the coefficient $F_s = 3$ is a judgment call. It could arguably be "1," or conceivably "1.5" or "2" or something else between 1 and 3.

F_s is essentially a "factor of safety," and the value $F_s = 3$ is possibly too conservative. As the quality and rigor of the grid convergence study increases, so does the conservatism of using the coefficient $F_s = 3$ in the definition of the GCI. However, consider practical complications such as rapidly varying coefficients from turbulent eddy viscosities or strong grid stretching, nonlinear systems, non-uniform behavior of various error metrics, experimental determination of spatially varying p (e.g., see de Vahl Davis, 1983), non-monotonic convergence (e.g., see Celik and Zhang, 1993, 1995). There can also be vagaries associated with defining r when power-law grid stretching is used (e.g., see Celik and Zhang, 1993, 1995), when regions are partitioned geometrically (as in Domain Decomposition or multi-block grid generation methods) giving non-smooth spatially varying r, and the surprisingly common practice of using $r_x \neq r_y$. Such complications, while not necessarily contradicting the ultimate applicability of Richardson Extrapolation (i.e., in the asymptotic range), do increase the uncertainty associated with the error estimate for practical engineering calculations. Likewise, if the grid convergence exercise is only performed for a representative "nearby" problem, uncertainty is increased. (See Cosner, 1995 for experienced testimony that "nearby" problems adequate to predict grid resolution requirements can be difficult to find.) Also, see discussion in Westerink and Roache (1995) and in Chapter 6 on the distinction between formal, actual asymptotic, and observed convergence rates. These considerations provide additional rationale for retaining "3" as the coefficient, in the sense of a "factor of safety," for reporting minimal two-grid convergence results.

As noted above, using the value $F_s = 3$ makes grid doubling with 2nd-order methods into the standard of comparison. This is not intended to make 2nd-order methods the goal, only the standard. (Like an IQ of 100, it is not meant to discourage genius.) It just means that for $p = 2$ and $r = 2$, we obtain GCI [fine grid] = ε. That is, it does not change what authors who use grid doubling with a 2nd-order method already have been reporting, namely ε.

Using the value $F_s = 1$ would make the GCI equal E_1, the error estimator obtained from Richardson Extrapolation. As noted earlier, since this is the best estimate we can make given only the information from calculations on two grids, we can only expect equal probability that the true answer is inside or outside of this band. Also, simple tests on the steady-state Burgers equation will quickly demonstrate that $F_s = 1$ is *not* usually conservative; see Table 5.9.1 below.

So the direct use of the Richardson Extrapolation Error Estimator E_1, i.e. GCI with $F_s = 1$, shows less than 50% probability of being conservative on this simple, model, 1D problem. Is <50% probability acceptable for an error band? I think not. Perhaps most damning, the use of $F_s = 1$ makes grid doubling with 1st-order methods into the standard of comparison, i.e. for $p = 1$ and $r = 2$, the GCI [fine grid] calculated using $F_s = 1$ would give GCI = ε. Clearly, we do not want 1st-order methods to be the standard of comparison! (See, e.g., Freitas, 1993.)

Case	p	Re	r	(mc, mf)	Behavior
1	2	10	2	20,40	Normal
2	2	10	1.1111	36,40	Normal
3	1	10	1.1111	36,40	Normal
4	2	50	1.5	100,150	Normal
5	2	1000	1.25	800,1000	Normal
6	1	1000	1.25	800,1000	Normal
7	2	5	2	500,1000	Normal
8	1	1000	1.1111	900,1000	Normal
9	2	30	3	100,300	Normal
10	2	50	2	50,100	Normal
11	2	50	1.1111	90,100	Normal
12	2	50	1.0101	99,100	Normal
13	1	50	2	50,100	Normal
14	1	50	1.1111	90,100	Normal
15	1	50	1.0101	99,100	Normal
16	1	0.1	2	20,40	all are conservative
17	1	0.5	2	20,40	coarse grid E_1 is also conservative
18	1	0.5	2	500,1000	Normal
19	2	−1	2	20,40	Normal
20	2	−10	2	20,40	Normal
21	1	−10	2	20,40	all are conservative
22	2	−100	1.1111	90,100	Normal
23	1	−100	1.1111	90,100	all are conservative
24	2	1000	2	1000,2000	Normal
25	1	1000	2	1000,2000	Normal
26	2	10	4	25,100	Normal
27	2	10	3.0303	33,100	Normal
28	2	10	2	50,100	Normal
29	2	10	1.1111	90,100	Normal
30	2	10	1.0101	99,100	Normal

TABLE 5.9.1. STEADY BURGERS EQUATION SOLUTIONS DEMONSTRATING THAT F_s = 1 IS NOT USUALLY CONSERVATIVE. The Richardson Error Estimator (E_1) and the Grid Convergence Index (GCI) are applied to the Steady-State Burgers Equation $-U\,U_x + U_{.} / Re = 0$. p = order of method. $Re > 0$ indicates boundary condItlons of $U(0) - 1$, $U(1) = 0$, suggestive of stagnation flow. $Re < 0$ indicates boundary conditions of $U(0) = 0$, $U(1) = 1$. r is the grid refinement ratio = mf / mc, where mf = number of fine grid cells, mc = number of coarse grid cells. Under the heading "Behavior," the entry "Normal" indicates the following pattern: compared to the exact solution, (a) the E_1 is not conservative, i.e. $|E_1|$ > |true error|, for both the coarse and fine grid estimators; and (b) the GCI is conservative, i.e. GCI < |true error| for both the coarse and fine grids.

A 50% "factor of safety" over the Richardson Error Estimator would be achieved with the value $F_s = 1.5$, or the naive value of $F_s = 2$ might prove to be a neat and reasonable compromise. But much experimentation would be required over an ensemble of problems to determine a near-optimum value and to establish the correspondence with statistical measures such as the 2σ band. Note that a true optimum would likely depend upon the family of numerical methods (e.g., medium-order FVM, high-order FDM or FEM, etc.) and upon the family of problems (e.g., turbulence, transonic, free surface, etc.).

All things considered, and after discussions solicited from many (on the order of 200) CFD practitioners, I had recommended in (Roache, 1994) the use of the value $F_s = 3$ in the definition of the GCI, even though I recognized and stated that it would be too conservative for high quality grid convergence studies. Of course, there is nothing to preclude an author from reporting *both* the GCI and the Richardson Error estimator E_1. As noted, the motivation for using $F_s > 1$ is that $F_s = 1$ is analogous to a 50% error band on experimental data, which is not adequate. For many reasons (see discussion above, and in Chapter 8) this is not unduly conservative when only two grids are used in the study. However, it is now clear that $F_s = 3$ is overly conservative for carefully performed grid convergence studies using three or more grid solutions to experimentally determine (or Verify) the observed order of convergence p (e.g., see the papers in Johnson and Hughes, 1995). For such high quality studies, a modest and more palatable value of $F_s = 1.25$ appears to be adequately conservative. However, for the more common two-grid study (often performed reluctantly, at the insistence of journal editors) I still recommend the value $F_s = 3$ for the sake of uniform reporting and adequate conservatism.

The 30 cases for Burgers equation in the above Table 5.9.1 were repeated for $F_s = 1.25$. These tests involved only 2 grids. There is no ambiguity about the order of convergence of the methods, with well-defined $p = 1$ or 2, but not all of the cases are clearly in the asymptotic regime, and some of the error estimates were fairly large, e.g. 20 % in Case 5. The question addressed was the following. Is the recommended $F_s = 1.25$ still overly conservative? The answer is, clearly not. Of the 30 cases, 18 were "Normal" (as defined in the Table legend) or more conservative, i.e. either the coarse or fine grid Richardson estimator itself was conservative, as well as the GCI's. This still leaves 12 out of 30 that are not reliably conservative. This is, of course, a small sample, but it does not indicate anything like the desired 2σ band (~1 in 20) which would only allow for 1 or 2 non-conservative cases out of the 30. The low order methods faired worse, as expected. Considering only the 25 stagnation-like flows ($Re > 0$), 12 of 15 cases (more like a 1σ band) were conservative for $p = 2$ (the 2nd-order method), whereas only 3 of 10 cases were conservative for $p = 1$. Furthermore, these 3 were the very low Re Cases (16,17,18).

All 30 cases were conservative for $F_s = 1.85$, and 26 of 30 were conservative for $F_s = 1.35$. The implication is that the 2σ band for this small sample would be achieved for roughly $F_s = 1.5$. However, I again emphasize that this is a toy problem: one-dimensional, steady state, unambiguous theoretical order of convergence, thorough iterative convergence, simplest possible boundary conditions, etc. It would seem that $F_s = 1.5$ is the minimum one could use reliably for 2 grid solutions. If one performs thorough grid convergence tests for a suite of representative problems of interest and establishes a convincing correlation, one could possibly convince the audience that $F_s = 1.5$ is akin to a 2σ band. But in the absence of such a thorough and specific study, I still recommend $F_s = 3$ for 2-grid studies, and reserve $F_s = 1.25$ only for carefully performed grid convergence studies using three or more grid solutions to experimentally determine the *observed* order of convergence p. (See further discussion in Chapter 8.)

Summary recommendations:

(a) use $F_s = 1.25$ for convergence studies with a minimum of three grids to experimentally demonstrate the observed order of convergence p on the actual problem,
(b) use $F_s = 3$ for two-grid convergence studies.

5.10. ADDITIONAL FEATURES OF GRID CONVERGENCE STUDIES FOR VERIFICATION OF CODES AND CALCULATIONS

The following sub-sections are applicable to the use of the GCI for Quantification of Uncertainty of actual calculations, and were originally presented in that context in Roache (1994). However, the discussion is equally applicable to *any* systematic grid convergence study, whether or not the GCI is used (and, if so, to any value for the "safety factor" F_s), and equally to Verification of Codes as well as Verification of Calculations.

5.10.1 Non-Integer Grid Refinement

Although it is generally assumed that grid doubling is preferable, it is argued here that, especially for the computer limitations frequently encountered in practice for multidimensional problems, it may be better to use a smaller change in grid resolution, say 10%. Consider a base grid, and refine or coarsen. If engineering intuition or studies on related ("nearby") problems have led to a good (economical, yet adequately accurate) grid selection for the base grid, then we are likely in the asymptotic range, but perhaps just barely. If we can afford to double the grid, we will certainly get more accurate answers, but the cost can be large. With an optimal numerical solution method, e.g. a good multigrid method, in which the computing cost is merely proportional to the number of unknowns, doubling a grid in three space dimensions and time will increase the cost over the base grid calculation by a factor of 16; if sub-optimal methods are used, the penalty is worse. If we coarsen the grid instead, the economics work for us, i.e. the coarse grid solution is only 1/16 as expensive as the base grid; however, the coarse grid solution may be out of the asymptotic range. This situation is especially evident in turbulent boundary layer calculations, wherein we need sublayer resolution ($y^+ < 1$) for the first grid point off the wall. (For example, see Shirazi and Truman, 1989 or Wilcox, 1993.) This applies only when the turbulence equations are integrated to the wall; different requirements apply if wall functions are used. (For example, see Celik and Zhang, 1993,1995 or Wilcox, 1993.)

Since the theory of generalized Richardson Extrapolation is valid for non-integer r, it is easier to use a small value (unless the computor is confident that the coarse grid with $r = 2$ will still be in the asymptotic range). However, there are practical limits to small r. For example, increasing the number of grid points by 1 in a base grid calculation of a 100×100 grid gives $r = 1.01$, and the theory is still valid. (Indeed, it would still be better than no grid convergence study at all.) But the results will now be obscured by other error sources, e.g. the "noise" of incomplete iterative convergence and machine round-off error, especially for accurate solutions. That is, as we reduce the change in the discretization error by using $r \rightarrow 1$, the leading truncation error term may be swamped by noise. As an intuitive engineering guess, a minimum 10% change ($r = 1.1$) is recommended. Of course, provided that the coarse grid is within the asymptotic range, it is intuitively obvious that the error estimates are more reliable for larger r, *for grid refinement.*

It is perhaps less obvious that the opposite is true for grid coarsening, i.e. when we keep the answers of the finer grid. In that case, it should be clear that there is *more* additional information, and therefore sharper error estimates, available for $r \sim 1$ than for $r >> 1$ (limited only by noise pollution from round-off and incomplete iteration errors). For example, consider a fine grid using 100 nodes. A coarse grid calculation using 90 nodes ($r = 1.1111...$) contains more information (and is more expensive) than a coarse grid calculation using 50 nodes ($r = 2$).

A reviewer of Roache (1994) expressed skepticism that one really can learn anything about grid convergence by changing the resolution by only 10%. A set of easily reproduced calculations was performed on the steady-state Burgers equation $-U\,U_x + U_{xx} / Re = 0$, with $U(0) = 1$ and $U(1) = 0$ (suggestive of stagnation flow), $Re = 10$, and $p = 2$, with a fine grid of 100 interior nodes and coarse grids from 25 to 99 nodes. The results are shown in Table 5.10.1.1. As seen, the error estimator for the one-dimensional "shear" $f = dU / dx$ at $x = 1$ obtained using the coarse grid of 90 nodes ($r = 1.1111$, or a factor of 0.9 coarsening) is 3.2 times *more* accurate than the error estimator obtained using the coarse grid of 50 nodes ($r = 2$, or a factor of 0.5 coarsening).

5.10.2 Independent Coordinate Refinement and Mixed Order Methods

The simplest mode in which to apply the Grid Convergence Index is to use a single parameter r to refine/coarsen the grid in all coordinates, space and time. However, there are often

N coarse	N fine	r	% error in $E_1 \times 10^3$
25	100	4	–9.5993
33	100	3.0303	–5.6013
50	100	2	–2.4713
90	100	1.1111	–0.7680
99	100	1.0101	–0.6351

Table 5.10.1.1. The improvement in the accuracy estimate from small r for grid coarsening. Steady-state Burgers equation $-U\,U_x + U_{xx} / Re = 0$ with $U(0) = 1$ and $U(1) = 0$ (suggestive of stagnation flow), $Re = 10$, $p = 2$. The error shown is the % error in the Richardson Extrapolated value (and therefore the error estimator E_1) for the one-dimensional "shear" dU / dx at $x = 1$. Note that the error estimate from $r = 1.1111$ (a factor 0.9 coarsening) is 3.2 times as accurate as the error estimate from $r = 2$ (a factor 0.5 coarsening).

good reasons for not doing this. In calculations of boundary layer flows (whether using boundary layer equations, full Navier-Stokes equations, or something intermediate) it is often the case that grid convergence is easy to establish in the longitudinal direction (being essentially dictated by the free-stream flow, which is not sensitive to Reynolds number) but is more problematical in the transverse direction, being sensitive to *Re*. Also, computer resource restrictions may prohibit grid doubling in each direction simultaneously, but allow it one coordinate at a time; e.g., see the convincing studies in Zha and Knight (1996). Also, in time- dependent problems, it is much easier to develop a code that is solution-adaptive in the time-step than in the spatial grid, so that time discretization errors might be independently controlled (e.g., see Roache, 1991, 1992a, 1993) and the systematic grid convergence test would be restricted to the spatial grid.

In such a case, the multidimensional theory indicates that the error estimates can be obtained orthogonally, by independent coordinate refinement, as is now demonstrated. The functions g_1, g_2, etc. in Equation 5.3.1 for the x direction now depend on y, and a constant (in x) term appears, but it may be verified that these do not affect the algebra of solving for g_1 , even with cross derivatives present from non-orthogonal grids, at least for $p = 1$ or 2. In each coordinate direction, Equation 5.6.1 or 5.7.2 is applied separately, with r_t, r_x, r_y, etc., and the resulting Grid Convergence Indexes are added.

$$\mathrm{GCI} = \mathrm{GCI}_t + \mathrm{GCI}_x + \mathrm{GCI}_y + \mathrm{GCI}_z + \ldots \qquad (5.10.2.1)$$

The additional error made by this "alternating direction Richardson Extrapolation," compared to refinement in all directions at once, is not merely heuristic but is ordered (it improves for fine grids and for $r \sim 1$) and small enough for the procedure to be practical. It is important to note that the procedure must be performed globally, i.e. with complete global solutions obtained for each refinement in independent coordinate directions; attempts to apply the extrapolation procedure by lines do not produce ordered or usable error estimates.

Consistent Richardson Extrapolation error estimators cannot be obtained from just two calculations (a coarse and a fine grid calculation) when different r are used in different coordinate directions, because there is no basis for separating out the directional contributions. In this case, a conservative GCI should be used, based on the smallest directional r. For example, in a two-dimensional steady flow calculated in a fine grid of 100×100 cells and a coarse grid of 50 $\times$ 75 cells, unless other theoretical considerations apply, we would have to conservatively attribute the change in solution to the more modest grid refinement, and use $r = 4/3$ to calculate the fine-grid GCI from Equation 5.6.1.

A similar situation occurs with mixed-order methods, e.g. the not uncommon situation of a method with 1st-order time accuracy and 2nd-order space accuracy. A conservative approach would be to use $p = 1$ in Equation 5.6.1, but a better estimate (and more appealing GCI) would be obtained using separate grid convergence studies in space and time, using $p = 1$ for the time contribution and $p = 2$ for the space contribution from Equation 5.6.1, and simply adding the results as in Equation 5.10.2.1. Alternately, one may use different r in different coordinates to compensate for different p. For time-dependent problems, Richards (1997) verified the theoretical order of convergence of an $O(\Delta t, \Delta x^2)$ method by a grid refinement study in which Δx is reduced by a factor of 2 and Δt by a factor of 4, giving a theoretical ratio of the errors on successive space-time grids of 4; his representative experimental values (in his Table I, i) are 3.76, 3.95, 3.97. Similarly convincing verification was obtained for his extension of the "Completed Richardson Extrapolation" (Roache and Knupp, 1993) to space and time. For example, the completed Richardson extrapolation applied to the $O(\Delta t, \Delta x^2)$ method should be $O(\Delta t^2, \Delta x^4)$. Reducing the Δx successively by factors of 3, and Δt by factors of 9, gives a theoretical error reduction by a factor of 81 for each successive grid refinement. Richards' representative experimental values (in his Table I, iv, a) are 78.8 and 80.1, a remarkable Verification.

For hybrid methods that shift locally to two-point upstream differencing for large cell *Re*, the conservative $p = 1$ must be used. However, hybrid methods are not only inaccurate, but often produce non-monotonic convergence even for gross properties like reattachment length (Barton, 1995) making it difficult to have confidence in the extrapolation process. Hybrid methods are not recommended (Barton, 1995; Leonard and Drummond, 1995).

For methods that use higher order stencils for advection than for diffusion, the error will be dominated asymptotically by the lower order term. For example, for solutions calculated by Leonard's ULTIMATE method (Leonard, 1991) which is 3rd-order for advection, the GCI would have to be reported conservatively using $p = 2$ in Equation 5.6.1 or 5.7.2. Note that the accuracy advantage of mixed order methods is not lost in this process. The 3rd-order accuracy for the (dominant) advection term will produce smaller errors than a 2nd-order method, so the *level* or *size* of error will be lower. However, the *slope* of the convergence asymptotically will be 2nd-order, so the extrapolation and error banding must be based upon $p = 2$.

5.10.3 Non-Cartesian Grids, Boundary Fitted Grids, Unstructured Grids, Adaptive Grids

The procedures for calculating GCI definitely apply to non-cartesian grids, with special considerations and caveats.

The Taylor series basis of Richardson Extrapolation applies to stretched orthogonal and non-orthogonal grids as long as the stretching is analytical. It is cleanest to apply in the transformed plane (ξ, η, ζ) where r is defined as above. The order of the extrapolation accuracy will now be affected by the order and iterative convergence of the grid generation equations. Shirazi and Truman (1989) found a surprising sensitivity of the error estimates to discretization of metrics and Jacobians, and to incomplete iterative convergence. For another example, if strong exponential source terms are used (e.g. near trailing edges of airfoils) which depend strongly on h (e.g., see Thompson et al, 1985) then a refined-grid generation will pollute the Richardson Extrapolation.

However, even if the grid generation equations are not converged, making the actual Richardson Extrapolation less dependable, it is still recommended that the uniform GCI be reported rather than the simple raw data of ε.

For the simple power-law stretching in one coordinate, commonly used in boundary layer codes, the ratio of successive grid increments h is constant, i.e.

$$h^{i+1} = sh^{i} \tag{5.10.3.1}$$

where s is the constant stretching factor. Nodal comparisons can be made during grid convergence without interpolation if the coarse and fine grids have nodes in common. This is accomplished during a grid doubling/halving by requiring (Ferziger and Peric, 1996) that the fine-grid stretching factor s_f be related to the coarse grid s_c by

$$s_f = \sqrt{s_c} \tag{5.10.3.2}$$

and by requiring the initial fine-grid spacing h_{f1} be related to the coarse grid h_{c1} by

$$h_{f1} = \frac{h_{c1}}{1+s_f} \tag{5.10.3.3}$$

The simpler strategy of inserting fine-grid points half way between the coarse-grid points introduces an additional truncation error term to differences "centered" (in logical space), but the additional error term is itself ordered. For a relatively coarse grid, the error is larger, but asymptotically the convergence rate is still 2nd-order (Ferziger and Peric, 1966).

Solution-adaptive grid generation codes may have their own internal *local* error estimators. Most often, solution adaptive grid generation of the redistribution or enrichment types (Thompson, et al, 1985; Knupp and Steinberg, 1993) is not based on any error estimator but on solution behavior (gradient, curvature, or simply resolution requirement) which is only loosely related to *local* error (which in turn is loosely related to the *global* error of interest herein). In such a case, the GCI reporting procedure recommended herein can be applicable if the solution-adaptive procedure is used only to obtain the base grid solution. This grid can then be changed non-adaptively, perhaps refined by a higher order interpolation or coarsened by simply removing every other point (as in Zingg, 1991,1992). Then the GCI of Equations 5.6.1 or 5.7.2 can be applied to this new grid. However, for *time-dependent* solution-adaptive grids, practical coding difficulties exist and it is not clear how to perform meaningful *global* error estimation nor uniform grid convergence reporting in this important situation.

Another difficulty occurs for unstructured grids. If the base grid is unstructured, the GCI procedure would still apply if one used a systematic method of grid refinement, that is, structured refinement of an unstructured grid. For example, refining each base grid triangle into four new triangles (by connecting the mid-points of the sides) gives $r = 2$ for use in Equation 5.6.1. However, if the refinement is also unstructured, as occurs in some algorithms and in user-interactive grid generation codes, there is no systematic and quantifiable grid refinement index like r to use in Equation 5.6.1. Such grid refinement FEM studies are customarily reported simply in terms of the total number of elements used in the coarse (N_2) and fine (N_1) grids. Use in Equation 5.6.1 of an

$$\text{effective } r = \left(\frac{N_1}{N_2}\right)^{1/D} \tag{5.10.3.4}$$

where D is the dimensionality of the problem, and reporting this GCI is clearly preferable to simply reporting ε. But it does not have the firm basis of a structured grid refinement, and may significantly underestimate or overestimate the accuracy, depending on whether the grid refinement algorithm (or the intuition of the interactive user) refined the grid in the critical areas or not. For example, well-placed refinement in a relatively small critical area combined with coarsening in a larger but unimportant area could lead to improvement in global accuracy with a *decrease* in the total number of cells. (For example, see Pelletier and Ignat, 1995.) This would render the effective grid refinement ratio r, and the Richardson Extrapolation procedure, meaningless.

For unstructured grid refinement and (structured or unstructured) grid adaption, I suppose it should be the burden of the algorithm developer to convince the reader/user that the local grid adaptivity process (sometimes based on a *local* error estimator) can be usefully correlated with a meaningful engineering *global* error estimate, which is the real interest. This is very difficult to accomplish convincingly for any problem which a fluid dynamicist would consider non-trivial (mainly because local truncation errors are advected downstream) but it can be done; see Schonauer et al (1981).

For an application of the GCI to unstructured grids with Finite Element Methods, using unstructured refinement, see Pelletier and Ignat (1995). They also show the correlation between the GCI results based on generalized Richardson Extrapolation with the theoretically distinct single-grid error estimators of the Zhu-Zienkiewicz (1990) class; see also Chapter 6, Sections 6.7 and 6.8. Pelletier and Ignat convincingly demonstrated the efficacy of their solution adaptive mesh generation algorithm; with a target increase of a factor of 3 in accuracy with the new mesh generation, they confirmed values *a posteriori* greater than 2. Hagen (1997) used the GCI theory with structured refinement of unstructured triangular grids in ocean calculations.

5.10.4 Shocks, Discontinuities, Singularities

In both his 1910 and 1927 papers, Richardson already considered the effect of singularities on the extrapolation procedure. These cases must be considered individually. If the form of the singularity is known *a priori*, it may be removed analytically. If unknown, its presence may be detected by checking to see if the asymptotic range has been reached (see below). Shocks and other discontinuities (e.g., contact surfaces) invalidate the Taylor series basis of Richardson Extrapolation, but unless the flow contains large numbers of complex shock patterns, the GCI procedure herein would still seem to have validity and be recommended, certainly an improvement over simply reporting the raw data for ε. As pointed out by Ferziger (1993), a more appropriate error measure here might be the shock position. Further experience with complex shocked flows is needed.

Blottner (1990) has shown how the concept of Richardson Extrapolation can be applied to systematically estimating the error due to artificial dissipation terms used in hypersonic shock calculations. The contribution of these terms to the GCI must also be calculated orthogonally to the other terms. If these terms are not estimated separately, the grid convergence tests will be polluted, since the (nonlinear) shock dissipation terms depend on h, and therefore the continuum problem being approximated changes from grid to grid. (This is the same difficulty that can appear with grid generation equations, noted above.). Blottner's results will be presented in Chapter 6, Section 6.11. (See also Kuruvila and Anderson, 1985.)

Also, the theory of Richardson Extrapolation is not applicable to nonlinear flux limiters, but again we expect these to be local applications, and still recommend the reporting of GCI over simply reporting the raw data for ε. But I would also recommend more detailed investigation, e.g. perhaps 3 grids (see below). The point is that the presence of shocks, other discontinuities or singularities can complicate grid convergence studies whether or not the GCI is used for reporting the results, so these complications do not constitute a criticism of the GCI.

5.10.5 Achieving the Asymptotic Range

The theory of Richardson Extrapolation, and therefore of the Grid Convergence Index, depends on the assumption that the Taylor series expansion (or at least, the definition of the order of the discretization implied by Equation 5.3.4) is valid asymptotically, and that the two grids are within the asymptotic range. For smooth elliptic problems, this is easy to achieve. (A 2nd-order accurate discretization of a Laplace equation with smooth boundary values is well behaved over virtually all discretizations.) For Reynolds numbers >> 1, it is more problematic, and more than two grid solutions are required. The methodology presented herein does allow for detecting this situation in a straightforward manner, provided that the order of the method, p, is uniform.

If an exact solution is known to a model problem, we can monitor

$$E_b = \frac{\text{error}}{h^p} \qquad (5.10.5.1)$$

as h is refined. Then the (approximate) constancy of E_p is a generally faithful verification of the order p and an indication that the asymptotic range is achieved. (For examples, see Chapter 3, Section 3.5, or Richardson, 1927; Steinberg and Roache, 1985; Roache et al, 1990; Blottner, 1990; Roache and Knupp, 1993.) In the practical case wherein the exact solution is not known, we perform at least three grid solutions and calculate two GCI, from fine grid to intermediate grid (GCI_{12}) and from intermediate grid to coarse grid (GCI_{23}). Then the (approximate) constancy of $E_p \approx \text{GCI}/(F_s\, h^p)$, or

$$\mathrm{GCI}_{23} = r^p \mathrm{GCI}_{12} \tag{5.10.5.2}$$

indicates that the asymptotic range has been achieved.

This indication that the asymptotic range has been achieved is usually faithful, in the author's experience (Roache, 1982; Roache et al, 1990; Roache and Knupp, 1993). But an exception (and unfortunately, an important one) occurs in problems with multiple scales of solution variation wherein a finer scale of the problem variation has been completely missed in the grid refinements. For example, in dual-continuum models of transport in porous media (WIPP PA Dept., 1992; Helton et al, 1995, 1996) the time scale for diffusion and storage in the material matrix blocks may be orders of magnitude less than the time scale for essentially advective transport in the fracture system. Time-step refinement may indicate no substantial change in the results (i.e., a false indication of convergence) if the time step is of the order of the advective time scale. Similar situations occur in turbulent boundary layer studies where some minimal viscous sublayer resolution is required (see Shirazi and Truman, 1989; Wilcox, 1993) and in chemically reacting flows which can have more time scales than species. Adaptive ODE solvers are good at detecting multiple time scales, but in multidimensional flows, at present there seems to be no substitute for an independent estimate (from theory or experiment) of the physical scales of interest.

Since so many journal authors are reluctant to perform even the most minimal grid convergence tests with two grids (Roache et al, 1986; Roache, 1990), it may seem unrealistic to recommend *three* grids as a matter of course. In fact, it is required to be sure that the calculations are in the asymptotic range, if this is not already inferred from experience with a "nearby" calculation as in, e.g., Nguyen and Maclaine-Cross (1988) or Blottner (1990). In any case, the GCI provides an easily achieved improvement over the simplistic reporting of raw data on ε.

G. de Vahl Davis (1983), in his classic benchmark calculations of a buoyancy-driven cavity, indicated local convergence rates of *less* than 1st-order for the relatively coarse grids used, even though the method was asymptotically 2nd-order accurate. In the absence of such meticulous work as that of de Vahl Davis, the reporting of a GCI based on the assumed $p = 2$ would be preferable to simplistic reporting the raw data of ε, but if there is any indication of less-than-theoretical convergence rates, the more conservative estimate with GCI evaluated from Equation 5.6.1 or 5.7.2 using $p = 1$ should be reported.

5.10.6 Extraction of the Observed Order of Convergence From Grid Convergence Tests

If an exact solution is known or manufactured (see Chapter 3), it is straight-forward to extract the order of convergence from results of a systematic grid convergence test using a minimum of two grid solutions. This serves to Verify a Code. However, it is also desirable to Verify the observed order for an actual problem, since the observed order of convergence depends on achieving the asymptotic range, which is problem dependent, and since the observed order may differ from the theoretical order, or from the order Verified for a test case, for a variety of reasons. (See discussion in Westerink and Roache, 1995 and in Chapter 6.) The following general method was presented in Roache (1995a), "Verification of Codes and Calculations." For an equivalent method with different notation (notably on the definition of the grid refinement factor) see Celik and Zhang (1995).

Blottner (1990) and others use graphical means, plotting the error on log paper and extracting the order from the slope. This procedure requires evaluation of the error itself, which is generally not known. If the finest grid solution is taken to be the reference value (unfortunately, often called the "exact" value, which it obviously is not) then the observed order will be accurate only for those grids far from the finest, and the calculated order approaching the finest grid will be indeterminate. Blottner (1990) improves on this by estimating the "exact" value by Richardson Extrapolation (see also Shirazi and Truman, 1989), but this procedure is somewhat ambiguous since the order is needed in order to perform the Richardson Extrapolation.

If the grid refinement is performed with constant r (not necessarily $r = 2$), the order can be extracted directly from three grid solutions, without a need for estimating the exact solution, following G. de Vahl Davis (1983, p. 254). With "1" being the solution on the finest grid in the present notation,

$$p = \ln\left(\frac{f_3 - f_2}{f_2 - f_1}\right) / \ln(r) \tag{5.10.6.1}$$

A generalization of this procedure, not restricted to constant r, is possible using the generalized theory of Richardson extrapolation. Equation (5.10.5.2) may be used to Verify an assumed order p. (It is not necessary to use the GCI itself.) One calculates

$$\alpha = \frac{\mathrm{GCI}_{23}^{fine}}{\mathrm{GCI}_{12}^{fine}} \tag{5.10.6.2}$$

If $\alpha \approx r^p$, then p is the observed order. However, Equation (5.10.6.2) requires r to be constant over the three grid set, and it cannot be used to calculate p directly since p is implicitly present in the GCI's. The more general procedure is to solve the equation

$$\frac{\varepsilon_{23}}{r_{23}^p - 1} = r_{12}^p \left[\frac{\varepsilon_{12}}{r_{12}^p - 1}\right] \tag{5.10.6.3}$$

for p. This is simple for r constant (not necessarily 2 or integer), giving

$$p = \ln\left(\frac{\varepsilon_{23}}{\varepsilon_{12}}\right) \tag{5.10.6.4}$$

But if r is not constant during the grid refinement, Equation (5.10.6.3) is transcendental in p. Usual solution techniques can be applied, e.g., direct substitution iteration, Newton-Raphson, even graphical. When considering the stability of the iteration, one should allow for observed $p < 1$. This can happen even for simple problems at least locally (de Vahl Davis, 1983) and in some cases the observed p is < 0. (Unfortunately, behavior far away from asymptotic convergence can be non-monotone.) Also, $r \sim 2$ will be easier to solve than $r \sim 1$, and $r >> 2$ is probably not of much interest. For well behaved synthetic cases which I have tested, direct substitution iteration with a relaxation factor $\omega \sim 0.5$ works well. With ρ = previous iterate for p, the iteration equation is

$$p = \omega\rho + (1-\omega)\frac{\ln(\beta)}{\ln(r_{12})} \tag{5.10.6.5a}$$

$$\beta = \frac{\left(r_{12}^{\rho}-1\right)}{\left(r_{23}^{\rho}-1\right)}\frac{\varepsilon_{23}}{\varepsilon_{12}} \tag{5.10.6.5b}$$

Note this form of the iteration gives the exact answer in one step for the case of $r = 2$.

Once p is known with some confidence, one may predict the next level of grid refinement r^* necessary to achieve a target accuracy, expressed as a target Error Estimate E_1 or GCI_1, call it GCI^*. With GCI_{23} being the value from Equation (5.6.1) for the previous two grids,

$$r^* = \sqrt[p]{\frac{GCI^*}{GCI_{23}}} \tag{5.10.6.6}$$

(If $F_s = 1$ is used in Equation 5.6.1, then the GCI = Error Estimate E_1.) This result, of course, depends only on the assumed definition of "order" of the discretization error, i.e. only on E_b = error / h^p, and not on the GCI theory itself.

5.10.7 Method of Characteristics and Spectral Methods

It is not clear how or if the GCI would be applicable to calculations obtained by the classic method of characteristics, as used in gas dynamics, because of the possibly discontinuous solutions and the irregular gridding. The various Modified Method of Characteristics (e.g., see references in Roache, 1992b or Chapter 13 of Roache, 1998) will produce more systematic grid refinement, but the concept of "order" is more tenuous for the Flux-Based MMOC (as evidenced by the fact that the accuracy *improves* as the Courant number increases above 1), and the GCI may not be applicable. Similarly, for spectral and pseudo-spectral methods, and certainly for spectral elements, different extrapolation procedures would be required. It is not known at this time how well the GCI would apply or could be extended.

5.10.8 Non-Smooth Property Variation and the GCI

In aerodynamics problems, one typically deals with smooth property variations over modest ranges, and often with constant property problems. In groundwater flow and transport calculations, uncertainty and range are much greater, and sensitivity studies are often performed with Monte Carlo techniques used to generate property variations of orders of magnitude, even from one grid block (finite volume) to the next (e.g., see WIPP PA Dept., 1992; Helton et al, 1995,1996). Geologic layering produces discontinuous variations in properties of several orders of magnitude.

In these situations, it is not advisable to use non-integer grid refinement parameters *r*, because additional errors would be introduced by interpolation of properties. This confusion would be aggravated by the common use of harmonic averaging for properties (e.g., Roache, 1991, 1992a, 1993). Likewise, grid coarsening is not advisable if a coarsened grid would not resolve the scale of the property variations (often the case for expensive two-phase flow calculations, e.g. WIPP PA Dept., 1992; Helton et al, 1995,1996). The only approach applicable is a brute-force grid refinement by a factor of 2, which avoids any necessity for interpolation of properties.

The use of the GCI still has two contributions to such problems: (a) including the effect of the order p of the method in Equation 5.6.1, and (b) economizing a consistent treatment of *further* grid refinement. That is, a third grid (second refinement) need not involve the expense of another grid doubling (to a quadrupling of the base grid resolution) but can be done on a tripled grid, and reported consistently with the GCI. The difference in computer time between calculating the sequence (base h_0, 1/2 h_0, 1/4 h_0) and the sequence (base h_0, 1/2 h_0, 1/3 h_0) can be significant. Consider an *optimal* method with base-case computer time = T_2 in 2D and T_3 in 3D, and time-step resolution increased in proportion to the spatial grid resolution. The quadrupling sequence in 2D costs $73T_2$ while the tripling sequence costs $36T_2$; in 3D, the costs are $273T_3$ and $98T_3$, respectively. These savings of a factor of 2 in 2D and somewhat less than 3 in 3D will be greatly amplified if *sub-optimal* direct solution methods are used.

5.10.9 Non-Smooth Property Variation and Geostatistical Realizations

A more fundamental question arises when geostatistical methods are used to generate particular realizations of grid-block property variations with specified statistical parameters. Only the statistical results are of interest, not the solutions of the individual realizations. The question is then, should the grid refinement studies be performed separately from the geostatistical realizations? That is, should the solution of the partial differential equations be converged on finer grids with the assumed continuum property variation fixed at a geostatistically generated coarse-grid distribution, or should the geostatistical generation also change as the grid is refined?

This is not an easy question. Although definition of a fixed continuum problem for the grid refinement studies is conceptually easier, it is clear that substantial computer savings could accrue to the combined convergence approach. The results from both approaches should be statistically consistent, though not identical. In either approach, the grid increments should be less than (be able to partially resolve) the correlation length of the property variation.

5.10.10 Stopping Criteria for Iterative Convergence

The stopping criteria for iterative convergence is similar to grid convergence testing in one important aspect: both must be properly normalized. As noted earlier, the raw ε, or change in the solution, has often been reported as a grid convergence criterion, but an (almost) arbitrarily small ε can be obtained for any problem with a small enough change in the grid size. The theory shows (as in Equation 5.4.2) that ε must be normalized by a quantity reflecting the influence of convergence rate p and the grid size reduction r. Analogously, for an iterative solution procedure in which the new iterate value f^{k+1} is under-relaxed by a factor ω, that ω must appear in the denominator of the stopping criteria. (For example, see Roache, 1972, p. 174 ff, 1975, 1995b; Roache and Ellis, 1975.) That is, many authors test something like

$$L_\infty = \max_{i,j} \left| \frac{f_{ij}^{n+1} - f_{ij}^{n}}{f_{norm}} \right| \leq tol \tag{5.10.10.1}$$

where n is the iteration level, f_{norm} is some suitable normalizing value, and tol is some convergence criterion like 10^{-4} or 10^{-6}. This test can be met meaninglessly, regardless of the actual convergence, in a relaxation procedure such as

$$f^{n+1} = \omega f^{new} + (1-\omega) f^{old} \tag{5.10.10.2}$$

by choosing ω small enough. A simple modification of the convergence test is the form

$$\frac{L_\infty}{\omega} \leq tol \tag{5.10.10.3}$$

For simple iteration schemes (such as SOR without immediate updating) it may be shown that this is algebraically equivalent to a criterion based on directly evaluating the residual.

Such iterative stopping criteria are still somewhat heuristic. What is intended is an estimate of the incomplete iteration error, defined as

$$\varepsilon_{iter}^{n} = f^{n+1} - f^{exact} \tag{5.10.10.4}$$

where, in the present context, f_{exact} is not the exact solution of the continuum equations but the exact solution of the algebraic (discretized) equations. Roughly,

$$f_{exact} \cong \lim_{n \to \infty} f^n \tag{5.10.10.5}$$

for a convergent iteration, the approximation being due only to the effects of machine round-off error. An estimate of the incomplete iteration error in terms of the iteration change is given by Ferziger (1988, 1993) as

$$\varepsilon^n_{iter} = f^{n+1} - f^{exact} \cong \frac{f^{n+1} - f^n}{\lambda_1 - 1} \qquad (5.10.10.6)$$

where λ_1 is the (assumed) *real* principal eigenvalue of the iteration matrix (defined by the iteration scheme), which can be estimated from the history of the iteration.

For a simple iterative solver like SOR applied to a discretized Poisson equation, the use of a relaxation factor ω less than the optimum value produces a real principal eigenvalue of the iteration matrix, making the normalizing of the stopping criteria easier to determine from the history of the iterative solution path. For SOR relaxation with larger than optimum values, and for more intricate iteration schemes, the principal eigenvalue may be complex. Ferziger and Peric (1996) present a method for normalizing the stopping criteria in this case, based on some assumptions and approximations that are justified by their test cases. These numerical experiments indicate that data from as many as 50 iterative steps must be used to produce a reliable normalization for the complex eigenvalue situation, but the amount of coding and the computational time penalty are insignificant compared to solving the discretized equations. This method is recommended for consideration, although Ferziger and Peric (1996) note that simpler methods based on monitoring either the reduction of the residual or the difference between successive iterates ε (as above in Equation 5.10.10.3) can also be used if normalized properly.

The same caveats that apply to discretization error convergence testing also apply to iterative convergence testing. Different variables, or the same variable evaluated at different locations, iteratively converge at significantly different rates.

5. 11 CONCLUSION

The Quantification of Uncertainty in CFD publications has been noticeably improving, with minimal grid refinement studies becoming more common. Unfortunately, it has not always been possible to uniformly interpret these studies.

In Roache (1994) it has been proposed that the results of systematic grid convergence studies be uniformly reported using the Grid Convergence Index of Equation 5.6.10 or 5.7.2, which is based upon a grid refinement error estimator derived from the theory of the generalized Richardson Extrapolation. While not answering all questions involved with Verification of a calculation, this method at least enforces some uniformity in the reporting and is based upon an objective asymptotic estimate of the grid convergence error, although the GCI is not a true error bound. Use of $F_s = 1.25$ should be restricted to convergence studies with a minimum of three grids that experimentally demonstrate the observed order of convergence p on the actual problem.

Since the GCI will often be less optimistic than the simplistic ε of Equation 5.4.3, especially for the all-too-popular 1st-order methods, some reluctance of authors may be anticipated.

Fortunately, the formulas are simple enough to be applied *a posteriori* by editors and reviewers. It is urged that they do so in the review process to continue improving the quality of CFD papers. To quote Ferziger (1993), "*... the frequently heard argument 'any solution is better than none' can be dangerous in the extreme. The greatest disaster one can encounter in computation is not instability or lack of convergence but results that are simultaneously good enough to be believable but bad enough to cause trouble.*"

REFERENCES FOR CHAPTER 5

AGARD (1988), *Validation of Computational Fluid Dynamics*, Lisbon, Portugal, 2–5 May 1988, NATO Advisory Group for Aeronautical Research and Development, AGARD CP 437. Volume 1: Symposium Papers and Round Table Discussion. Volume 2: Poster Papers, December 1988.

Anderson, W. K. and Batina, J. T. (1988), "Accurate Solutions, Parameter Studies, and Comparisons for the Euler and Potential Flow Equations," in AGARD (1988), Paper No. 14.

Barton, I. E. (1995), "A Numerical Study of Flow over a Confined Backward-Facing Step," *International Journal for Numerical Methods in Fluids*, Vol. 21, pp. 653–665.

Blottner, F. G. (1990), "Accurate Navier-Stokes Results for the Hypersonic Flow over a Spherical Nosetip," *Journal of Spacecraft and Rockets,* Vol. 27, No. 2, March–April 1990, pp. 113–122.

Caruso, S. C., Ferziger, J. H., and Oliger, J. (1985), "Adaptive Grid Techniques for Elliptic Flow Problems," Report No. TF–23, Thermosciences Division, Stanford University, Stanford, California, 1985.

Celik, I., Chen, C. J., Roache, P. J., and Scheuerer, G. (1993), Eds., *Symposium on Quantification of Uncertainty in Computational Fluid Dynamics,* FED Vol. 158, ASME Fluids Engineering Division Summer Meeting, Washington, D. C., June 20–24, 1993.

Celik, I. and Zhang, W-M. (1993), "Application of Richardson Extrapolation to Some Simple Turbulent Flow Calculations," in Celik et al (1993), pp. 29–38.

Celik, I. and Zhang, W-M. (1995), "Calculation of Numerical Uncertainty Using Richardson Extrapolation: Application to Some Simple Turbulent Flow Calculations," *ASME Journal of Fluids Engineering*, Vol. 117, September 1995, pp. 439–445.

Conte, S. D. and de Boor, C. (1965), *Elementary Numerical Analysis: An Algorithmic Approach*, McGraw-Hill, New York, 1965, p. 313.

Cosner, R. R. (1995), "CFD Validation Requirements for Technology Transition," AIAA Paper 95–2227 and the oral presentation, AIAA 26th Fluid Dynamics Conference, San Diego, California, 19–22 June 1995.

de Vahl Davis, G. (1983), "Natural Convection of Air in a Square Cavity: A Bench Mark Numerical Solution," *International Journal for Numerical Methods in Fluids,* Vol. 3, No. 3, pp. 249–264.

Ferziger, J. H. (1981), *Numerical Methods for Engineering Application*, John Wiley & Sons, New York.

Ferziger, J. H. (1988), "A Note on Numerical Accuracy," *International Journal for Numerical Methods in Fluids*, Vol. 8, pp. 995–996.

Ferziger, J. H. (1993), "Estimation and Reduction of Numerical Error," *FED Vol. 158, Symposium on Quantification of Uncertainty in Computational Fluid Dynamics,* ASME Fluids Engineering Division, Summer Meeting, Washington, D.C., June 20–24, 1993, pp. 1–8.

Ferziger, J. H. and Peric, M. (1996), "Further Discussion of Numerical Errors in CFD," *International Journal for Numerical Methods in Fluids*, Vol. 23, pp. 1263–1274.

Freitas, C. J. (1993), "Editorial," *Journal of Fluids Engineering*, Vol. 115, No. 3, September 1993, pp. 339–340.

Hagen, S. C. (1997), "Finite Element Grids Based on a Localized Truncation Error Analysis," Ph.D. Dissertation, Department of Civil Engineering and Geological Sciences, University of Notre Dame, September 1997.

Helton, J. C., Anderson, D. R., Roache, P. J., et al (1995), "Effect of Alternative Conceptual Models in a Preliminary Performance Assessment for the Waste Isolation Pilot Plant," *Nuclear Engineering and Design*, Vol. 154, 1995, pp. 251–344.

Helton, J. C., Anderson, D. R., Roache, P. J., et al (1996), "Uncertainty and Sensitivity Analysis Results Obtained in the 1992 Performance Assessment for the Waste Isolation Pilot Plant," *Reliability Engineering and System Safety*, Vol. 51, 1996, pp. 251–53–100.

Johnson, R. W. and Hughes, E. D. (1995), Eds., *Quantification of Uncertainty in Computational Fluids Dynamics–1995,* Joint JSME–ASME Fluid Mechanics Meeting, ASME FED Volume 213, Hilton Head, South Carolina, August 14–18, 1995.

Kessler, R., Peric, M. and Scheuerer, G. (1988), "Solution Error Estimation in the Numerical Predictions of Turbulent Recirculating Flows," in AGARD (1988), pp. P9–1 to 12.

Kline, S. J. and McClinstock, F. A. (1953), "Describing Uncertainties in Simple-Sample Experiments," *Mechanical Engineering*, Jan. 1953, pp. 3–8.

Knupp, P.M. and Steinberg, S. L. (1993), *Fundamentals of Grid Generation*, CRC Press, Boca Raton, Florida.

Kuruvila, G. and Anderson, J. D. (1985), "A Study of the Effects of Numerical Dissipation on the Calculation of Supersonic Separated Flows," AIAA Paper 85–0301, January 1985.

Gourlay, R. and Morris, J. Ll. (1968), "Deferred Approach to the Limit in Non-Linear Hyperbolic Systems," *Computing Journal*, Vol. 2, 1968, pp. 95–101.

Leonard, B. P. (1991), "The ULTIMATE Conservative Difference Scheme Applied to Unsteady One-Dimensional Advection," *Computational Methods in Applied Mechanics and Engineering*, Vol. 88, 1991, pp. 17–74.

Leonard, B. P. and Drummond, J. E. (1995), "Why You Should Not Use 'Hybrid', 'Power Law' or Related Exponential Schemes for Convective Modelling–There Are Much Better Alternatives," *International Journal for Numerical Methods in Fluids*, Vol. 20, 1995, pp. 421–442.

Nguyen, T. V. and Maclaine-Cross, I. L. (1988), "Incremental Pressure Drop Number in Parallel-Plate Heat Exchangers," *ASME Journal of Fluids Eng*ineering, Vol. 110, March 1988, pp. 93–96.

Pelletier, D. and Ignat, L. (1995), "On the Accuracy of the Grid Convergence Index and the Zhu-Zienkiewicz Error Estimator," in Johnson and Hughes (1995), pp. 31–36.

Richards, S. A. (1997), "Completed Richardson Extrapolation in Space and Time," *Communications in Numerical Methods in Engineering*, Vol. 13, No. 7, pp. 573–582.

Richardson, L. F. (1910), "The Approximate Arithmetical Solution by Finite Differences of Physical Problems Involving Differential Equations, with an Application to the Stresses in a Masonry Dam," *Transactions of the Royal Society of London, Series A*, Vol. 210, pp. 307–357.

Richardson, L. F. (1927), "The Deferred Approach to the Limit," *Transactions of the Royal Society of London, Series A.*, Vol. 226, pp. 299–361. "Part I. Single Lattice." "Part II. Interpenetrating Lattices" by J. A. Gaunt.

Roache, Patrick J. (1972), *Computational Fluid Dynamics*, First Edition, Hermosa Publishers, P. O. Box 9110, Albuquerque, New Mexico 87119-9110, USA. http://www.hermosa-pub.com/hermosa. Revised Printing 1976. See also Roache (1998).

Roache, P. J. (1975), "The LAD, NOS, and Split NOS Methods for the Steady-State Navier-Stokes Equations," *Computers and Fluids*, Vol. 3, 1975, pp. 179–195.

Roache, P. J. and Ellis, M. A. (1975), "The BID Method for the Steady-State Navier-Stokes Equations," *Computers and Fluids*, Vol. 3, 1975, pp. 305–320.

Roache, P. J. (1982), "Scaling of High-Reynolds-Number Weakly Separated Channel Flows," in *Numerical and Physical Aspects of Aerodynamic Flows,* T. Cebeci, Ed., Chapter 6, Springer Verlag, NY, pp. 87–98.

Roache, P. J., Ghia, K. N. and White, F. M. (1986), "Editorial Policy Statement on the Control of Numerical Accuracy," *ASME Journal of Fluids Engineering*, Vol. 108, No. 1, p. 2.

Roache, P. J. (1990), "Need for Control of Numerical Accuracy," *Journal of Spacecraft and Rockets,* Vol. 27, No. 2, March–April 1990, pp. 98–102.

Roache, P. J., Knupp, P. M., Steinberg, S., Blaine, R. L. (1990), "Experience with Benchmark Test Cases for Groundwater Flow," ASME FED Vol. 93, *Benchmark Test Cases for Computational Fluid Dynamics*, I. Celik and C. J. Freitas, Eds., Book No. H00598, June 1990, pp. 49–56.

Roache, P. J. (1991), "Computational Fluid Dynamics Algorithms and Codes Developed for WIPP Site Simulations," in *Computational Mechanics,* Cheung, Y. K., Lee, J. H. W. and Leung, A. Y. T., Eds., *Proceedings of the Asian Pacific Conference on Computational Mechanics,* A. A. Balkema, Rotterdam, Vol. 2, Hong Kong, 11–13 December 1991, pp. 1325–1335.

Roache, P. J. (1992a), "Computational Fluid Dynamics Algorithms Developed for WIPP Site Simulations," in Russell, T. F. et al, eds., *Proceedings of the IX International Conference on Computational Methods in Water Resources,* Denver, Colorado, Elsevier Applied Science, New York, June 9–12, 1992, pp. 375–382.

Roache, P. J. (1992b), "A Flux-Based Modified Method of Characteristics," *International Journal for Numerical Methods in Fluids,* Vol. 15, pp. 1259–1275.

Roache, P. J. and Knupp, P. M. (1993), "Completed Richardson Extrapolation," *Communications in Numerical Methods in Engineering,* Vol. 9, pp. 365–374.

Roache, P. J. (1993), "The SECO Suite of Codes for Site Performance Assessment," *Proceedings of the 1993 International High-Level Radioactive Waste Management Conference,* Las Vegas, Nevada, April 26–30, 1993.

Roache P. J. (1994), "Perspective: A Method for Uniform Reporting of Grid Refinement Studies," *ASME Journal of Fluids Engineering,* Sept. 1994, Vol. 116, pp. 405–413. See also Roache, P. J. (1993), "A Method for Uniform Reporting of Grid Refinement Studies," in Celik et al (1993), pp. 109–120. See also Roache, P. J. (1993), *Proceedings of the AIAA 11th Computational Fluid Dynamics Conference*, Part 2, Orlando, Florida, 6–9 July 1993, pp. 1057–1058.

Roache, P. J. (1995a), "Verification of Codes and Calculations," AIAA Paper 95–2224, 26th AIAA Fluid Dynamics Conference, San Diego, California, June 19–22, 1995. Also, *AIAA Journal*, May 1998, to appear.

Roache, P. J. (1995b), *Elliptic Marching Methods and Domain Decomposition*, CRC Press, Boca Raton, Florida.

Roache, P. J. (1997), "Quantification of Uncertainty in Computational Fluid Dynamics," *Annual Review of Fluid Mechanics*, Vol. 29, 1997, pp. 123–160.

Roache, P. J. (1998), *Fundamentals of Computational Fluid Dynamics*, Hermosa Publishers, P. O. Box 9110, Albuquerque, New Mexico 87119-9110, USA. http://www.hermosa-pub.com/hermosa. To appear.

Schonauer, W., Raith, K., and Glotz, G. (1981), "The Principle of the Difference of Difference Quotients as a Key to the Self-Adaptive Solution of Nonlinear Partial Differential Equations," *Computer Methods in Applied Mechanics and Engineering*, Vol. 28, No. 3, pp. 327–359.

Shirazi, S. A. and Truman, C. R. (1989), "Evaluation of Algebraic Turbulence Models for PNS Predictions of Supersonic Flow Past a Sphere-Cone," *AIAA Journal,* Vol. 27, No. 5, pp. 560–568.

Steinberg, S. L. and Roache, P. J. (1985), "Symbolic Manipulation and Computational Fluid Dynamics," *Journal of Computational Physics*, Vol. 57, pp. 251–284.

Thompson, J. F., Warsi, Z. U. A., and Mastin, C. W. (1985), *Numerical Grid Generation Foundations and Applications,* North-Holland, New York.

Westerink, J. J. and Roache, P. J. (1995), "Issues in Convergence Studies in Geophysical Flow Computations," in Johnson and Hughes (1995), pp. 1–6.

Wilcox, D. C. (1993), *Turbulence Modeling for CFD,* DCW Industries, Inc., La Cañada, California.

Williams, M. (1989), "A Truncation Error Elimination Method for Steady-State Navier-Stokes Calculations," AIAA Paper 89–1934, 1989.

WIPP PA Dept. (1992), *Annual Performance Assessment for the Waste Isolation Pilot Plant, December 1992, Vol. 1: Preliminary Comparison with 40 CFR 191, Subpart B Vol. 2: Technical Basis,* Performance Assessment Department, Sandia National Laboratories, Albuquerque, New Mexico.

Zha, G-C. and Knight, D. (1996), "Three-Dimensional Shock/Boundary-Layer Interaction Using Reynolds Stress Equation Turbulence Model," *AIAA Journal*, Vol. 34, No. 7, July 1996, pp. 1313–1321.

Zhu, J. Z. and Zienkiewicz, O. C. (1990), "Superconvergence Recovery Technique and *A Posteriori* Error Estimates," *International Journal for Numerical Methods in Engineering*, Vol. 30, 1990, pp. 1321–1339.

Zingg, D. W. (1991), "Viscous Airfoil Computations Using Richardson Extrapolation," AIAA Paper 91–1559–CP, AIAA 30th Aerospace Sciences Meeting, Reno, Nevada, January 6–9, 1992.

Zingg, D. W. (1992), "Grid Studies for Thin-Layer Navier-Stokes Computations of Airfoil Flowfields," *AIAA Journal*, Vol. 30, No. 10, October 1993, pp. 2561–2564. See also AIAA Paper 92–0184.

6

APPLICATIONS OF SYSTEMATIC GRID CONVERGENCE STUDIES AND THE GRID CONVERGENCE INDEX

6.1 INTRODUCTION

This chapter presents applications of systematic grid convergence studies and the Grid Convergence Index (GCI) in both Verification of Codes and Verification of Calculations. The examples are taken from various articles in the open literature to exemplify good work and to illustrate the details of sensitivity for particular calculations. Such Verification is performed external to the code and therefore is applicable to commercial codes, with or without access to the source code (which would be dangerous). It is the easiest to apply of all error estimation or Verification methods (it just requires methodical plodding) and is the most reliable and flexible.

6.2 TWO FURTHER EXAMPLES OF (PARTIAL) CODE VERIFICATION IN GROUNDWATER FLOW

In Chapter 3, I presented a detailed example of systematic Verification of code (produced by Symbolic Manipulation) using the Method of Manufactured Solutions to obtain the non-trivial analytical "benchmark" solution. This methodology is very general. Whether or not the benchmark solution is "manufactured" or more realistic, the Code Verification involves the same procedures. So long as the benchmark solution exercises all the aspects of the discretization, the Code Verification is convincing.

Following are two further examples of (partial) Code Verification via systematic grid convergence studies, both taken from Roache et al (1990). In this paper, we described some our of experiences with formulating and applying several benchmark test cases in groundwater hydrology CFD problems. The problems were all single phase, Darcy law flows. Three problem categories were considered:

1. full 2-D groundwater hydrology flow codes tested on a steady problem with constant conductivity in stretched cartesian coordinates;
2. Fortran subroutines produced by computer Symbolic Manipulation for the stencil array evaluation of tensor conductivity in general non-orthogonal 2-D and 3-D coordinates;
3. particle tracking in 2-D and 3-D.

Freedom from coding errors, consistency of the discretization, and order of convergence were Verified. The first two exercises, described below, demonstrate successful Code Verification and successful error detection. The third example demonstrates an unsuccessful "false negative" test, and will be presented in Chapter 8.

6.2.1 Darcy Flow in Stretched Orthogonal Coordinates

Four codes were considered in this part of the study: SWIFT II (Reeves, et al, 1986), HST3D (Kipp, 1987), MODFLOW (McDonald and Harbaugh, 1988), and SECO_FLOW (Roache et al, 1995). SWIFT II is a code developed by Intera Technologies for Sandia Laboratories. HST3D and MODFLOW were developed by the USGS (United States Geological Survey). SECO_FLOW (for **S**andia-**ECO**dynamics) is a suite of codes (including 2-D and 3-D SECO_TRACKER and SECO_TRANSPORT) developed at Ecodynamics for Sandia Laboratories. SWIFT II and HST3D use primitive variables and allow for varying fluid compositions (e.g. brine, heat) whereas MODFLOW and the early version of SECO_FLOW allowed for only a pure component. (A version developed later included variable density.) SWIFT II and MODFLOW use a Marker-And-Cell (MAC, or Arakawa "C") staggered grid (e.g., see Roache, 1972), HST3D uses a non-staggered grid (i.e., collocated variables) and SECO_FLOW has options for either. All four use fully implicit (backward Euler) time differencing, and all four appear from the problem formulations and code descriptions to be uniformly 2nd-order accurate in space and 1st-order accurate in time.

The simplest formulation of the governing equations for this problem gives a single parabolic equation in terms of hydraulic (piezometric) head *h*.

$$S_s \frac{\partial h}{\partial t} = \nabla \cdot (k \nabla h) \tag{6.2.1.1}$$

The codes considered all use stretched cartesian (i.e. planar orthogonal) coordinates, so the conductivity *k* may be tensor, provided that the principal axes of *k* are aligned with *x* and *y*; this is effectively a scalar conductivity assumption, since it is equivalent to a simple rescaling of *x* or *y*. All four use or have options to use harmonic averaging for the conductance (a combination of the physical variable conductivity and discretization terms), but for this first set of tests, only the constant *k* was verified. (Variable *k* tests, including full tensor *k* with discontinuities, will be described following.)

SWIFT II and HST3D treat much more complicated problems, so that many of their calculations are null for the simple problem described here. However, MODFLOW and SECO_FLOW are also fairly complex codes, owing not to the complexity of the governing PDE itself, but to the options for harmonic averaging and many *modeling* issues such as definition of aquifer properties, locally confined (artesian) or unconfined aquifer conditions, simulation of rivers, recharge, well schedules, inactive regions, etc. (All the SECO codes have the feature of using separate grids to define aquifer properties and to perform calculations; this is highly advantageous for performing grid convergence studies.) All of these options introduce significant nonlinearities (step function dependencies on the dependent variable *h*) and coding complexities. Also, SECO_FLOW has both regional and local area grid solution capability (i.e. an elementary domain decomposition approach) and can treat discontinuous boundary definition of the general Robin type, whereas HST3D allows Dirichlet or non-homogeneous Neumann, and the other codes all use only homogeneous Neumann boundary conditions.

A benchmark (partial Verification) test case for steady flow was formulated as follows. With constant aquifer properties on a square domain of $(0, a) \times (0, a)$ and no wells, the boundary conditions are as follows. (The variable "*h*" = head for MODFLOW and SECO_FLOW, or pressure for SWIFT II and HST3D.) At $y = a$, the Dirichlet condition applies as

$$h(x,a) = c\left(\frac{\pi x}{a} + 1\right) \tag{6.2.1.2}$$

$$c = \cosh(\pi) = 11.59... \tag{6.2.1.3}$$

and on the other three sides of the domain, the homogeneous Neumann condition applies as

$$\frac{\partial h}{\partial n} = 0. \tag{6.2.1.4}$$

The analytic solution is

$$h(x,y) = \cos\left(\frac{\pi x}{a}\right)\cosh\left(\frac{\pi y}{a}\right) + c \qquad (6.2.1.5)$$

The domain chosen was 40 km × 40 km, but these values, like the constant aquifer properties, scale out of the results for the test problem. The analytic solution was used to set initial conditions at all points.

SECO_FLOW uses this discretized analytic solution to set discrete boundary conditions, which means boundary values and values at the first interior cells. Thus, at boundaries with a homogeneous Neumann condition, the two-point difference equation for $\partial h / \partial n$ is not set to zero, but to the discrete values from the exact solution, which only approach zero as the grid is refined.

Two problem sets are presented. Table 6.2.1.1 presents results for a uniform staggered (MAC) grid, and Table 6.2.1.2 for a stretched MAC grid. The Tables show maximum error ERR_MAX, and COEFF_MAX = ERR_MAX · (IL – 1 + mac)^2, where mac = 1 for the MAC grid, or mac = 0 for the collocated grid option. For a uniformly 2nd-order accurate solution, the value of COEFF_MAX should become roughly a constant as the grid is refined.

The results in Table 6.2.1.1 display the expected 2nd-order accuracy. The maximum error occurs well away from boundaries, e.g., in the 40 × 40 grid, at (I, J) = (7, 34).

Table 6.2.1.2 presents the same type results, but for a stretched grid. The grid stretching used in x concentrated (arbitrarily) the stretching at x = 20 km in the 40 km × 40 km grid, with a power-law stretching with b = 0.5. This gives a ratio of maximum to minimum x-spacing of 1.72 in the 5 × 5-cell grid, and 6.33 in the 80 × 80-cell grid. The grid stretching used in y concentrated the stretching at y = 23 km with b = 0.4. This gives a ratio of maximum to minimum y-spacing of 2.97 in the 5 × 5-cell grid, and 5.92 in the 80 × 80-cell grid.

Again, the results display the expected 2nd-order accuracy, and the maximum error occurs well away from boundaries, e.g., in the 40 × 40 grid, at (I, J) = (35, 34). Asymptotic behavior is reached more slowly, and the maximum truncation error is larger, for the stretched grid than for

IL × JL	ERR_MAX	COEFF_MAX
5 × 5	1.129E–03	2.82E–02
10 × 10	4.142E–04	4.14E–02
20 × 20	1.390E–04	4.14E–02
40 × 40	2.274E–05	3.64E–02
80 × 80	4.902E–06	3.14E–02

Table 6.2.1.1. Convergence of SECO_FLOW Uniform Grid. (From Table 1 of Roache et al, 1990.)

IL × JL	ERR_MAX	COEFF_MAX
5 × 5	5.906E–04	1.48E–02
10 × 10	5.354E–04	5.35E–02
20 × 20	4.242E–04	1.70E–01
40 × 40	1.704E–04	2.73E–01
80 × 80	4.101E–05	2.62E–01

Table 6.2.1.2. Convergence of SECO_FLOW Stretched Grid. (From Table 2 of Roache et al, 1990.)

the uniform grid, as is to be expected for a smoothly varying solution. This is the cost for the increased resolution near the grid concentration point (20 km, 23 km).

A second similar test problem, which we had previously used on SWIFT II tests, was also exercised on SECO_FLOW. At $y = a$, the Dirichlet condition applies as

$$h(x,a) = c \cdot \cos\left(\frac{0.5\pi x}{a}\right) \tag{6.2.1.6}$$

$$c = \cosh\left(\frac{\pi}{2}\right) = 2.509... \tag{6.2.1.7}$$

At $x = a$, the Dirichlet condition applies as

$$h(a,y) = 0 \tag{6.2.1.8}$$

and on the other two sides of the domain, the homogeneous Neumann condition applies as

$$\frac{\partial h}{\partial n} = 0. \tag{6.2.1.9}$$

The analytic solution for this second problem is

$$h(x,y) = \cos\left(\frac{\pi x}{2a}\right)\cosh\left(\frac{\pi y}{2a}\right). \tag{6.2.1.10}$$

This problem was considered for SECO_FLOW only because it caused difficulties for SWIFT II. Although the solution behavior is not significantly more difficult than the previous problem, it does contain Dirichlet boundary conditions on two adjacent boundaries. This causes a local error in SWIFT II near that corner. No such difficulty arises with SECO_FLOW, as expected. In the 80 × 80 uniform grid, ERR_MAX = 6.265E–06 and COEFF_MAX = 4.01E–02, which are comparable to the errors for the first problem. Also, the maximum error occurs at (I, J) = (22, 41) indicating no difficulty near the corner, in contrast to the SWIFT II results, which were clearly indicative of a larger error. The problem was not unknown to the code authors; the SWIFT II user manual indicates correctly that Dirichlet conditions at a corner cell, where different Dirichlet values apply at each side, will be over-written so that only the second entered value is active. Thus, the two adjacent faces of a corner cell cannot have different Dirichlet values. That is, due to coding peculiarities (attributable to its age), SWIFT II needs to have

$$h\left(\frac{a-dx}{2},a\right)=h\left(a,\frac{a-dy}{2}\right). \qquad (6.2.1.11)$$

What we could not explain was that, even with the first problem (involving 3 Neumann and 1 Dirichlet boundaries), neither SWIFT II nor MODFLOW displayed 2nd-order convergence rates in the resolution ranges tested. Both of these use MAC grids. HST3D, which uses a collocated variable grid, did display 2nd-order convergence rates (as did SECO_FLOW above, also using MAC grids). However, for a related problem with specified non-zero flux boundary conditions, HST3D proved to have an error by a factor of 2 in the magnitude of the computed flux boundary value, only for a steady state calculation. The pressure solution itself was correct, but the post-processing evaluation of boundary flux was in error. It turned out that this coding error could easily be averted by changing the time differencing scheme selected for the steady state calculation. (The coding error was easily corrected by the author.)

6.2.2 Darcy Flow with Tensor Conductivity in Non-Orthogonal Coordinates

The tests described above Verified SECO_FLOW and HST3D for the simple (and unrealistic, for groundwater modeling) cases of constant properties. The next tests involved a much more general problem, and exemplify the debugging that can be accomplished with this approach of systematic grid convergence testing.

Modules for the SECO_FLOW codes were developed that use 2-D and 3-D non-orthogonal coordinates and allow for tensor conductivity. The Fortran source codes for the stencil loading were produced using computer Symbolic Manipulation, rather than hand coding. (For example, see Roache and Steinberg, 1984, Steinberg and Roache, 1985, 1986a, 1986b). These code subroutines were verified by solving several test problems, the most difficult of which involves an ellipse shaped region on the unit square with a discontinuous jump in tensor conductivities across the boundary.

A driver code was written to test the code and theory of automatic generation of Symmetric Finite Difference Stencils in Generalized 3-D Coordinates (Steinberg and Roache, 1990). The goal of the exercise was to verify 2nd-order accuracy of the finite difference approximations to the continuum problem, the latter consisting of any second-order symmetric elliptic operator with tensor coefficients applied to a continuous function on an arbitrary connected domain in the plane. A discontinuous coefficient tensor model was included to simulate the case of abrupt changes in the permeability of adjacent geological formations.

The basic capabilities of the driver code were, therefore, to generate

i. a set of transformations from logical space to physical space in the plane,
ii. a symmetric elliptic operator, and
iii. a set of boundary conditions based on the choice of solution function.

With these inputs, the stencil-loader code was called to generate the symmetric stencils. A simple point SOR solver then used the stencils to obtain the discrete solution. Finally, the discrete solution was compared to a known exact solution to obtain the global truncation error.

The irregular domain produces a general non-orthogonal grid, the most general problem for which the stencil-loader was designed. A number of other domains (e.g., rectangles, parallelograms, trapezoids) were also used during the debugging phase of this exercise, but were superfluous in the final Verification. Likewise, separate 2-D tests were run, but only the 3-D results are shown herein. The test problems were designed by P. Knupp.

The Transformation

Two transformations were used.

A. Unit Cube:

$$\begin{aligned} x(\chi,\xi,\zeta) &= \chi \\ y(\chi,\xi,\zeta) &= \xi \\ z(\chi,\xi,\zeta) &= \zeta \end{aligned} \tag{6.2.2.1}$$

B. Irregular Domain:

$$\begin{aligned} x(\chi,\xi,\zeta) &= \chi + \frac{1}{2}\sin\left[\frac{\pi}{4}\cdot\xi\right] \\ y(\chi,\xi,\zeta) &= \xi + \frac{1}{2}\sin\left[\frac{\pi}{8}\cdot(\chi+\zeta)\right] \\ z(\chi,\xi,\zeta) &= \zeta + \frac{1}{2}\sin\left[\frac{\pi}{8}\cdot(\chi+\xi+\zeta)\right] \end{aligned} \tag{6.2.2.2}$$

Using the bound $|\cos(x)| \le 1$, it is possible to show that the Jacobian of this transformation is strictly positive everywhere. The domain looks basically like a distorted cube.

The Operator

Three possible choices of coefficients define the Operator.

A. Laplace Operator.

$$\begin{aligned} &k11(x,y,z) = 1. \\ &k12(x,y,z) = k21(x,y,z) = 0. \\ &k13(x,y,z) = k31(x,y,z) = 0. \\ &k22(x,y,z) = 1. \\ &k23(x,y,z) = k32(x,y,z) = 0. \\ &k33(x,y,z) = 1. \end{aligned} \tag{6.2.2.3}$$

B. Arbitrary Elliptic Operator with Continuous Coefficients.

$$\begin{aligned} &e1(x,y,z) = 1 + 2x^2 + y^2 + z^2 \\ &e2(x,y,z) = 1 + x^2 + 2y^2 + z^2 \\ &e2(x,y,z) = 1 + x^2 + y^2 + 2z^2 \end{aligned} \tag{6.2.2.4}$$

$$\mathbf{D} = \begin{bmatrix} e1 & 0 & 0 \\ 0 & e2 & 0 \\ 0 & 0 & e3 \end{bmatrix} \tag{6.2.2.5}$$

$$\mathbf{P} = \begin{bmatrix} \cos(u)\cdot\cos(v) & \sin(u) & \cos(u)\cdot\sin(v) \\ -\sin(u)\cdot\cos(v) & \cos(u) & -\sin(u)\cdot\sin(v) \\ -\sin(v) & 0 & \cos(v) \end{bmatrix} \tag{6.2.2.6}$$

with $0 < u$, $v < 2\pi$, and

$$\mathbf{K} = \begin{bmatrix} k11 & k12 & k13 \\ k21 & k22 & k23 \\ k31 & k32 & k33 \end{bmatrix} \tag{6.2.2.7}$$

Then **D** has positive eigenvalues. If we let

$$\mathbf{K} = \mathbf{P}^{-1}\mathbf{D}\mathbf{P} \tag{6.2.2.8}$$

then **K** has the same eigenvalues. Further, since **P** is an orthogonal matrix, **K** is symmetric. The elements of **K** are as follows.

$$\begin{aligned} k11(x,y,z) = e1\cdot\cos(u)\cdot\cos(u)\cdot\cos(v)\cdot\cos(v) \\ +e2\cdot\sin(u)\cdot\sin(u)\cdot\cos(v)\cdot\cos(v) \\ +e3\cdot\sin(v)\cdot\sin(v) \end{aligned} \tag{6.2.2.9}$$

$$k12(x,y,z) = k21(x,y,z) = (e1-e2)\cdot\cos(u)\cdot\sin(u)\cdot\cos(v) \tag{6.2.2.10}$$

$$\begin{aligned} k13(x,y,z) &= k32(x,y,z) \\ &= \cos(v)\cdot\sin(v)\cdot\left[e1\cdot\cos(u)\cdot\cos(u)+e2\cdot\sin(u)\cdot\sin(u)-e3\right] \end{aligned} \tag{6.2.2.11}$$

$$k22 = e1\cdot\sin(u)\cdot\sin(u)+e2\cdot\cos(u)\cdot\cos(u) \tag{6.2.2.12}$$

$$k23(x,y,z) = k32(x,y,z) = (e1-e2)\cdot\cos(u)\cdot\sin(u)\cdot\sin(v) \tag{6.2.2.13}$$

$$\begin{aligned} k33(x,y,z) = e1\cdot\cos(u)\cdot\cos(u)\cdot\sin(v)\cdot\sin(v) \\ +e2\cdot\sin(u)\cdot\sin(u)\cdot\sin(v)\cdot\sin(v) \\ +e3\cdot\cos(v)\cdot\cos(v) \end{aligned} \tag{6.2.2.14}$$

For the numerical calculations, we took

$$u = \frac{5\pi}{12} \quad \text{and} \quad v = \frac{2\pi}{9} \tag{6.2.2.15}$$

C. Elliptic Operator with Discontinuous Coefficients.

The test case of discontinuous tensor conductivity was established by defining two regions, *I* and *II*, with continuous tensor conductivity in each but with a discontinuity at the boundary. To model the geologic situation, normal fluxes across a geological boundary are assumed continuous:

$$\left[\mathbf{K}\frac{\partial f}{\partial n}\right]_I = \left[\mathbf{K}\frac{\partial f}{\partial n}\right]_{II} \tag{6.2.2.16}$$

$$\begin{aligned} \mathbf{K}\frac{\partial f}{\partial n} &= \begin{bmatrix} n1 & n2 & n3 \end{bmatrix} \begin{bmatrix} k11 & k12 & k13 \\ k21 & k22 & k23 \\ k31 & k32 & k33 \end{bmatrix} \begin{bmatrix} fx \\ fy \\ fz \end{bmatrix} \\ &= n\mathbf{K}\nabla f \end{aligned} \tag{6.2.2.17}$$

The unit cube domain was divided into the two regions *I* and *II* wherein *I* was an ellipse embedded in the unit cube.

$$I = \{(x, y, z) \text{ in } U |\ r < 2\} \tag{6.2.2.18}$$

$$II = \{(x, y, z) \text{ in } U |\ r > 2\} \tag{6.2.2.19}$$

$$r = \sqrt{(1+x)^2 + 12y^2 + 12z^2} \tag{6.2.2.20}$$

On region *I*, we let

$$\mathbf{D} = \begin{bmatrix} 1.5 & 0 & 0 \\ 0 & 1.0 & 0 \\ 0 & 0 & 0.75 \end{bmatrix} \tag{6.2.2.21}$$

with $u = \frac{5 \cdot \pi}{12}$ and $v = \frac{2 \cdot \pi}{9}$. Then

$$\begin{aligned} k11 &= 0.916361, \quad k12 = 0.095755, \\ k13 &= 0.139593, \quad k22 = 1.466506, \\ k23 &= 0.080348, \quad k33 = 0.867133. \end{aligned} \tag{6.2.2.22}$$

On region *II*, we let

$$\mathbf{K}_{II} = \rho \cdot \mathbf{K}_{I} \tag{6.2.2.23}$$

with $\rho > 1$. For our tests, $\rho = 2$. We then took as the solution $f(x, y) = F(r)$, so that

$$F_I(r) = \rho \cdot \ln(r) \tag{6.2.2.24}$$

$$F_{II}(r) = \rho \cdot \ln(2) + 0.5 \cdot \ln(r-1) \tag{6.2.2.25}$$

and the flux continuity condition reduces to

$$[\mathbf{K} \quad \mathbf{F}_r]_I = [\mathbf{K} \quad \mathbf{F}_r]_{II}. \tag{6.2.2.26}$$

This is automatically satisfied by the choice of $F(r)$. Furthermore, $F(r)$ is continuous and differentiable (even at $r = 2$), as is needed for F is to represent pressure or head.

The Boundary Conditions

The general Robin boundary condition is

$$\alpha \cdot \frac{\partial f}{\partial n} + \beta \cdot f = \gamma \tag{6.2.2.27}$$

for each (continuum) point on the boundary of domain. For each of the four boundaries, the user selected from the following choices.

A. All Dirichlet: $\alpha = 0, \beta = 1$

B. All Neumann: $\alpha = 1, \beta = 0$ (6.2.2.28)

C. All Robin: $\alpha = 1, \beta = 1$

D. Random: $0 < \alpha, \beta < 1$

(Choice D uses α and β varying randomly along a boundary.)

The γ was then computed, based on the user's choice of solution function, f.

A. $f = 1$ on all of domain ($g = 0$ for Laplace Operator) (6.2.2.29)

B. $f = \sin(\pi \cdot x) \cdot \sin(\pi \cdot y) \cdot \sin(\pi \cdot z)$ (6.2.2.30)

C. $f = F(r)$ for discontinuous tensor (6.2.2.31)

The source term g is also computed from f, using $g = Lf$. The values of g and f are computed at cell centers, while γ is computed on boundary cell faces. For the discontinuous tensor case, $F(r)$ given in (6.2.2.24–25) must be selected.

Interface Conductivity Evaluation

Finally, the user can define PDE coefficients (i.e. the conductance, a combination of the physical variable conductivity and discretization terms) at the cell interfaces either by linear or harmonic averages. For a uniform grid, harmonic averaging is equivalent to defining k at the cell face $(i + 1/2)$ as

$$k(i+1/2) = \frac{2k(i)k(i+1)}{k(i)+k(i+1)} \tag{6.2.2.32}$$

as opposed to the more obvious linear average

$$k(i+1/2)=\frac{k(i)+k(i+1)}{2} \tag{6.2.2.33}$$

Harmonic averaging is commonly used in groundwater flow codes.

Debugging Successes

Several bugs in the 2-D codes were uncovered during the course of this Code Verification procedure.

i. The coefficient averaging procedure was not defined on the physical boundary of the problem, giving zero coefficient values there. This was easily remedied by changing the way in which the metric coefficients were computed in the stencil-loader.
ii. An error in the point-SOR routine at the corner stencils prevented 2nd-order convergence there.
iii. Two of the boundary loops used to compute the right-hand-side arrays in the stencil-loader were indexed over "*i*" when they should have been indexed over "*j*," causing incorrect zeros in the right hand side array.
iv. The corner stencil formula, e.g. at the lower left corner (i, j), was modified from

$$f(i,j)=\frac{1}{2}\left[f(i,j+1)+f(i,j-1)\right] \tag{6.2.2.34}$$

to

$$f(i,j)=f(i+1,j)+f(i,j-1)-f(i+1,j+1) \tag{6.2.2.35}$$

to preserve 2nd-order convergence there. (This also required adding a few lines of code setting the right-hand-side arrays to zero during the computation of the corner stencils).

With the 2-D code fixes built into the *n*-dimensional Symbolic Manipulation code, the 3-D tests uncovered only one additional bug.

v. The right-hand-side arrays on the edges of the cube were not properly initialized to zero (being outside the needed loops).

When these items were corrected, the 2-D and 3-D stencil-loader routines were Verified to produce 2nd-order accurate stencils for problems on general domains using non-orthogonal grids, provided smooth PDE coefficients were used. Solution function (B), Equation 6.2.2.30 was used in all the runs shown herein.

The full set of 3-D problems is described in Table 6.2.2.1. Linear averaging of the conductivities was used except in Run 10. Numerical results are given here only for Runs 1, 8, 9 and 10.

Run	Domain	Operator	B.C.
1	Unit Cube	Laplace	Dirichlet
2	Irregular	Laplace	Dirichlet
3	Unit Cube	Continuous	Dirichlet
4	Irregular	Continuous	Dirichlet
5	Unit Cube	Laplace	Robin
6	Irregular	Laplace	Robin
7	Unit Cube	Continuous	Robin
8	Irregular	Continuous	Robin
9	Unit Cube	Discontinuous	Dirichlet
10	Unit Cube	Discontinuous–harmonic average	Dirichlet

Table 6.2.2.1. Conductivity Problem Configurations. (From Table 3 of Roache et al, 1990.)

N	xerr	at	xerr $\cdot N^2$
$5 \times 5 \times 5$	0.03356	(3, 3, 3)	0.839
$10 \times 10 \times 10$	0.00796	(6, 6, 6)	0.796
$20 \times 20 \times 20$	0.00202	(11, 11, 11)	0.808
$40 \times 40 \times 40$	0.00051	(21, 21, 21)	0.816
$5 \times 5 \times 40$	0.02224	(3, 3, 21)	–
$5 \times 40 \times 5$	0.02226	(3, 21, 3)	–
$40 \times 5 \times 5$	0.02226	(21, 3, 3)	–

Table 6.2.2.2. Run 1. (From Table 4 of Roache et al, 1990.)

The symmetry of the solution and boundary conditions made Run 1, shown in Table 6.2.2.2, ideal to check the case $\Delta x \neq \Delta y \neq \Delta z$. As seen in the Table, the truncation error is the same no matter which direction is refined, and the maximum error occurs in the 21st-cell in the direction having 40 cells. Therefore, we can safely conclude there are no coding errors related to $\Delta x \neq \Delta y \neq \Delta z$.

In another experiment, Run 1 was repeated with the PDE coefficients set to zero in the ring of cells outside the physical boundary to obtain some idea of the effect of extrapolating these coefficients. The results showed that the accuracy was degraded to roughly 1st-order.

Run 8 in Table 6.2.2.3 shows the 2nd-order convergence of the continuous operator with Robin boundary conditions on an irregular domain, exercising all the metric terms in the 3-D boundary fitted coordinate transformation.

N	xerr	xerr · N^2
5	0.06970	1.743
10	0.02119	2.119
20	0.00559	2.236
40		

Table 6.2.2.3. Run 8. (From Table 5 of Roache et al, 1990.)

N	xerr	xerr · N^1
5	0.05395	0.27
10	0.02808	0.28
20	0.01442	0.29
40	–	–

Table 6.2.2.4. Run 9: Linear Averaging for K. (From Table 6 of Roache et al, 1990.) (Note the unity exponent in the last column.)

N	xerr	xerr · N^1
5	0.05906	0.29
10	0.02079	0.21
20	0.01301	0.26
40	0.00688	0.28

Table 6.2.2.5. Run 10: Harmonic Averaging for K. (From Table 7 of Roache et al, 1990.) (Note the unity exponent in the last column.)

Runs 9 and 10 in Tables 6.2.2.4 and 6.2.2.5 show the convergence of the discontinuous coefficient problem on the unit cube, for linear and harmonic averaging of conductivities.

The theoretical basis for harmonic averaging (e.g. see McDonald and Harbaugh, 1988) is well established in 1-D. We found 2nd-order convergence in 1-D provided that the discontinuity stayed on a cell interface as the grid was refined, and that harmonic averaging was used; linear averaging produced only 1st-order convergence. However, our nonorthogonal grid multidimensional results in Tables 6.2.2.4–5 above are not impressive. For a discontinuous k with ratio $\rho = 2$ (which is mild by geologic standards), 3-D harmonic averaging is slightly worse than linear averaging at $N = 5$, and slightly better at higher resolution. Both certainly give a consistent discretization, but both are only 1st-order accurate. (Note the unity exponent in the last column.) The order of convergence was *not* restored to 2nd-order by aligning the grid with the discontinuity.

Since the publication of Roache et al (1990), the basis for this section, Shashkov and Steinberg (1995,1996) have shown how to correctly generalize the concept of harmonic averaging to 2-D. The solution is elegant but surprisingly expensive. (See also Shashkov, 1997 and Hyman et al, 1997.)

The above tests Verify the accuracy of the spatial differencing. The time differencing was separately Verified by a simple test of the decay of a single Fourier component, the solution being obtained by elementary separation of variables (Roache et al, 1995). The separation of Verification for spatial and temporal differencing is justified only by the knowledge that the algorithms did not mix space and time derivatives (unlike algorithms such as Approximate Factorization, for example). One could argue that this is cheating on the Verification process, since it is not, strictly speaking, treating the code as a "black box." As noted elsewhere, this approach amounts to partitioning the option matrix of the code, and greatly reduces the complexity of the option combination problem, if it can be justified.

This partitioning is further made palatable, even to one who is not familiar with the code structure, by the modest claim of only $O(\Delta t)$ accuracy. In fact, it is difficult to imagine a non-contrived coding error that could produce reasonable time solutions and not be at least $O(\Delta t)$ accurate. As noted in Chapter 3 in the Section 3.14 titled "Warnings: What the Method Does Not Verify," it is easy to make coding mistakes that do not strictly follow the algorithm as intended, yet do not affect the order of time accuracy. For example, in the fully implicit (backward time) differencing algorithm in the presently considered codes, the intention was to set the boundary conditions at the advanced time level $(n + 1)$. If a coding mistake is made that results in some boundary values being lagged, i.e. set at time level n, stability might be affected, but the order of convergence would still test at $O(\Delta t)$. The *size* of the error might be (slightly) affected, compared to the intended algorithm, but the *order* would not be changed. On the other hand, $O(\Delta t^2)$ algorithms, especially those like Approximate Factorization that mix space and time differences, can easily be corrupted by coding mistakes to $O(\Delta t)$. More scrupulous Verification is required, as in Salari and Blaine (1996); see also Section 6.24.

6.3 ISSUES IN CALCULATION VERIFICATION

The above examples involved accuracy Verification of Codes. The following examples involve Verification of individual Calculations, and assume the use of an already Verified code.

This initial discussion on issues in calculation Verification is taken from Westerink and Roache (1995), "Issues in Convergence Studies in Geophysical Flow Computations."

In performing systematic grid convergence studies, there are a variety of issues which lead to a level of uncertainty in the estimate which must be taken into consideration. These issues include the following.

- Higher order truncation term interaction even in the asymptotic range
- Non-asymptotic truncation term interaction
- Time-space truncation error interaction and related superconvergence
- Effect of parameter convergence

The wavenumber distribution of the response will also affect observed convergence rates, and the response spectra will include flow features artificially generated by the numerics. The computations may also experience nonuniform behavior of various error norms. All these possible factors, and the others noted earlier in Chapter 5, introduce uncertainty into the Verification process based on Richardson Extrapolation or other methods, and provide further rationale for the safety factor $F_s > 1$ in the Grid Convergence Index.

6.3.1 Formal, Actual and Observed Convergence Rates

It is a common feature of discrete solutions to PDE's to encounter convergence rates for complicated applications which are less than formal rates. It is assumed that the formal convergence rate is indicated by the leading order space and/or time truncation error terms. However, even in the asymptotic range (where discrete space and time steps tend to zero), formal convergence rates may never be achieved, leading to the definition of the *actual asymptotic* convergence rate. Even neglecting the well-recognized problems with computer round-off error, this *actual asymptotic* convergence rate may be different from the *formal* convergence rate when, for example, the formal analysis misses interaction between the leading order and subsequent order truncation terms in the error series (or when other approximations are required). This can come about due to the form of the truncation terms, which consists of products of the space step to a power and higher order derivatives of the response function. When gradients of the response function increase with increasing grid refinement, as they do in many nonlinear and/or complicated flows, the higher order terms in the truncation series continue to compete with the leading order truncation term.

Finally, the convergence rate actually *observed* in numerical experiments may be different from *either* the *formal* or *actual asymptotic* convergence rates, simply due to grid resolution not being adequate to achieve the asymptotic range. When the coarse grid Richardson Error Estimator does not equal (approximately) the fine grid Richardson Error Estimator computed for the same grid, this is an indication that the formal convergence rate is not being achieved. It is in fact common for the fine grid error estimator to be less than the coarse error estimator for the same grid, indicating that the observed convergence rate is less than the formal rate. A more complete discussion of the variety of factors that can influence the observed convergence rates will be given in Chapter 8, Section 8.1.

6.4 AN EXAMPLE OF THE EFFECTIVE GRID REFINEMENT RATIO

As an example of the effect of grid refinement ratio on the reporting of grid convergence tests, consider the following Figure 6.4.1 for local Nusselt numbers (normalized heat transfer coefficients) taken from the second-order accurate Finite Element solutions by Comini et al (1995).

The presentation of results looks converged almost to the plotting accuracy, perhaps 0.1%. However, the grid refinement is not a grid doubling, as noted in the legend. To calculate an

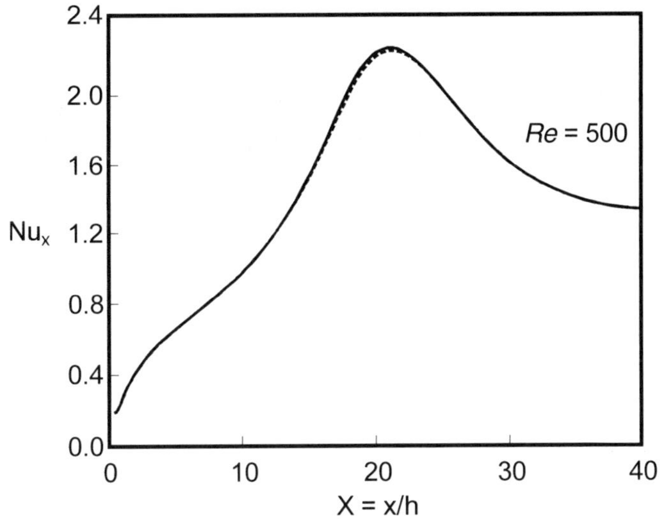

Figure 6.4.1. Grid dependence of calculated distributions of local Nusselt numbers. Pr = 0.7, *Re* = 500. Solid lines represent results from a mesh with 1024 parabolic elements and 3273 nodes. Dashed lines represent results from a mesh with 1504 parabolic elements and 4725 nodes. (From Figure 4 of Comini et al, 1995.)

effective grid refinement ratio r as suggested in Chapter 5 for this unstructured grid, we could use either the ratio of (parabolic) elements, giving

$$r_{elements} = \left(\frac{1505}{1024}\right)^{1/2} = 1.21 \tag{6.4.1}$$

or the ratio of nodes,

$$r_{elements} = \left(\frac{4725}{3273}\right)^{1/2} = 1.20 \tag{6.4.2}$$

which are equivalent within the approximation of the empiricism of the concept of effective r for unstructured grid refinement. With $p = 2$ for the parabolic elements used by Comini et al, this gives

$$\text{GCI} = \frac{3\varepsilon}{r^p - 1} = \frac{3\varepsilon}{1.2^2 - 1} = 6.64\varepsilon \tag{6.4.3}$$

With the small $\varepsilon \sim 0.1\%$ this still indicates a level of grid convergence that is totally acceptable, especially considering the conservatism of the "safety factor" of $F_s = 3$ in the GCI; yet, it is considerably less converged than suggested by the original reporting. Even if we use $F_s = 1$, so that the reported GCI equals the Richardson Error Estimator, corresponding roughly to only a 50% confidence band, we would obtain

$$\mathrm{GCI}(F_s = 1) = 2.21\varepsilon \tag{6.4.4}$$

somewhat less impressive than just ε. If results similar to Figure 6.4.1 had been obtained with only first-order methods, the discrepancy in the reporting would be marked. For $F_s = 3$, we would obtain

$$\mathrm{GCI} = 14.63\varepsilon \tag{6.4.5}$$

and for $F_s = 1$,

$$\mathrm{GCI} = 4.88\varepsilon \tag{6.4.6}$$

again showing the difficulty of judging grid convergence with first-order methods.

6.5 A BENCHMARK PROBLEM FOR DRIVEN CAVITY FLOW

The driven cavity problem (Roache, 1972) is geometrically simple, and has been used for decades as an easily reproducible problem. Ghia et al (1982) produced a benchmark solution that is often cited for comparison purposes. Although no error estimate *per se* was included, the results can still be used with confidence for debugging purposes. Because of its wide utility for benchmark comparisons, we reproduce the solution here in Table 6.5.1.

6.6 A BENCHMARK PROBLEM FOR FREE CONVECTION

As common as the driven cavity problem is, it has a disadvantage, in that the moving lid introduces a singularity. In the upper corners, the velocity is discontinuous, and the vorticity is unbounded. (This does not prevent the vorticity from converging on the lid away from the corners; see, e.g., Roache, 1975,1995). De Vahl Davis et al (1979) devised a comparison problem specifically to remove the singularity but maintain the geometric simplicity of the driven cavity. Non-trivial motion is induced by a heated vertical wall. While requiring that the code include buoyancy terms, this problem is more realistic than the pure driven cavity, and contains no singularities. (A non-realistic mathematical model that eliminates the singularity could use a prescribed lid velocity varying from 0 at the corners to 1 in the middle, and would serve equally well for code Benchmarking, but it is not as attractive because it is physically unrealizable.) De Vahl Davis and Jones (1983) compared several contributed solutions for this problem, which showed a disturbing range of uncertainty, especially for the early finite element solutions.

De Vahl Davis (1983) produced a classic Benchmark solution of this free convection problem for comparison. He used systematic grid convergence and Richardson Extrapolation very carefully, with an experimental determination rather than an assumption of the *local* order of convergence. Seven years later, Hortmann et al (1990) recalculated this problem with modern multigrid algorithms and higher computer power. The increased resolution (up to 320 × 320 for most of the calculations, and 640 × 640 for one) of course provided more accurate answers and

129-Grid Pt. No.	y	Reynolds Number *Re* 100	400	1000	3200	5000	7500	10,000
129	1.00000	1.00000	1.00000	1.00000	1.00000	1.00000	1.00000	1.00000
126	0.9766	0.84123	0.75837	0.65928	0.53236	0.48223	0.47244	0.47221
125	0.9688	0.78871	0.68439	0.57492	0.48296	0.46120	0.47048	0.47783
124	0.9609	0.73722	0.61756	0.51117	0.46547	0.45992	0.47323	0.48070
123	0.9531	0.68717	0.55892	0.46604	0.46101	0.46036	0.47167	0.47804
110	0.8S16	0.23151	0.29093	0.33304	0.34682	0.33556	0.34228	0.34635
95	0.7344	0.00332	016256	0.18719	0.19791	0.20087	0.20591	0.20673
80	0.6172	−0.13641	0.02135	0.05702	0.07156	0.08183	008342	0.08344
65	0.5000	−0.20581	−0.11477	−0.06080	−0.04272	−0.03039	−0.03800	0.03111
59	0.4531	−0.21090	−0.17119	−0.10648	−0.86636	−0.07404	−0.07503	−0.07540
37	0.2813	−0.15662	−0.32726	−0.27805	−0.24427	−0.22855	−0.23176	−0.23186
23	0.1719	−0.10150	−0.24299	−0.38289	−0.34323	−0.33050	−0.32393	−0.32709
14	0.1016	−0.06434	−0.14612	−0.29730	−0.41933	−0.40435	−0.38324	−0.38000
10	0.0703	−0.04775	−0.10338	−0.22220	−0.37827	−0.43643	−0.43025	−0.41657
9	0.0625	−0.04192	−0.09266	−0.20196	−0.35344	−0.42901	−0.43590	−0.42537
8	0.0547	−0.03717	−0.08186	−0.18109	−0.32407	−0.41165	−0.43154	−0.42735
1	0.0000	0.00000	0.00000	0.00000	0.00000	0.00000	0.00000	0.00000

129-Grid Pt. No.	x	Reynolds Number *Re* 100	400	1000	3200	5000	7500	10,000
129	1.0000	0.00000	0.00000	0.00000	0.00000	0.00000	0.00000	0.00000
125	0.9688	−0.05906	−0.12146	−0.21388	−0.39017	−0.49774	0.53858	−0.54302
124	0.9609	−0.07391	−0.15663	−0.27669	−0.47425	−0.55069	0.55216	−0.52987
123	0.9531	−0.08864	−019254	−0.33714	−0.52357	−0.55408	0.52347	−0.49099
122	0.9453	−0.10313	−0.22847	0.39188	−054053	−0.52876	−0.48590	−0.45863
117	0.9063	−0.16914	−0.23827	0.51550	−0.44307	−0.41442	−0.41050	−0.41496
111	0.8594	−0.22445	−0.44993	−0.42665	−0.37401	−0.36214	−0.36213	−0.36737
104	0.8047	−0.24533	−0.38598	−0.31966	−0.31184	−0.30018	−0.30448	−0.30719
65	0.S000	0.05454	0.05186	0.02526	0.00999	0.00945	0.00824	0.00831
31	0.2344	0.17527	0.30174	0.32235	028188	0.27280	0.27348	0.27224
30	0.2266	0.17507	0.30203	0.33075	0.29030	0.28066	0.28117	0.28003
21	0.1563	0.16077	0.28124	0.37095	0.37119	0.35368	0.35060	0.35070
13	0.0938	0.12317	0.22965	0.32627	0.42768	0.42951	0.41824	0.41487
11	0.0781	0.10890	0.20920	0.30353	0.41906	0.43648	0.43564	0.43124
10	0.0703	0.10091	0.19713	0.29012	0.40917	0.43329	0.44030	0.43733
9	0.0625	0.09233	0.18360	0.27485	039560	0.42447	0.43979	0.43983
1	0.0000	0.00000	0.00000	0.00000	0.00000	0.00000	0.00000	0.00000

Table 6.5.1. Results for *u*- and *v*-velocity components for driven cavity flow, from Ghia et al (1982).

tighter error bands, but the original estimates of de Vahl Davis were confirmed. "For example, [de Vahl Davis] estimated the bounds for [dimensionless heat transfer rate, Nusselt number] to be 0.2%, 0.3% and 1.0% for Rayleigh numbers 10^4, 10^5, and 10^6 respectively; the difference to the present results is 0.08%, 0.06% and 0.2%." Because of complexities of normalizing schemes, etc., the paper by Hortmann et al (1990) should be consulted directly for its dependable Benchmark comparisons.

6.7 LAMINAR PLANE JET IMPINGING ON A HEATED FLAT PLATE

Pelletier and Ignat (1995) developed a simple analytical solution for the temperature field for an incompressible planar laminar jet impinging on a heated flat plate. They used the solution to calibrate and evaluate their solution adaptive unstructured grid generation methods, and to compare of their Zhu-Zienkiewicz type error indicators with the GCI extended to unstructured grids. (See details in the original paper.) The Reynolds number was set to 50, and their analytical solution is as follows.

$$u = 4x^2 e^{-\alpha^2 a}(1 - 2\alpha^2 a) \tag{6.7.1}$$

$$v = -8xye^{-\alpha^2 a}(1 - 2\alpha^2 a) \tag{6.7.2}$$

$$p = \frac{1}{1 + C(x^2 + y^2)} \tag{6.7.3}$$

$$T = \exp\left[\frac{\gamma x}{\delta_0(1 + \beta y)}\right] \tag{6.7.4}$$

with the following parameters: $\alpha = x^2 y$, $a = 5000$, $\gamma = 5$, $\beta = 4$, $\delta_0 = 0.01$.

6.8 A *K*-ε MODEL OF A FREE SHEAR LAYER

In the same paper, Pelletier and Ignat (1995) developed a simple analytical solution for an incompressible free shear layer using the *k*-ε turbulence model. As in the previous section, they used this solution to compare their Zhu-Zienkiewicz type error indicators with the GCI extended to unstructured grids. (See details in the original paper.) The solution should be of considerable interest for turbulence modelers, since it displays most of the characteristics of an experiment of Patel (1973) yet is convenient to encode. A similar approach should be used to develop analytical Benchmarks for variants of the *k*-ε model, and for the Wilcox (1993) *k*-ω model and variants.

$$u = \left(\frac{u_1 + u_2}{2}\right) - \left(\frac{u_1 - u_2}{2}\right) erf\left(\frac{\sigma\, y}{x}\right) \tag{6.8.1}$$

$$v = v_0 + \left(\frac{u_1 - u_2}{2}\right)\frac{1}{\sigma\sqrt{\pi}}\exp\left[-\left(\frac{\sigma y}{x}\right)^2\right] \tag{6.8.2}$$

$$B = \exp\left[-\left(\frac{\sigma y}{x}\right)^2\right] \tag{6.8.3}$$

$$k = k_0(c_k + B) \tag{6.8.4}$$

$$\varepsilon = \frac{\varepsilon_0}{x}(c_k + B) \tag{6.8.5}$$

$$\mu_T = \mu_{T0}x(c_k + B) \tag{6.8.6}$$

$$p = 0, u_1 = 1.0, u_2 = 0.0, \sigma = 13.5 \tag{6.8.7}$$

$$k_0 = \frac{343}{75{,}000}u_1(u_1 - u_2)\frac{\sigma}{\sqrt{\pi}} \tag{6.8.8}$$

$$\varepsilon_0 = \frac{343}{22{,}500}C_\mu u_1(u_1 - u_2)^2\frac{\sigma^2}{\pi} \tag{6.8.9}$$

$$\mu_{T0} = \frac{343}{250{,}000}\rho u_1 \tag{6.8.10}$$

$$Re_l = \frac{\rho u_1 L}{\mu_l} = 10^4 \tag{6.8.11}$$

6.9 TRANSONIC AIRFOIL CALCULATIONS

Zingg (1991,1992) applied the Richardson Error Estimator to airfoil lift and drag calculations in body-fitted grids. Zingg's work demonstrates the necessity of grid convergence testing even when experimental data are available. In 4 of 7 cases, experimental agreement was better with *coarse* grid calculations than with fine (Zingg, 1992).

The results of the Viscous Transonic Airfoil Workshop (Holst, 1987) indicate that "while turbulence modeling is a significant source of error in the computation of airfoil flows, particularly for separated flows, a large portion of the error is often associated with inadequate numerical resolution and an insufficient distance to the outer boundary of the grid" (Zingg, 1991). Zingg's work convincingly brought these errors well within engineering accuracy. He used Pulliam's well-established viscous airfoil code ARC2D (Pulliam, 1986) and the algebraic

Baldwin-Lomax turbulence model with the thin-layer approximation to the full Reynolds-Averaged Navier-Stokes equations on structured non-orthogonal "C-type" grids. Zingg presented results for 8 different flow cases and two fine base grids, each base grid being coarsened two times by removal of alternate cells in each direction; thus, $r = 2$ and $p = 2$ in the Richardson extrapolations. Here, I will give the lift and drag results for just two flow cases and the "A7A" base grid of 497 × 194 points, with 401 of the 497 points on the body and 48 in the wake. Points were clustered at the leading and trailing edges with a spacing of 10^{-4} chords; the normal spacing at the body was 10^{-7} chords. The most coarse grids used still had adequate resolution of the viscous sublayer (with the largest $y^+ = 0.2$ for the first grid point, less than the commonly used value of $y^+ = 1$). The outer boundary distance was 12 chords.

The first result presented here is for *subsonic* flow, Zingg's Case 3, a NACA 0012 airfoil at $M = 0.16$ and angle of attack $\alpha = 12°$. The lift and drag coefficients are given in Table 6.9.1.

For this case, trailing edge separation was predicted on the upper surface (at about 96% of chord for the fine grid calculation.) Second order accuracy is Verified by the reduction in % changes by factors of 4 as the grid is refined by factors of 2, from coarse to medium, and medium to fine, correct to the appropriate level of precision. This is a difficult case as judged by the drag calculation, which is 5.2% in error even for the fine grid of 497 × 194 points, compared to the best RE (**R**ichardson **E**xtrapolation) results (using the fine and medium grids). As Zingg notes, "Richardson Extrapolation is very effective for this case," as can be judged by evaluating the results of RE from the medium and coarse grids, labeled "RE m-c" in the Table 6.9.1. These give accuracy roughly equal to the finest grid for lift, and about a 9× improvement for drag ($\%C_D = 0.6$ compared to 5.2). Zingg notes the favorable economics involved with grid coarsening (see also Chapter 3); a grid coarsened by removal of every other point gains a factor of 4 in number of points, and (roughly) another factor of 2 in number of time steps, so RE costs about an additional 1/8 or 12.5% of computer time. Applying these ratios to this case, the medium grid calculation requires about 1/8 of the fine grid calculation, and the coarse grid calculation requires another $1/8^2$. So the medium-coarse RE requires about 9/64 the time of a single fine grid calculation.

Grid	C_L	C_D	% C_L	% C_D
Fine	1.3081	0.01327	−0.3	5.2
Medium	1.2957	0.01526	−1.3	21.0
Coarse	1.2563	0.02299	−4.3	82.3
RE m-c	1.3088	0.01268	−0.3	0.6
RE f-m	1.3122	0.01261	–	–

Table 6.9.1. Lift and drag coefficients for NACA 0012 airfoil at $M = 0.16$ and $\alpha = 12°$. The "RE f-m" results were obtained by **R**ichardson **E**xtrapolation from the fine and medium grid solutions, and "RE m-c" from the medium and coarse grids. The %'s are changes from the RE results. (From Table 3 of Zingg, 1991.)

Allowing a little slack for additional arithmetic, the Richardson Extrapolation method on medium and coarse grids produces about the same error as the fine grid on lift and about 9× better on drag, plus it provides an error estimate (whereas the fine grid calculation alone does not), at about 15% of the cost!

The economics are improved in 3-D by another factor of 2, so RE m-c costs about (1/16 + $1/16^2$) = 17/256 or about 7% of the cost of a single fine grid solution. The cost could further be reduced by using coarse grid solutions as initial guesses for the finer grids.

Note, however, as discussed in Chapter 3, that we really require 3 grid solutions to be certain that we are in the asymptotic range, and to evaluate the observed order of convergence p. Zingg found that the 3-grid solutions were well-behaved for the subsonic cases such as the above. But for transonic cases, he found that "the streamwise point spacing of the [coarse] grid is not sufficient, and shocks are excessively smeared. As a result, the extrapolated solutions [from the medium and coarse grids] generally do not reduce the error." Accordingly, the coarse grid results are not included herein for the *transonic* Case 5, a NACA 0012 airfoil at $M = 0.55$ and $\alpha = 8.34°$, shown in Table 6.9.2.

For this case, "there is a strong shock on the upper surface at about 20% chord. There is a small separated flow region after the shock and the flow finally separates at about 90% chord." The fine grid solution is good, predicting lift and drag to within 0.4% and 0.2% respectively of the RE solution. [The GCI's for lift and drag, using the very conservative $F_s = 3$, would be 1.2% and 0.6%.] This is a high quality numerical solution for a strong shock transonic flow. Zingg notes that "the second-order artificial dissipation required at shocks scales with the grid spacing to first order. Therefore, Richardson Extrapolation can only be used if the global influence of the second-difference dissipation is small. This is likely if shocks are weak or well-resolved." [Alternately, separate Richardson extrapolation can be performed separately for the dissipation, as in Blottner (1990) and Kuruvila and Anderson (1985); see Section 6.11.] Generally, Zingg (1991) found that the error in pressure drag is larger than the error in friction drag. See the original paper for these and other additional results, and comparison with experiments. (See also comments in Chapter 10.)

Grid	C_L	C_D	% C_L	% C_D
Fine	0.9977	0.03484	−0.4	0.2
Medium	0.9868	0.03509	−1.4	0.9
RE f-m	1.0013	0.03476	–	–

Table 6.9.2. Lift and drag coefficients for NACA 0012 airfoil at $M = 0.55$ and $\alpha = 8.34°$. The "RE f-m" results were obtained by **R**ichardson **E**xtrapolation from the fine and medium grid solutions. The %'s are changes from the RE results. (From Table 5 of Zingg, 1991.)

6.10 ORDERED ESTIMATION OF FAR-FIELD BOUNDARY ERROR

In a later paper (Zingg, 1992) on airfoil calculations, Zingg systematically varied the position of the outer computational boundary, i.e. the "far-field" boundary. Inspection of his data indicates that Richardson Extrapolation can be applied to the estimation of far-field boundary errors, with the error proving to be 1st-order in the inverse of distance to the boundary, as now shown.

Zingg's (1992) Case 3 was a subsonic NACA 0012 airfoil at $M = 0.16$, $\alpha = 12°$, $Re = 2.88\times10^6$, with transition at 0.01 chords on the upper surface and 0.95 on the lower. From inspection of the data in his Table 8, "Results of outer boundary position studies," it appeared that the errors for Case 3 were approximately ordered in $1/L_b$, the distance to the outer boundary. Calculations proved to be remarkably consistent with this assumption. The Case 3 data are reproduced below in Table 6.10.1. The $\%C_L$ and $\%C_D$ were calculated as differences from the 96 chord solution.

Note that both $\%C_L$ and $\%C_D$ approximately doubled going from 24 to 12 chords, consistent with the assumption of boundary error $BE \sim 1 / L_b$ where L_b is the distance from the airfoil to the outer boundary. If we make this assumption, extrapolate to $L_b = \infty$, and re-calculate $\%C'_L$ and $\%C'_D$ as differences from this extrapolated solution (as one does with the mesh spacing results), the behavior even more accurately fits the model of $BE \sim 1 / L_b$, as now shown.

If we were doing Richardson Extrapolation from $L_b = 12$ and 24 with the assumption that $BE \sim 1/L_b^p$ and $p = 1$, it would look like the usual grid extrapolation with grid doubling and $p = 1$, i.e.

$$C_L \sim C_{L24} + (C_{L24} - C_{L12}) \tag{6.10.1}$$

However, since we are doing the analog of a grid quadrupling when we go from $L_b = 24$ to 96, the extrapolation now becomes

$$C'_L \sim C_{L96} + 1/3(C_{L96} - C_{L24}). \tag{6.10.2}$$

L_b	C_L	C_D	% C_L	% C_D
96 chords	1.3111	0.01397	–	–
24	1.3121	0.01376	0.08	−1.5
12	1.3139	0.01349	0.2	−3.4

Table 6.10.1. Effect of outer boundary position on lift and drag coefficients of transonic airfoil calculations. % C_L and % C_D are differences from the 96 chord boundary values. (From Zingg, 1992, Case 3, Table 8.)

L_b	C_L	C_D	$\%C'_L$	$\%C'_D$
96 chords	1.3111	0.01397	0.02289	−0.4986
24	1.3121	0.01376	0.09917	−1.9943
12	1.3139	0.01349	0.23649	−3.9174

Table 6.10.2. Effect of outer boundary position on lift and drag coefficients of transonic airfoil calculations. $\%C'_L$ and $\%C'_D$ are differences from the extrapolated boundary values.

(That is, the factor of 1/3 appears either from the common doubling with $p = 2$, or from quadrupling with $p = 1$.) Applying this gives

$$C'_L \sim 1.3111 + 1/3(1.3111 - 1.3121) = 1.3108 \tag{6.10.3}$$

$$C'_L \sim 0.01397 + 1/3(0.01397 - 0.01376) = 0.01404 \tag{6.10.4}$$

Recalculating $\%C'_L$ based on the differences between C_L and C'_L (rather than C_{L96}), and similarly for $\%C'_D$, gives the following Table 6.10.2. (I have kept more significant figures than justified in order to show the asymptotic behavior.)

For the assumed $BE \sim 1 / L_b$, the theoretical ratios R of successive $\%C'_L$ and $\%C'_D$ would be 4 from 96 to 24 chords, and 2 from 24 to 12 chords. The actual values are shown below in Tables 6.10.3–4.

L_b	$\%C'_L$	R actual	R theory
96 chords	0.02289		
24	0.09917		
		4.332	4
12	0.23649		
		2.38470	2

Table 6.10.3. Extrapolations for Boundary Error in C_L.

L_b	$\%C'_D$	R actual	R theory
96 chords	−0.4986		
24	−1.9943		
		3.9998	4
12	−3.9174		
		1.9643	2

Table 6.10.4. Extrapolations for Boundary Error in C_D.

As in the case of Zingg's grid size calculations, the extrapolations for C_L are a little noisy, since those calculations are already very accurate. The extrapolations for the less accurate C_D are remarkably consistent with the assumption $BE \sim 1 / L_b$. The need for the unrealistic retention of so many significant figures is now evident; they are required to indicate any departure from the theory! The ratios are exact for anything less than 4 significant figures.

These new error estimates are slightly larger, which is to be expected since the original reference was $L_b = 96$ to approximate $L_b = \infty$ and the convergence is monotone. The original estimate of 1.5% error for $\%C_D$ at $L_b = 24$ now becomes $\%C'_D = 2.0\%$.

6.11 ARTIFICIAL DISSIPATION EFFECTS

Blottner (1990) has used the same procedure to estimate effects of artificial dissipation terms in hypersonic flow calculations. Blottner's study used the boundary fitted coordinate Navier-Stokes code NS3D in the axisymmetric option to calculate hypersonic ($M = 5$) laminar flow over a spherical nose tip at $Re = 1.8875\times10^6$. The code is 2nd-order in space and time, and incorporates both an implicit 2nd-order and an explicit 4th-order artificial dissipation De. The form of De used is

$$De = \frac{S}{J}\left[\Delta\xi^4 \frac{\partial^4 Q}{\partial\xi^4} + \Delta\zeta^4 \frac{\partial^4 Q}{\partial\zeta^4}\right] \tag{6.11.1}$$

in which S is a user-defined weight (a smoothing parameter), J is a normalizing Jacobian, Q represents each of the conservation variables, and ξ and ζ are the streamwise and cross-stream general coordinates.

Since both the implicit and explicit artificial dissipations are ordered terms, either or both of these could be evaluated simply as part of an overall grid convergence study. However, it is also of interest to separately evaluate the explicit artificial dissipation De, since the effects of this term can be extrapolated to $De \to 0$ for a *fixed* grid as well. For a particular grid, we obtain solutions with two values of S, and extrapolate the solution as follows.

$$\begin{aligned} Q(S \to 0) &= Q_1 + S_1 \times D \\ D &= -\frac{Q_1 - Q_2}{S_1 - S_2} \end{aligned} \tag{6.11.2}$$

Table 6.11.1 from Blottner (1990) shows this extrapolation for three different grids, each a multiple of two, for the pressure and heat transfer rate at one representative point, on the body surface and 90° away from the stagnation point. (The iterative convergence criteria was tight, $L_2 \leq 10^{-8}$.)

Grid Cells	S	$P_{wall}/P_{freestream}$	q_{wall}
22 × 24	0	1.62258	7.97707
	5	1.61099	7.76188
	10	1.60598	7.97235
	20	1.58938	7.96763
	40	1.56158	7.86627
44 × 48	0	1.57535	8.27713
	5	1.57453	8.27711
	10	1.57437	8.28781
	20	1.57339	8.29849
	40	1.57137	8.31405
	80	1.56760	8.33694
88 × 96	0	1.56433	8.42643
	10	1.56427	8.42615
	20	1.56421	8.42587
	40	1.56409	8.42526

Table 6.11.1. Influence of explicit artificial dissipation parameter *S* on surface pressure and heat flux *q* [kW/m^2]. (From Blottner, 1990, Table 2.)

Note the double convergence involved. For each grid, the extrapolation gives the corresponding solution for $S \to 0$, but the values are significantly different for each grid. For the coarse grid, there is some departure from linearity for the effect of S (see Blottner, 1990, Figure 2). For the finest grid, extrapolation to $S \to 0$ proves to be unnecessary, since the pressure is already converged to 5 significant figures for $S = 20$ and 10. This is consistent with the fact that the explicit artificial dissipation is also an ordered approximation. See also Kuruvila and Anderson (1985) and Zingg (1992). Salari and Roache (1990) show that the convergence as artificial dissipation is reduced in a Approximate Factorization solutions of a compressible flow code is well behaved even at low Mach numbers ($M = 0.1$); see also Section 6.28 below. Blottner notes that "This result indicates that the [explicit] dissipation term introduces a significant error for the coarse grid solution, but the fine grid solution [for pressure] is only slightly in error." However, "it appears that the heat flux requires a finer grid before this type of analysis can be used effectively."

[Note also the dangers of compensating errors in the pressure solution. The pressure solution on the coarse grid with large dissipation ($S = 40$) is 1.56158, which differs by only 0.2% from the best fine grid solution extrapolated to $S = 0$, which is 1.56433. (Some intermediate value of about $S = 39$ would give equality.) Such a spot check, without systematic convergence studies, could be dangerously misleading, especially since the errors for the heat flux do *not* compensate, but add.]

The paper of Blottner (1990) contains many more details, including Richardson Extrapolation for the grid convergence.

6.12 SINGLE AND DUAL POROSITY CONTAMINANT TRANSPORT: SOURCE TERM LOCATION

Grid resolution studies of radionuclide transport in fractured porous media were performed by Salari et al (1995) using a finite volume code, SECO_TRANSPORT. Transport calculations were performed for the grid convergence test using both single and dual-porosity models. (All the SECO codes have the feature of using separate grids to define aquifer properties and to perform calculations; this is highly advantageous for performing grid convergence studies.) The velocity field for the transport calculations was provided by an algebraic relation, removing any additional errors that might influence transport convergence results. The Grid Convergence Index was used to examine the convergence results. The grid convergence results for radionuclide transport calculations in porous media proved to be significantly influenced by the *plausible but improper* modeling of the source in a finite volume grid during the grid refinement. A source term could be modeled as a point source, or can have some physical dimensions. The distinction was shown to have major consequences for grid convergence tests.

6.12.1 Transport Code

The SECO_TRANSPORT code developed by K. Salari (Salari et al, 1992, Roache 1993; Salari and Blaine, 1996) is one of the suite of SECO (**S**andia–**ECO**dynamics) codes that perform groundwater flow, transport, and particle tracking. SECO_TRANSPORT computes multiple component solute transport in (possibly) fractured porous media using single porosity or dual porosity models. It allows for radioactive decay and generation of daughter products.

The "dual porosity" model is a conceptualization of the geologic system as a matrix of porous material with fractures. When no (significant) fractures are present, all flow and transport takes place in the rock matrix. When (significant) fractures are present, flow in the fracture system dominates flow in the matrix system; one may safely neglect *flow* in the matrix, but not *transport* in the matrix. The contaminant disperses from the fracture system through the interface (possibly including a skin resistance due to clay lining) and into the matrix system, where it also decays, reacts chemically, etc. As a pulse of contaminant passes a point in the fracture system, some contaminant concentration will diffuse into the matrix, then be released again into the fracture fluid (usually water) after the advective pulse has passed. One *could* try to model the system of fractures as a single-porosity system, but the behavior is qualitatively altered. (It is a question of scale.) The mathematical behavior of a dual porosity system vs. a single porosity system is distinct enough to be used to interpret field pumping experiments.

The matrix block equation of the dual porosity model can treat the matrix material and a clay lining. For the fracture-matrix system, transport in the fracture is produced by advection and the related hydrodynamic dispersion/diffusion, while transport in the matrix block is assumed to be dominated by molecular diffusion. Thus, the governing partial differential equations contain advection-dispersion (in the fracture system only), adsorption, source, and decay terms.

These equations are solved using an efficient Approximate Factorization procedure with implicit boundary conditions. The code uses an implicit Total Variation Diminishing (TVD) scheme with three-level time differencing and directional splitting for improved accuracy and execution time. The algorithms are 2nd-order accurate in space and time. In our applications, the flow field for transport is usually obtained from the SECO_FLOW code (Roache, 1993) or equivalent.

The SECO_TRANSPORT code on a 80 × 80 mesh is about 35 times faster than the classical approach (using direct banded solvers) and requires about 6 times less memory. The solution produced is more accurate and less diffusive for low-high Peclet number cases compared to the usual 2-point upstream differencing calculation.

6.12.2 Problem Definition for Source Term Location

To run the SECO_TRANSPORT code, a flow field is needed and is usually generated by the SECO_FLOW code. In Salari et al (1995), the intention was to conduct the study independent of the flow code so that the convergence of the flow field is decoupled from the transport. A non-trivial 2-D flow field representative of an active study was constructed using analytical relations as follows. Initially, a coarse grid representation of the desired velocity field was established, followed by crudely fitting the components of this field, u and v, with analytical functions. The velocity fields generated by these functions were examined and the following functions were chosen by R. Blaine to represent the flow field:

$$u = a_1 + a_2 x^2 \ln(x) + a_3 y^2 \ln(y) \tag{6.12.2.1}$$

$$v = \exp\left(b_1 + b_2 x^2 \ln(x) + b_3 y^3\right) \tag{6.12.2.2}$$

where $0 \le x \le 1$, $0 \le y \le 1$, and the coefficients are given in Table 6.12.2.1.

For the test problem the coordinates were scaled by 5000 m and the velocity components were scaled by 2.5E–10 m/s to obtain representative values for physical domain and velocity magnitude. The maximum divergence of the flow field scaled by maximum velocity is about 1.0E–5 1/m which is adequate for the test problems. (The interest was in adequately representing the structure of the representative velocity field, rather than in using a truly divergence-free velocity field.) Since the flow field is defined analytically, the velocity field can easily be constructed on an arbitrary grid. This is an advantage for the grid convergence study.

For the grid convergence test, we used 3 uniform grids ($\Delta x = \Delta y$) of 40 × 40, 80 × 80, and 160 × 160 cells. The Δx's for each grid are 125, 62.5, and 31.25 meters. The *point* source Q is

i	1	2	3
a_i	−0.046969054	−0.114209345	−0.613161177
b_i	−1.374151646	−6.544653318	−1.327978144

Table 6.12.2.1. Coefficients for Equations 6.12.2.1–2.

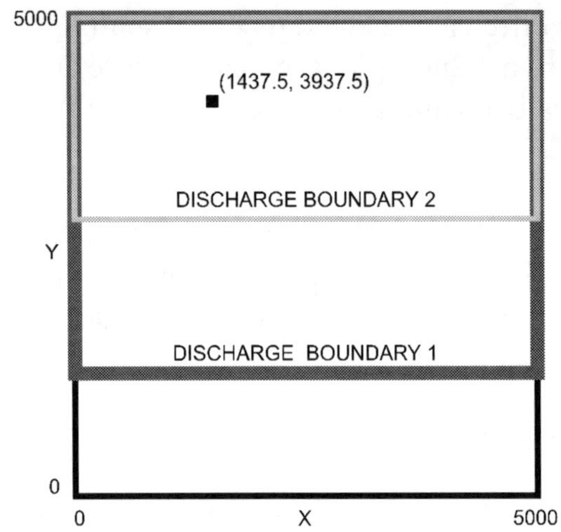

Figure 6.12.2.1. Groundwater Transport Domain Including Location of the Point Source. (From Figure 1 of Salari et al, 1995.)

assumed to be located at 1437.5, 3837.5 meters and there are two rectangular discharge boundaries as shown in Fig. 6.12.2.1. The integrated discharges from these boundaries are used in the convergence study.

For both model problems, single and dual-porosity, the parameters are as follows.

- matrix porosity = 0.16
- matrix tortuosity = 0.0001
- fracture tortuosity = 1.0
- fracture spacing = 3.5 m
- fracture aperture = 0.0035 m
- longitudinal dispersivity = 100.0 m
- transverse dispersivity = 10.0 m
- transported actinide ^{233}U
- decay constant = 1.39E–13 1/sec
- free water molecular diffusion coefficient = 1.7E–10 m^2/sec.

See Salari et al (1992) for details of the continuum equations. The time for the transport is 4500 years starting at 1000 years ending at 5500 years. Three different Δt's were used: 22.5, 11.25, and 5.625 years, which correspond to 1000, 2000, and 4000 time-steps respectively.

The time-dependence of the source function for the transport calculation was constructed using a half sine function over the pulse duration of ~3200 years with maximum amplitude of 1.0E–15 kg/s. The source release began at time 1005 years. The total mass injected (^{233}U) over ~3200 years was 3.1832E–5 kg.

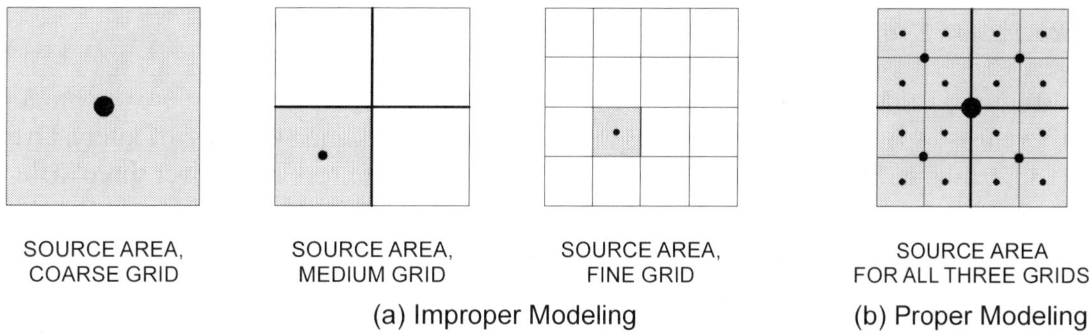

Figure 6.12.2.2. Effective Change in Source Location or Size as Grid is Refined. (a) Improper Modeling of the Point Source. (b) Proper Modeling of the Point Source (From Figure 2 of Salari et al, 1995.)

One of the important aspects of setting up this convergence study was to relate the physical location of the source Q to the different grids. For a finite difference calculation, the computed properties exist at node points. This allows for a physical location to be preserved through arbitrary *integer* grid refinements. However, this is not the case for a finite volume grid where the computed properties are located at cell centers, and grid doubling will not preserve a physical location. (In order to retain the location of a cell center during grid refinements, the grid size would have to increase by a factor of 3.) Therefore, for grid doubling, a finite volume grid poses the following difficulties.

Assume the source location corresponds to a cell center on the coarse grid, so that the grid doubling splits this cell into four cells. Now, one has to decide which of the four new cells can be associated with the source. It should be apparent that this is an arbitrary choice and, in any case, that the continuum location is now positioned at the *edge* of a cell, rather than at the center. Since the computational representation covers the cell, the effective position is always located at the center of the cell, consistent with a second-order representation. Thus, the grid doubling with a finite volume representation effectively moves the point source by $\Delta x/2$ and $\Delta y/2$, introducing a first-order error in the location of the point source. Further grid doubling would introduce a similar choice. This process is shown in Figure 6.12.2.2a and is labeled as *improper* modeling of the source.

Intuition might suggest that such a small change in effective location would only affect results in the immediate neighborhood of the source, but would not affect rate of convergence of global results such as integrated discharge. Our results will show that this is not the case.

In order to remedy this shortcoming of the finite volume grid one has to physically preserve the initial cell associated with the source. This means that as the grid is refined, for example from one coarse cell to 4 medium cells and to 16 fine cells, the total injected mass should be equally distributed among the smaller cells. As a result, the area and location of the initial coarse cell are preserved as well as the total mass injected. This is shown schematically in Figure 6.12.2.2b and is labeled the *proper* modeling of the source. As we show below, only the latter approach will produce the correct 2nd-order convergence behavior.

6.12.3 Results on Source Term Location

The study was conducted for transport calculations using single and dual-porosity models on three different *x-y* spatial grids. Only the *x-y* spatial grid convergence tests are of interest herein, but we also examined time-step convergence, verifying adequate resolution over three different time steps. For brevity and focus, these tests are not reported herein; the results presented were obtained with the smallest Δt. Likewise, for the dual porosity calculations; the matrix block was discretized with 15 stretched cells into the dual porosity direction. The minimum cell size (which occurs at the fracture-matrix interface) was 0.001 (nondimensional). Since the diffusion rate into the matrix is slow, there was no need to further refine the mesh at the fracture-matrix interface for finer fracture grids. In any case, the primary results of the spatial *x-y* grid convergence results presented herein would hold for any reasonable time and dual-porosity resolution.

Figures 6.12.3.1–2 show the results for the single and dual porosity calculations for all three grids. The ^{233}U concentration contours for single-porosity are shown at time = 4150 years and dual-porosity results and the breakthrough curves for the single and dual porosity calculations are shown at time = 5000 years. In both single and dual porosity calculations there are noticeable differences in the concentration contours going from the coarse to the medium grids; however, the concentration contours remain relatively the same between the medium and the fine grid solutions. The breakthrough curves for both transport models do not show any visible variation through the grid refinements.

Tables 6.12.3.1–2 present the convergence results for dual-porosity calculations for discharge surfaces 1 and 2. In this case, the source is modeled properly by preserving the initial cell size and location. The theory indicates that the ratio of successive fine-grid GCI's should equal 4 for a grid doubling with a 2nd-order method in the asymptotic range. As the ratios of GCI's in the Tables show, we have attained 2nd-order convergence for the integrated discharge for both surfaces.

Tables 6.12.3.3–4 present the same dual-porosity calculations as shown previously in Tables 6.12.3.1–2, with the exception that the source is now improperly modeled with a single cell on all grids. From the behavior of the ratio of GCI's, it is clear that the convergence is no longer 2nd-order. In fact, examination of the integrated discharge column shows that the convergence is not even monotone.

Tables 6.12.3.5–6 present the results for the single-porosity calculations for both discharge surfaces and the proper modeling of the source. The solutions are 2nd-order accurate as shown by the ratio of the GCI's.

Tables 6.12.3.7–8 present the same single-porosity calculations as shown previously in Tables 6.12.3.5–6, with the exception that the source is now improperly modeled with a single cell on all grids. Again, the results are similar to the dual-porosity calculations in that they are no longer 2nd-order accurate.

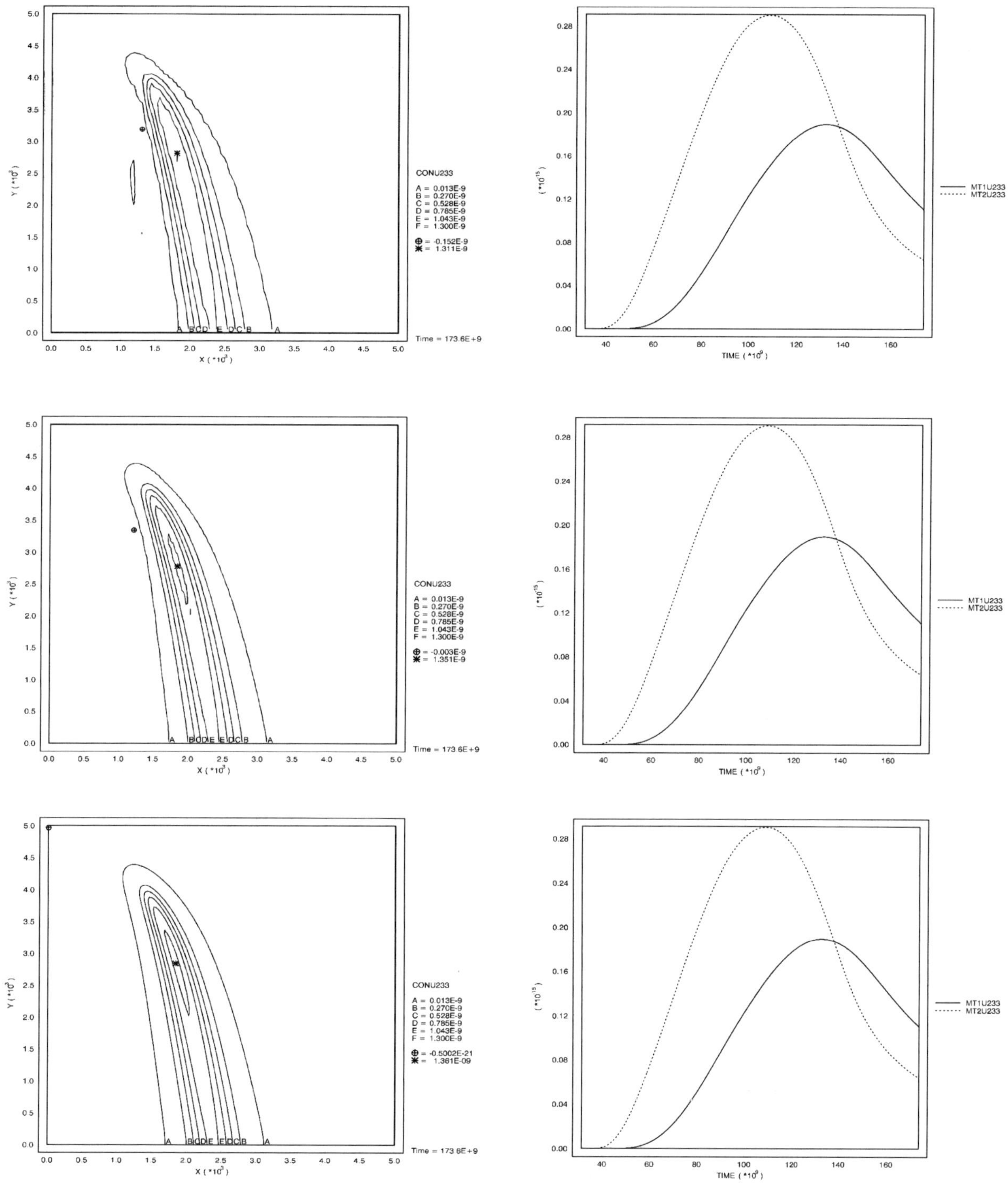

Figure 6.12.3.1. Dual-Porosity Calculations with Proper Source Modeling. Concentration contours and breakthrough curves for all three grids at time = 5000 years. MT1 and MT2 are the discharge boundaries 1 and 2 shown in Figure 6.12.2.1. (From Figure 3 of Salari et al, 1995.)

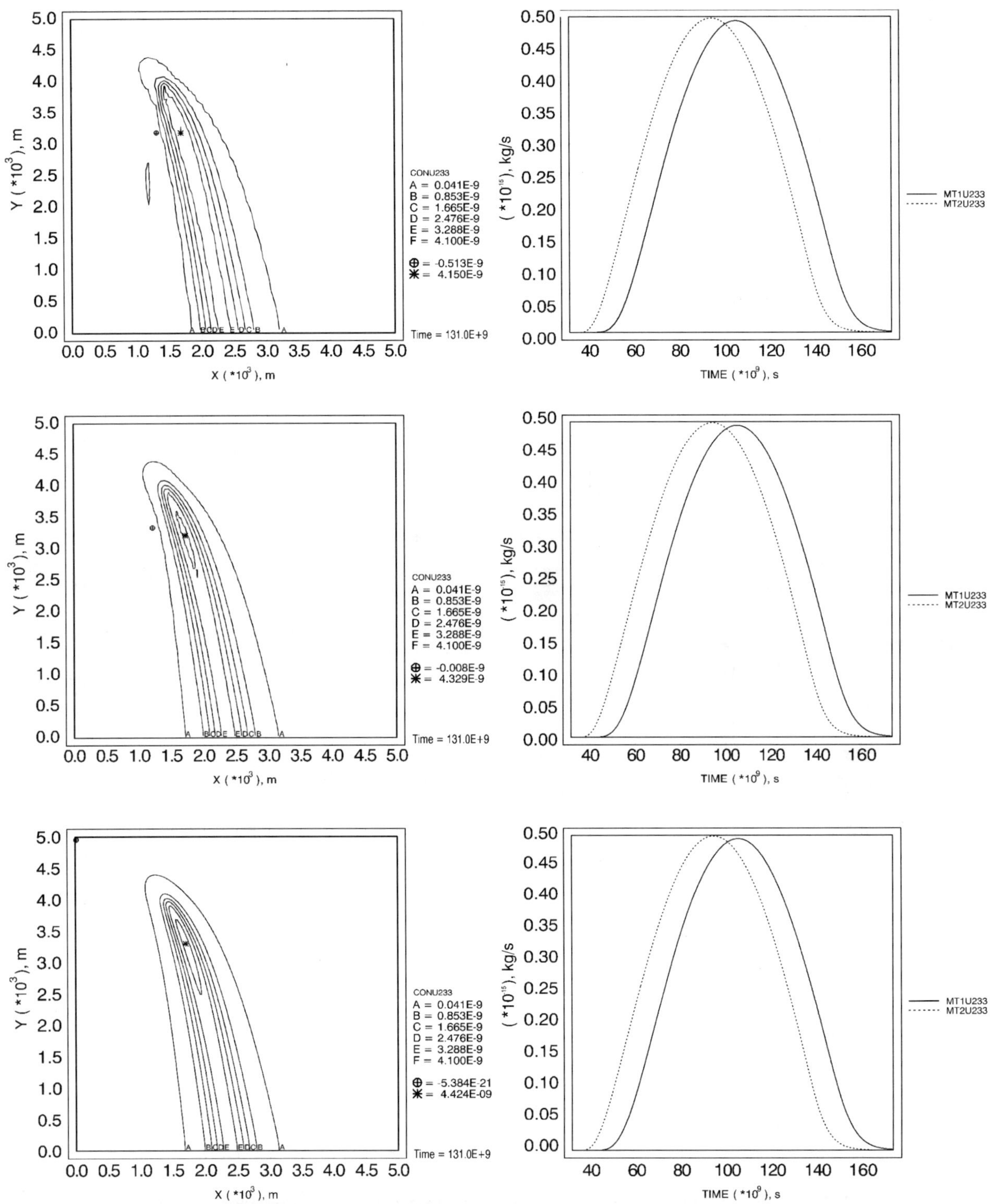

Figure 6.12.3.2. Single-Porosity Calculations with Proper Source Modeling. Concentration contours at time = 4150 years and breakthrough curves at time = 5000 years for all three grids. MT1 and MT2 are the discharge boundaries 1 and 2 shown in Figure 6.12.2.1. (From Figure 4 of Salari et al, 1995.)

Grid	Integrated Discharge	GCI_{fine}	GCI ratio
40 × 40	2.2110467E–5	–	–
80 × 80	2.2115032E–5	0.021 %	–
160 × 160	2.2116186E–5	0.0052 %	3.95

Table 6.12.3.1. Grid convergence results, dual-porosity, *proper* source modeling, surface 1. (From Table 2 of Salari et al, 1995.)

Grid	Integrated Discharge	GCI_{fine}	GCI ratio
40 × 40	1.4098546E–5	–	–
80 × 80	1.4102589E–5	0.028 %	–
160 × 160	1.4103605E–5	0.0072 %	3.98

Table 6.12.3.2. Grid convergence results, dual-porosity, *proper* source modeling, surface 2. (From Table 3 of Salari et al, 1995.)

Grid	Integrated Discharge	GCI_{fine}	GCI ratio
40 × 40	2.2110467E–5	–	–
80 × 80	2.2278868E–5	0.75 %	–
160 × 160	2.2200042E–5	0.36 %	2.08

Table 6.12.3.3. Grid convergence results, dual-porosity, *improper* source modeling, surface 1. (From Table 4 of Salari et al, 1995.)

Grid	Integrated Discharge	GCI_{fine}	GCI ratio
40 × 40	1.4098546E–5	–	–
80 × 80	1.4145352E–5	0.33 %	–
160 × 160	1.4127100E–5	0.13 %	2.54

Table 6.12.3.4. Grid convergence results, dual-porosity, *improper* source modeling, surface 2. (From Table 5 of Salari et al, 1995.)

Grid	Integrated Discharge	GCI_{fine}	GCI ratio
40 × 40	3.0177976E–5	–	
80 × 80	3.0173064E–5	0.016 %	–
160 × 160	3.0171838E–5	0.0041 %	4.01

Table 6.12.3.5. Grid convergence results, single-porosity, *proper* source modeling, surface 1. (From Table 6 of Salari et al, 1995.)

Grid	Integrated Discharge	GCI_{fine}	GCI ratio
40 × 40	2.7125801E–5	–	–
80 × 80	2.7120203E–5	0.021 %	–
160 × 160	2.7118813E–5	0.0051 %	4.03

Table 6.12.3.6. Grid convergence results, single-porosity, *proper* source modeling, surface 2. (From Table 7 of Salari et al, 1995.)

Grid	Integrated Discharge	GCI_{fine}	GCI ratio
40 × 40	3.0177976E–5	–	–
80 × 80	3.0219098E–5	0.14 %	–
160 × 160	3.0196543E–5	0.075 %	1.81

Table 6.12.3.7. Grid convergence results, single-porosity, *improper* source modeling, surface 1. (From Table 8 of Salari et al, 1995.)

Grid	Integrated Discharge	GCI_{fine}	GCI ratio
40 × 40	2.7125801E–5	–	–
80 × 80	2.7141090E–5	0.056 %	–
160 × 160	2.7131694E–5	0.035 %	1.63

Table 6.12.3.8. Grid convergence results, single-porosity, *improper* source modeling, surface 2. (From Table 9 of Salari et al, 1995.)

The level of numerical accuracy attained in these calculations is far beyond that ordinarily expected of groundwater transport predictions. The high resolution is necessary to unambiguously Verify the principal results of sensitivity to the representation of the source term. If the discharge boundaries were farther from the source, the asymptotic convergence results would have been corrupted by computer round-off error. In fact, it is somewhat surprising (and encouraging) that the asymptotic convergence rates were in such good agreement with theory even in Table 6.12.3.1, in which the variation in integrated discharge occurred only in the 5-th significant figure. However, the results would be more pronounced if the integrated discharge boundary were closer to the source and/or if the grid were more coarse. It will be worthwhile (and involves little effort) to use the proper modeling of the source Q described herein in practical modeling work.

6.12.4 Summary on Source Term Location Problem: 1st-Order Performance with a 2nd-Order Code

This study (Salari et al, 1995) showed that the grid convergence results for radionuclide transport calculations in porous media are significantly influenced by the improper spatial modeling of the source Q in a finite volume grid during the grid refinement. The most obvious approach degrades the experimental order of accuracy attained grid convergence tests to 1st-order. The proposed solution is to preserve the initial area and location of the source. The implementation of this method is quite simple and effective, as has been shown in our examples, and will obviously be applicable to source term representations for problems other than transport in porous media.

Perhaps the point of this study of most general interest is that the *code* itself had previously been *rigorously Verified* to be 2nd-order in space and time. (See the synopsis of the Verification and Confirmation exercises from Salari and Blaine, 1996 given below in Section 6.24.) Indeed, the observed convergence rate in the present example with proper source term modeling confirms the 2nd-order accuracy. However, even with a correct code, improper *modeling* (not coding) caused degradation to 1st-order accuracy of the *particular calculation* (rather than the code). By no stretch of imagination could this failure be construed as a fault of the code builder or the code itself; the failure arose in the *use* of the code.

6.13 CONVERGENCE BEHAVIORS FOR MIXED-ORDER METHODS

The previous example from Salari et al (1995) illustrates an important aspect of the convergence behaviors for mixed-order methods. Although the code itself is demonstrably uniformly 2nd-order accurate, the inappropriate modeling of the source term Q caused an observed (an indeed, asymptotic) 1st-order convergence rate. The "method," including the inappropriate modeling for Q, is now a mixed order method, like one obtained by using 1st-order upstream discretization on advection terms with 2nd-order on diffusion terms, or like one obtained using uniformly 2nd-order on interior points and 1st-order discretization of boundary conditions. Since the magnitude of the error is ultimately dominated by the lowest order terms, one may ask what, if any, is the advantage of using higher order accuracy whenever possible, if any 1st-order discretization requires $p = 1$ to be used in the extrapolation and error estimation/banding?

The answer is shown in Figure 6.13.1 for hypothetical examples, qualitatively comparing the convergence behavior of three hypothetical methods. Method A is a uniformly 2nd-order method, as shown by the slope of ~2 in the log/log plot of error vs Δx. Method B is a hybrid method, in which the advection term is a weighted combination of 1st-order and 2nd-order discretization, the (adaptive) weighting depending on the cell Reynolds number. The weighting of the 1st-order discretization goes to zero as the grid is refined, so that ultimately the convergence rate would be 2nd-order; however, over the observed (and usually, the practical) range of Δx, some intermediate rate of convergence $1 \leq p_{hybrid} \leq 2$ is observed. Method C is a 4th-order method except for some locally 1st-order treatment, e.g. the inappropriate Q described above, or an outflow boundary condition, or a small (non-adaptive) weighting of a 1st-order discretization of advection used to "filter" a marginally stable method (e.g., see Roache and Dietrich, 1988).

Over the range shown, hypothetical Method C is more accurate than A, in the sense that its error is smaller. Yet its convergence rate is slower, as evidenced by its slope of ~1. Note that even though C is more accurate than A, the extrapolated solution (and therefore its error

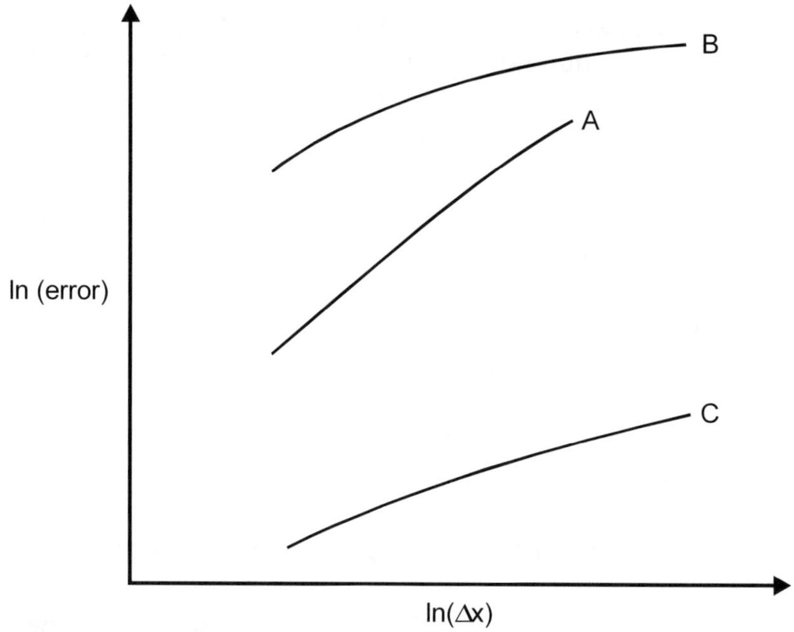

Figure 6.13.1. Qualitative convergence rate and error magnitude for three hypothetical methods. A is a uniformly 2nd-order method. B is a hybrid 1st- to 2nd-order method. C is a 4th-order method for the dominant terms, with a minor term or boundary condition discretized with only 1st-order accuracy.

estimation/banding) is correctly obtained using $p = 1$ for B and $p = 2$ for A. Likewise, the extrapolated solution for the hybrid Method B is best obtained empirically using the observed rate p_{hybrid}.

6.14 GRID CONVERGENCE OF ZERO DRAG COEFFICIENT

When inviscid equations are used to model airfoils in shock-free flow, the drag should be zero. This known exact value can be used to monitor grid convergence, as in Jameson and Martinelli (1996). This variable is not only convenient, but it is deserving of attention because of its engineering importance, as noted by the authors. "In aeronautical applications the accurate prediction of drag is particularly important, and an error as large as 0.0005 is significant since the total drag coefficient of the wing of a transport aircraft (including friction, vortex and shock drag) is in the range of 0.0150." Their calculations using the H-CUSP scheme are zero to 4 significant figures on a 160×32 mesh, as shown in the following Table 6.14.1, which also demonstrates the roughly 2nd-order convergence rate.

For non-zero drag coefficient problems, grid convergence estimation requires at least three grid solutions to estimate p, as noted previously. Jameson and Martinelli also give examples of time-dependent calculations for incompressible flow over an oscillating cylinder, obtaining a value of p for the maximum (over the oscillation cycle) drag coefficient MD (not the original notation) in three grids. In a transparent notation,

Mesh	RAE 2822 Airfoil Mach 0.50, $\alpha = 3°$	NACA 0012 Mach 0.50, $\alpha = 3°$	Korn Airfoil Mach 0.75, $\alpha = 3°$
40×8	0.0062	0.0047	0.0098
80×16	0.0013	0.0008	0.0017
160×32	0.0000	0.0000	0.0000

Table 6.14.1. Drag Coefficient on a Sequence of Meshes using the H-CUSP Scheme.
(From Table 2 of Jameson and Martinelli, 1996.)

$$p_D = \log_2\left[\frac{MD_{65\times33} - MD_{129\times65}}{MD_{129\times165} - MD_{257\times129}}\right] = 1.84 \tag{6.14.1}$$

"which is close to 2." Better Verification is obtained for incompressible vortex shedding from a stationary half-cylinder at Reynolds Number = 250, for maximum drag coefficient *MD*, maximum lift coefficient *ML*, and Strouhal number (a dimensionless shedding frequency) *S*.

$$p_S = \log_2\left[\frac{S_{65\times65} - S_{129\times129}}{S_{129\times129} - S_{257\times257}}\right] = 2.27 \tag{6.14.2}$$

$$p_L = \log_2\left[\frac{ML_{65\times65} - ML_{129\times129}}{ML_{129\times129} - ML_{257\times257}}\right] = 1.83 \tag{6.14.3}$$

$$p_D = \log_2\left[\frac{MD_{65\times65} - MD_{129\times129}}{MD_{129\times129} - MD_{257\times257}}\right] = 2.01 \tag{6.14.4}$$

"which are indeed close to 2." Note that Strouhal frequency is not a difficult quantity to predict physically nor to converge mathematically, because it is essentially an inviscid phenomenon and therefore fairly insensitive to viscous modeling, boundary layer resolutions, etc. (These calculations were all carried out with the outflow boundary located 16 diameters downstream, presumably based on some unreported sensitivity calculations.) The paper (Jameson and Martinelli, 1996) also contains an excellent example (in its Figures 30 and 31) of the radical improvement in convergence for the drag polar produced by graduating from a 2nd-order time-accurate method to a 3rd-order method.

6.15 ANOMALOUS RESULT POSSIBLY DUE TO GRID STRETCHING

Steffen et al (1995) used the Grid Convergence Index (GCI) to report a grid convergence study on a calculation of incompressible flow in a rectangular nozzle using a well-Verified code with a *k*-ε turbulence model. They used GRIDGEN, a commercial code grid generation package (Steinbrenner and Chawner, 1992) to generate a 10 block, structured, generalized coordinate mesh with cells clustered towards the nozzle walls and exit plane. The base grid contained

450,000 cells, and grid refinement and coarsening of 10% was used to generate 3 grids, with the motivation of keeping the solution in the asymptotic range (see discussion in Chapter 5). The GCI values were calculated with $r = 1.1$ and $p = 2$ for two functional quantities: (a) the axial location at which the jet centerline velocity has decayed to 1/2 the value at the exit plane, and (b) the axial location at which the jet half-velocity-width along the nozzle minor axis has spread from 0.5 to 1.5.

The results were fairly convincing for functional (a). In their notation, the coarse-to-medium grids GCI(c-m:a) = 10.3%, and the medium-to-fine GCI(m-f:a) = 7.64%. These notations correspond to GCI_{23} and GCI_{12} in Equation (5.10.5.2) of Chapter 5, repeated here as

$$\mathrm{GCI}_{23} = r^p\ \mathrm{GCI}_{12} \tag{6.15.1}$$

Recall that approximate satisfaction of this equation indicates that the asymptotic range has been achieved. With $r^p = 1.1^2 = 1.21$, the above values give

$$10.3\% \approx 1.21 \times 7.64\% = 9.24\% \tag{6.15.2}$$

Considering the inherent difficulties of grid smoothness across boundaries of 10 grid blocks, this is reasonable satisfaction (~10%), and indicates that all 3 grid calculations are in the asymptotic range for functional (a).

However, the results for functional (b) were anomalous, with GCI(c-m:b) = 3.44% and GCI(m-f:b) = 4.81%, an incorrect trend. The authors note that these results were "located in a region of the computational mesh where stretching could have affected the accuracy. This deserves further study before a solid conclusion can be reached." It would appear likely that the results are adequately converged for both functionals (a) and (b), as the authors believe, but that the assumption of observed $p = 2$ is not Verified due to grid stretching problems of the type discussed in Chapter 5, including subtleties associated with multi-block grid generation methods. No matter what the final conclusion proves to be, this sort of candid and painstaking reporting of grid convergence tests builds confidence effectively.

6.16 NON-SMOOTH PROPERTY VARIATION: GLOBAL ERROR NORMS

The general issues of non-smooth property variation were discussed in Chapter 5. Here, we present results of a particular study, and special considerations for global error norms.

Ruge (1995) has performed a thorough study of grid convergence and an evaluation of the Grid Convergence Index (GCI) for 1-D two-phase flow in porous media with non-smooth property variation. The 1-D convergence study was performed to support 2-D studies of a geologic waste repository (WIPP PA Dept., 1992; Helton et al, 1995, 1996). The 2-phase Darcy flow code uses mixed 1st- and 2nd-order spatial differencing and adaptive 1st-order time differencing, common for this class of problems. Beginning with a base case of 61 spatial cells, four levels of spatial refinement (by factors of 2, 4, 8 and 16 from the base case) were used, and the effect of timestep refinement (with the same nominal refinement factors) was also examined. In preliminary 2-D studies, Ruge found that the spatial error dominated, so that error estimates could be examined in terms of refinement of spatial step Δ alone; however, for the more extreme refinement possible in the 1-D problem, this was not true. The nominal time-step refinement factors sometimes were overcome by the solution-adaptive time-stepping algorithm, producing higher than nominal refinement. This caused some inconsistency with the temporal order inferred, compared to runs wherein the nominal (scheduled) time step refinements were followed.

This study focused on global rather than local convergence, and Ruge found that the estimates of convergence could depend strongly on the quantities compared and the norms used. A major factor was the cell alignment of faces with the boundaries of the computational domain. As already noted above in Section 6.12, in this grid configuration, doubling the number of cells requires the *location* of the center of a single cell to shift by Δ/2 (See Figure 6.12.2.2.). Combined with non-smooth property variation, this can confuse grid convergence calculations. It was necessary to avoid any kind of interpolation across the coarse grid cell faces, since discontinuities in the problem coefficients can interfere with accuracy and affect error estimates. For this reason, spatial comparisons were made only *within* cells, and comparisons between two levels of discretization were made only on the coarser level, i.e. the finer mesh solution was transferred to the coarser mesh, where the non-smooth properties were defined to be piecewise constant. This allowed 2nd-order interpolation between fine level cell centers (within a coarse grid cell, so that material properties are constant) to obtain fine-grid values at the coarse level cell centers. Comparisons were further complicated by the fact that the 1-D problem of interest represented a highly stretched quasi-radial mesh, so that uniform subdivision in the coordinate did not produce uniform subdivision of volumes.

Consequently, initial attempts at grid convergence studies clearly showed that it was necessary to carefully define the norms used for measuring solution differences over space and time. With S^k the space of grid functions f^k defined on the level k refinement grid, so $f^k \in S^k$, and v the volume of the ith cell, Ruge used the following volume-weighted discrete approximation to the spatial L_2 norm.

$$\left\| f^k \right\|_{S^k} = \left(\frac{\sum_i v_i \left(f_i^k \right)^2}{\sum_i v_i} \right)^{1/2} \tag{6.16.1}$$

The time norm was defined similarly, with the simplification that the solutions being compared were all stored at the same sequence of timesteps, regardless of what sequence they were calculated on (further time step refinements being skipped over) so that the level need not be specified for t. With f being assumed piecewise linear on each time interval, the normalized L_2 norm in time was defined as follows.

$$\|f\|_t = \left(\frac{1}{t_n - t_0} \sum_{i=1}^{n} \frac{f_i^2 + f_i f_{i-1} + f_{i-1}^2}{3} \right)^{1/2} \quad (6.16.2)$$

The time norm was applied to the space norm to obtain the full norm, as follows.

$$\|f\|_k = \left\| \left(\|f\|_{S^k} \right) \right\|_t \quad (6.16.3)$$

The norms as defined do not commute, but Ruge found that the order in which the norms were applied to obtain the full norm did not have much effect. The same norm order was used consistently in all the comparisons. Generally, consistency is important to the convergence study. For example, "the norm of the difference between two functions can depend heavily on the level on which it is measured, so that when norms of several differences are involved in a GCI calculation, for example, all functions involved should be weighted to some common level." (Ruge, 1995)

With the dependent variables of interest being pressure and saturation, Ruge found the temporal order $q = 1$ in most cases. The spatial orders of convergence for the global measures of pressure were found to be approximately $p = 2$ at all refinement levels; however, the $p = 2$ result is largely an artifact of the highly stretched quasi-radial mesh, which makes most of the spatial region covered to be in virtually uniform flow. Near the modeled waste repository, where spatial variations were significant, the convergence order was only $p = 1$. For the saturation, $p \cong 2$ on the coarser levels, but decreases to $p \cong 1$ on the finer levels. [This would indicate that the coarser levels of mesh refinement are *not* in the asymptotic range of convergence, contrary to Ruge's statement.] The methods used are mixed order, so that use of the more conservative *a priori* values $p = 1$ and $q = 1$ in the GCI calculation was expected to be more conservative. Indeed, Ruge found that the GCI generally was about 4 times the actual error, compared to the factor $F_s = 3$ in the GCI definition (see Chapter 3). [Note, however, that the actual 2-D calculation of engineering interest involved much more coarse resolutions than Ruge's 1-D study, so the additional conservatism of $F_s = 3$ is probably appropriate.] Thus, Ruge (1995) concludes that "the GCI is performing well in our case."

6.17 DISCRETE VORTEX METHODS

Discrete vortex methods can present particular difficulties for grid convergence testing because of sensitivities. In discrete vortex methods, there is no grid. Rather, the discretization involves the selection of a finite number of vortices. The velocity field at every point needed is calculated from the Biot-Savart law, summing over all the vortices. The velocity field is calculated at each location of the vortex centers, and the velocity field is then used to move the vortices in a Lagrangian calculation. Implicit algorithms are difficult to implement efficiently, and virtually all the published methods use fully explicit differencing. Stability requires the addition of a viscous core to the inviscid vortex, which has a singularity at the center. Although physical vortices have a viscous core of size ordered by the Stokes radius, the modeled vortices can have much larger cores without destroying the accuracy. In fact, it is necessary for stability that each vortex stay within the viscous core of a neighbor; if vortices move apart so this condition would be violated, the vorticity is locally redistributed between old vortices and new vortices "injected" into the simulation. Hugo (1995) has shown that this condition corresponds to a physically intuitive requirement that the flow field at a neighboring vortex in a one-dimensional simulation is not changed, i.e. if a vortex is injected outside the core, the induced velocity at neighbors changes, evidently leading to instability.

The operation count for a fully viscous flow field (such as the driven cavity problem at low *Re*) is prohibitive. The efficiency of the method comes from the fact that vortex resolution is not required in the inviscid regions. For example, in the calculation of a shear layer, vortices are only required in the shear layer, not in the free stream region. However, the rapid increase in computer time with the number of vortices N makes convergence testing difficult. The time step convergence is straightforward. Because of the requirement for the viscous core, and the (usually) non-physical modeling of the core, it is not quite clear to what system of equations the method would converge in the limit of $N \to \infty$. Convergence issue are further complicated when the modeling is statistical, e.g. aero-optical phase distortion due to turbulence fluctuations; see Jumper et al (1994), Jumper and Hugo (1995), Hugo (1995).

6.18 OBSERVED CONVERGENCE RATES FOR EULER EQUATIONS WITH SHOCKS

This section summarizes results of Engquist and Sjogreen (1996) as presented in Yee and Sweby (1996) on the observed convergence rate for discontinuous (shock) solutions of the Euler equations (for inviscid, compressible flow) using high-resolution shock-capturing schemes such as TVD (Total Variation Diminishing) and ENO (Essentially Non-Oscillatory) schemes. The critical lesson and mathematical distinction is that the Euler equations are a *system* of hyperbolic conservation laws, and the convergence rate behavior unfortunately cannot be inferred from the study of a model *scalar* equation (i.e., for a single variable).

For a shock-containing solution of a scalar nonlinear conservation law, the characteristics point into the shock. TVD, ENO and similar schemes going back to the original FCT, or Flux-Corrected Transport algorithm of Boris (see Book et al, 1975) adapt to a low order scheme just at a shock, since the concept of high-order is meaningless across a discontinuity. According to the linear theory of Kreiss and Lundqvist (1968), dissipative schemes like these damp out the errors propagating backwards against the direction of the characteristics. Thus, it is reasonable to expect that the locally large errors at the shock will stay in the immediate vicinity of the shock. In numerical experiments with formal pth order methods , one usually obtains $O(\Delta^p)$ convergence away from the shock.

For systems of equations, complications arise because other families of characteristics exist, and it becomes possible that the large error at the shock can propagate out into the entire post-shock (downstream) region by following a characteristic that emerges from the shock. An example is given for quasi-1D nozzle flow (using a model system of equations that exhibit shocks; see Yee and Sweby, 1996 for details) for an area distribution given by

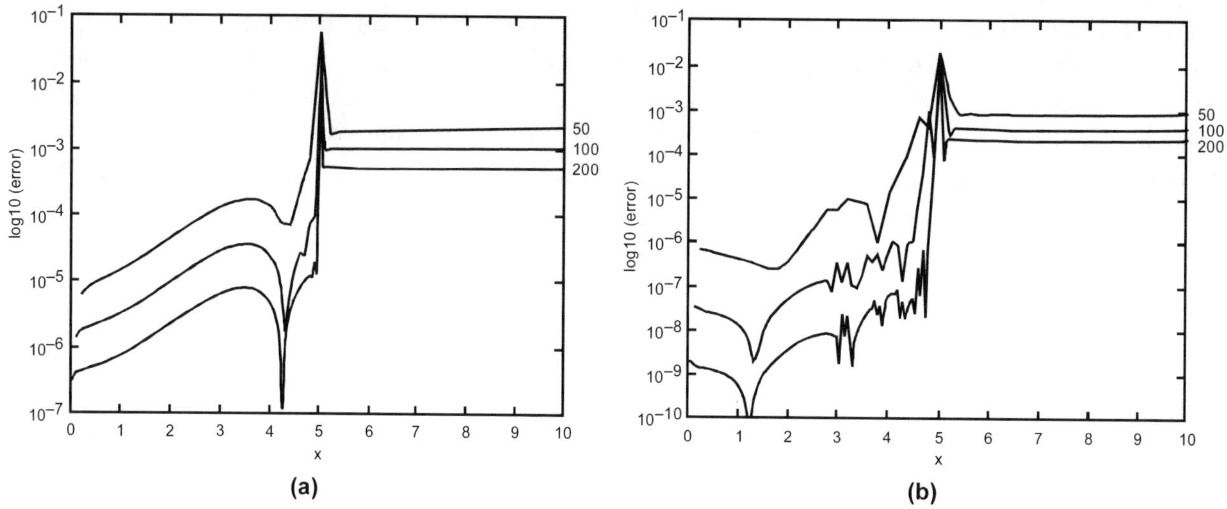

Figure 6.18.1. Effect of Grid Resolution on Error in Momentum for a Quasi-1-D Model System of Equations Exhibiting a Shock. (a) 4th-order ENO method. (b) 2nd-order TVD method. (From Yee and Sweby, 1996, Figure 4.3.)

$$A(x) = 1.398 + 0.347 \tanh(0.8x - 4) \qquad (6.18.1)$$

The exact steady-state solution for the model equations used has a shock at $x = 5$. Figure 6.18.1 shows the error in momentum obtained using a formally 4th-order ENO methods (part a) and a formally 2nd-order TVD method (part b), for three grids containing 50, 100 and 200 points.

The apparently constant jumps in error between the three grid solutions is due to the log scale used; when the observed convergence rate p is calculated, the result for the ENO method (between the 100 and 200 point grids) is 3.9 before the shock, in excellent agreement with the formal rate of $p = 4$, However, the observed rate is only $p = 1$ downstream of the shock, indicating that the 1st-order error at the shock is being propagated downstream. Likewise for the TVD method, the observed rate upstream of the shock is $p = 2.2$, somewhat better than the formal rate of $p = 2$, but downstream of the shock it deteriorates to $p = 1.1$. Similar results are obtained for 2-D Euler solutions.

In these example calculations, the exact solution of the model problem is known, so 2 grids suffice to determine the observed convergence rate p. For non-model problems wherein the exact solution is not known, it would be necessary to perform at least 3 grid solutions to calculate p, and 4 to verify that the observed p is constant, i.e. that the calculation is indeed within the asymptotic range. Unless such thorough testing is performed, the above calculations on model problems would dictate that error estimation and banding should be done assuming the post-shock $p = 1$.

6.19 COMPLETED RICHARDSON EXTRAPOLATION

The calculations in Roache and Knupp (1993) Verifying the method of Completed Richardson Extrapolation provide a good example of the sensitivity of the grid convergence testing. See details in Chapter 3, Section 3.12.2.

6.20 TRUNCATION ERROR IN ELLIPTIC GRID GENERATION

Methods of generating curvilinear, boundary-fitted grids by solving elliptic equations are highly developed; e.g., see Thompson et al (1985) or Knupp and Steinberg (1993). The seminal paper in this area was that of Thompson, Thames and Mastin (1974) which used the following grid generation equations.

$$\alpha x_{\xi\xi} - 2\beta x_{\xi\eta} + \gamma x_{\eta\eta} + J^2\left(P x_\xi + Q x_\eta\right) = 0 \tag{6.20.1a}$$

$$\alpha y_{\xi\xi} - 2\beta y_{\xi\eta} + \gamma y_{\eta\eta} + J^2\left(P y_\xi + Q y_\eta\right) = 0 \tag{6.20.1b}$$

where

$$\alpha = x_\eta^{\ 2} + y_\eta^{\ 2} \tag{6.20.1c}$$

$$\beta = x_\xi x_\eta + y_\xi y_\eta \tag{6.20.1d}$$

$$\gamma = x_\xi^{\ 2} + y_\xi^{\ 2} \tag{6.20.1e}$$

$$J = x_\xi y_\eta - x_\eta y_\xi \tag{6.20.1f}$$

(J is the Jacobian of the transformation.)

These are PDE's for the physical (x, y) coordinates in the transformed coordinates (ξ, η) and are solved in the (ξ, η) "logical" plane, with easy boundary conditions. (Note that no interpolation of boundary position is required.) An earlier method (for triangle grids) due to Winslow (1967) is recovered when the source terms P and Q are taken to be zero. It was claimed that this *homogeneous* elliptic method of grid generation has the theorctical property that the grid lines should not cross, i.e. the grid transformation generated should automatically be "one-to-one and onto." However, the present author showed by simple hand calculations (exact solutions) on a trivial 3×3 example problem that the method could easily be made to fail (Roache and Steinberg, 1985). For a horseshoe shaped region (geometrically similar inner and outer ellipses, with size ratio of 2), as shown in Figure 6.20.1, the 3×3 grid produces only one interior point. The position of this point falls on the boundary for a critical aspect ratio $A_C = 3\sqrt{2} \cong 2.1$. For aspect ratios smaller than this, the degenerate 1-point grid generation is acceptable. For aspect ratios larger than this, the point falls outside the region, i.e. the grid lines are crossed.

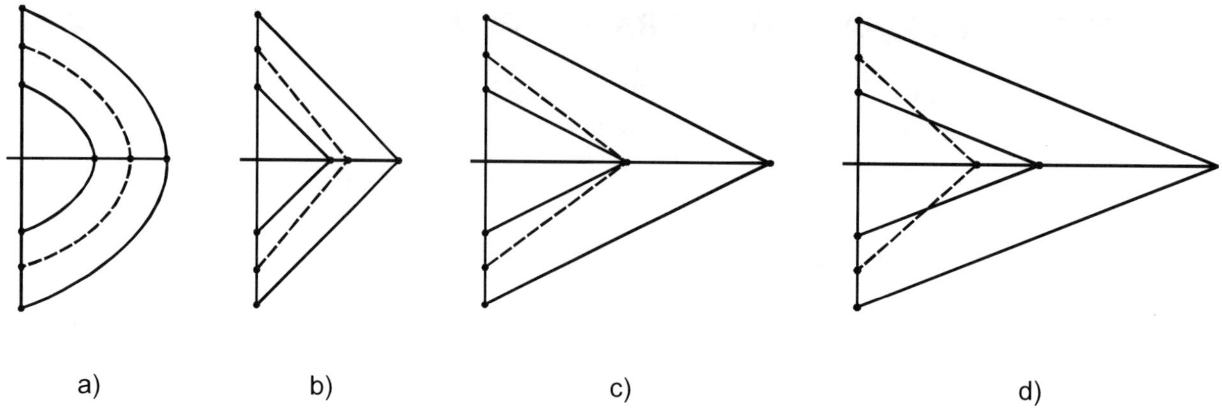

Figure 6.20.1. Horseshoe shaped region for elliptic grid generation, showing the grid folding failure of the homogeneous grid generation equations for the degenerate 3 × 3 grid. *A* is the aspect ratio of the ellipses, and the ratio of outer to inner ellipses = 2. (From Roache and Steinberg, 1985, Figure 2.)

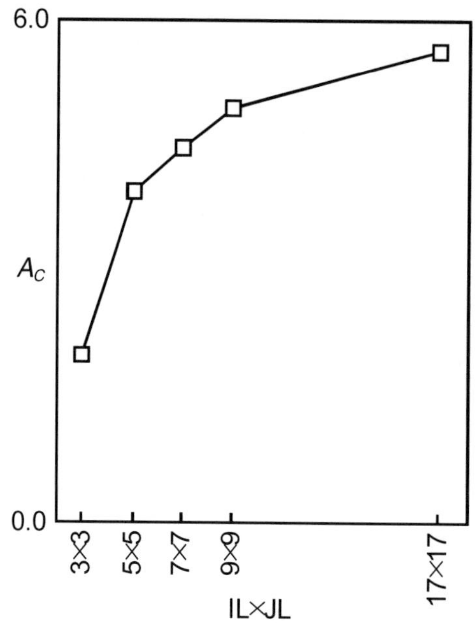

Figure 6.20.2. Variation of the critical aspect ratio A_C with mesh size. The theory would indicate *lim* ($\Delta \rightarrow 0$) $A_C = \infty$. (From Roache and Steinberg, 1985, Figure 3.)

Furthermore, crude grid resolution studies indicated that the critical aspect ratio did not diminish very rapidly with increased resolution, as shown in Figure 6.20.2.

It was far from clear whether or not this failure of the homogeneous elliptic grid generation equations indicated an error in the theorems of Thompson et al (1974). The question was later essentially resolved (with some reservations) by Knupp and Luczak (1995) who showed (a) that the failure was sensitive to the boundary parameterization, and (b) that the failure was due to truncation error. That is, in the limit of $\Delta\xi$, $\Delta\eta \rightarrow 0$, a correct grid is generated. (However, the phenomenon remains a problem at finite grid resolution.)

6.21 ONE DIMENSIONAL MOVING ADAPTIVE GRID PROBLEMS

Salari and Steinberg (1994) have developed a flux-corrected transport (FCT) algorithm suitable for moving, solution-adaptive grids for difficult 1-D shock propagation problems. They discuss limitations of the algorithm, as well as its significant successes. Clearly, such algorithms are much more efficient than fixed grid methods, yet there are difficulties associated with Verification of Calculations.

6.22 GCI APPLICATION IN SOLUTION-ADAPTIVE GRIDS WITH NON-INTEGER GRID REFINEMENT

Lotz et al (1995, 1997) present an exemplary application of the Grid Convergence Index in solution-adaptive grids with non-integer grid refinement. The subject problem was the aerodynamics calculation for viscous transonic flow over supercritical airfoils with blunt trailing edges (more specifically, divergent trailing edge or DTE airfoils) designed to create a "rooftop" pressure distribution on the upper lifting surface. The Briley-MacDonald (1977, 1980) Approximate Factorization method was used with explicit artificial dissipation, and turbulent boundary layers were modeled with a mixing-length model; the objective was to quantify the numerical uncertainties (i.e., Verification of the Calculation) rather than Validation of the turbulence model, etc. The GCI's were evaluated for several quantities: three aerodynamic coefficients, shock location, physical extent of the recirculation wake behind the blunt trailing edge, and distributions of the surface pressure coefficients.

Initial grid distributions were established with increased mesh density in the anticipated shock locations, then new grids were adapted to the initial solution. Then the new grids were refined for the grid convergence study. Three grid resolutions were used. The advantages of non-integer grid refinement (see discussion in Chapter 5) are evident; over all three grids, the overall grid refinement ratio is only $r_{13} \sim 1.68$ (compounded of $r_{12} \sim 1.27$ and $r_{23} \sim 1.32$). The relatively constant GCI over both refinements indicates that all three grids are reasonably in the asymptotic range (except in the wake). Interestingly, the distribution of *local* GCI for the local pressure coefficients was almost uniform around the airfoil, except in the shock -boundary layer interaction region and at the point where turbulence transition was fixed. [This uniformity suggests that the adapted grid distribution is appropriate, i.e. near-optimum.]

The evaluation by Lotz et al (1997) of the numerical uncertainties indicates good convergence for lift, but only fair convergence for moment and drag, and indicates that drag calculations are more sensitive to numerical error. This is partially in contrast to Salari and Roache (1990), who also found lift to be the easiest to converge, but who found moment errors to be more slowly converging than drag. However, a related result in a similar vein is shown in Table 1 of Lotz et al (1997). Although the value of the GCI is larger for drag than moment, the variation in GCI (from coarse-medium GCI to medium-fine GCI) is larger for moment than drag.

This would suggest that the relative error for moment is less than that for drag, even though the asymptotic behavior is not as good; this is perhaps unexpected, but not really contradictory. In any case, both the calculations of Lotz et al (1997) and Salari and Roache (1990) on viscous airfoil grid convergence studies, as well as countless other studies, make it clear that convergence of any functional cannot be assumed *a priori* to guarantee the same quantitative level of convergence for any other functional (as discussed earlier in Chapters 4 and 5; see also Chapter 7).

These studies (Lotz et al, 1995, 1997) clearly show the power and utility of the GCI in providing conservative error bands without the computational expense of successive grid doublings (see discussion in Chapter 5).

6.23 HIGH QUALITY GRID RESOLUTION STUDIES LEADING TO A SAFETY FACTOR OF 1.25

The previously cited studies by Lotz et al (1995, 1997) and other studies, notably those in the ASME publication edited by Johnson and Hughes (1995), including Chang and Haworth (1995), Salari et al (1995), Pelletier and Ignat (1995), Celik and Karatekin (1995), Leonard and MacVean (1995), and Ruge (1995), all led to the suggestion made in Chapter 5 that the value of the "factor of safety" used in the GCI (Equations 5.6.1, 5.7.2–3) be adopted as $F_s = 1.25$, but only for such high quality studies with a minimum of 3 grids to experimentally ascertain the observed rate of convergence p. As noted in Chapter 5, Section 5.9, for the more common 2-grid study, I still recommend the value $F_s = 3$ for the sake of uniform reporting and adequate conservatism.

6.24 TRANSPORT CODE VERIFICATIONS USING THE GCI: PARTITIONING THE OPTION MATRIX

The SECO_TRANSPORT code developed by K. Salari (Salari et al, 1992, Roache 1993; Salari and Blaine, 1996) has been previously described in Section 6.12.1. The complexity of the options of the code create the difficulty described in Chapter 2, Section 2.18, which for simplicity is repeated here, with emphasis added.

"Another concern in Code Verification is the number of user options in a code, especially general-purpose commercial CFD codes. This is a genuine practical problem, but does not nullify claims of Verification; it just limits those claims. The exponentially expanding complexity of the option tree does not nullify the definition of Verification of Code; it simply qualifies the definition. 'Code Verification' is restricted to that combination of options claimed to be Verified. There is a gray area here, as one might expect, in the judgment of the independence of options. Some knowledge of algorithm and code structure may be necessary to infer the reasonableness *of simplifications of option interactions (essentially, partitioning of the full matrix of option interactions)*. The easy answer is the rigid one of categorical "no." This would usually appeal to the rigid SJ personality type (see Chapter 11) that gives Quality Assurance a bad name. A more intelligent and economical approach is possible, bearing in mind that subtle and unanticipated option interactions have occurred, especially before the acceptance of structured coding and modularity."

We address here the "simplifications of option interactions" which essentially constitutes a partitioning of the full matrix of option interactions. We will rely on a knowledge of algorithm and code structure to guide the partitioning, but then "cover our bets" after the Verification by further Confirmation exercises to further build confidence that the partitioning is correct.

The partitioning is applied to the following aspects of the code.

- Verification of the Geologic Matrix Equations for Diffusion, Single Species
- Verification of the Fracture Equations for Advection, Dispersion, Diffusion at low Peclet Numbers, single Species
- Verification of the Coupling between Matrix and Fracture Equations
- Verification of the Radioactive Decay Equations (which are the same in matrix and fracture systems) for Multiple Species
- Verification of the TVD Algorithms (6 options) for Fracture Equations at high Peclet Numbers, Single Species
- Confirmation Exercises on a Commonly Used Analytic Solution, single Species

Code features to be Verified must include

- Time-Dependent and Spatially-Variable Velocity and Dispersivity Field
- Both Constant and Time-Dependent Boundary Conditions
- Time and Space Order of Convergence
- Discharge calculations on pre-defined closed boundaries, involving quadratures.

For many PDE codes, time and space convergence can be partitioned (e.g., the SECO_FLOW codes as discussed in Section 6.2). With the Approximate Factorization algorithm used, spatial convergence for steady state solutions can be established without considering the time differencing options (in theory), but the time accuracy is coupled through the mathematics of the Approximate Factorization, so this major option partitioning is not available for this code. The Verification was made possible by the development by P. M. Knupp of two benchmark analytical solutions for the fracture equations (see Appendix in Salari and Blaine, 1996). The first is a solution for a spatially variable dispersivity field with a time-dependent and spatially variable velocity field, but with time-constant boundary conditions. The second is a solution for time-dependent boundary conditions, a test necessitated again by the subtlety of the AF algorithm. A separate analytical solution was used for the geologic matrix equations, with spatial property variation and time-dependent boundary conditions. The time-dependent solutions were Verified separately for two 2nd-order time differencing options: 3-point backward and trapezoidal time differencing.

With the fracture and matrix equations separately Verified, the implicit coupling of the two was Verified with the previously developed solution by Tang et al (1981) for constant properties.

The justification of this partitioning (i.e., not testing the coupling with spatially variable properties) depends on knowledge of the algorithm and code structure. Then the multiple species transport is Verified against the analytical solution of Lester et al (1981). No spatial variability enters into these equations. The solution is for 3 species in a chain. The partitioning assumption here involves knowledge that the code is written for N species, and has been exercised for $N = 9$; it hardly seems necessary to test all possible numbers of species.

The only aspect of discharge calculations on pre-defined closed boundaries to be Verified involves quadratures. Knowing that these do not depend on other options of the code allows partitioning of this feature. However, since time-and space-dependent velocity fields are involved, the Verification uses the first benchmark problem. I believe this to be a high standard.

The Verification is performed with both uniform and smoothly stretched grids. As is well known, extreme grid stretching will affect accuracy of an individual calculation, but the code is Verified with moderate stretching, knowing that there is nothing special about the stretching function. (One cannot Verify a code for an infinity of stretching functions, nor an infinity of velocity fields, property variations, etc.)

Likewise, the 6 different TVD flux limited options (for eliminating oscillations in shock-like high Peclet number solutions) are partitioned from the rest of the code, since the coding does not interact with dispersion, matrix equations, radioactive decay, etc.

Needless to say, only a sampling of the Verification exercises can be presented here, in Table 6.24.1 and Figure 6.24.1. As noted in Chapter 3, such exercises require that the scientist-engineer "think like an accountant" for a while.

A later version of the SECO_TRANSPORT code included the 3-D *moving* free surface boundary conditions (developed for the single porosity case only). It was difficult to achieve 2nd-order accuracy in space and time with a moving free surface and a general non-orthogonal moving coordinate system, but it was convincingly Verified using two Manufactured Solutions in Roache et al (1996).

After these and many other Verifications of the partitioned features of the SECO_TRANSPORT code, the authors justifiably claimed Code Verification. This Verification was followed by further Confirmation exercises.

Although not logically necessary (if one really believes the Verification) Confirmation exercises can be useful for further confidence building in the user/customer. However, it is *highly* recommended that the code builder (or whoever may be Verifying the code) stand firm on distinguishing between Verification, which at least in principle is accomplished once and for all (like a theorem), and Confirmation, which can go on forever, depending on the whims of users who may be mathematically semi-literate.

Grid	Δt	L_2 error	Max. error	GCI	GCI ratio
10 × 10	0.04	3.3742E–3	6.8663E–3		
20 × 20	0.02	8.2240E–4	1.7046E–3	205.360%	
40 × 40	0.01	2.0385E–4	4.2610E–4	49.779%	4.13
80 × 80	0.005	5.0512E–5	1.0590E–4	12.340%	4.03
160 × 160	0.0025	1.2426E–5	2.6112E–5	3.065%	4.03

a. Single porosity fracture equations, time-dependent flow, trapezoidal time differencing, Van Leer's MUSCL limiter, uniform grid. The solutions are 2nd-order accurate, as shown by the GCI ratio ≅ 4. (From Salari and Blaine, 1996, Table 11.)

Grid	Δx	Δt	L_2 error	GCI	GCI ratio
20 × 20	0.05	0.25	7.697E–3		
40 × 40	0.025	0.125	1.954E–4	46.540%	
80 × 80	0.0125	0.0625	4.921E–4	11.847%	3.92
160 × 160	0.00625	0.03125	1.234E–4	2.988%	3.96

b. Single porosity fracture equations, time-dependent flow, 3-point backward time differencing, central differences in space (no limiter), uniform grid. The solutions are 2nd-order accurate, as shown by the GCI ratio ≅ 4. (From Salari and Blaine, 1996, Table 14.)

Grid	Δs	L_2 error	Max. error	GCI	GCI ratio
20 × 20	0.04	4.7240E–3	1.1464E–2		
40 × 40	0.02	1.0251E–3	2.3426E–3	244.200%	
80 × 80	0.01	2.4657E–4	5.6313E–4	51.398%	4.75
160 × 160	0.005	6.0886E–5	1.3810E–4	12.259%	4.19
320 × 320	0.0025	1.5147E–5	3.4211E–5	3.020%	4.06

c. Matrix equations only, steady-state calculations, non-uniform grid (s is the spatial increment in logical space). The solutions are 2nd-order accurate, as shown by the GCI ratio ≅ 4. (From Salari and Blaine, 1996, Table 16.)

Grid	Δs	L_2 error	Max. error	GCI	GCI ratio
20 × 20	0.04	4.7240E–3	1.1464E–2		
40 × 40	0.02	1.0251E–4	2.3426E–3	244.200%	
80 × 80	0.01	2.4657E–4	5.6313E–4	51.398%	4.75
160 × 160	0.005	6.0886E–5	1.3810E–4	12.259%	4.19
320 × 320	0.0025	1.5147E–5	3.4211E–5	3.020%	4.06

d. Discharge calculations. Single porosity fracture equations, steady- state flow, central space differencing (no limiter), non- uniform grid. The solutions are 2nd-order accurate, as shown by the GCI ratio ≅ 4. (From Salari and Blaine, 1996, Table 16.)

Tables 6.24.1. Sampled Verification tests of the SECO_TRANSPORT 2D code using Manufactured Solutions. (See Salari and Blaine, 1996, for parameters and details.)

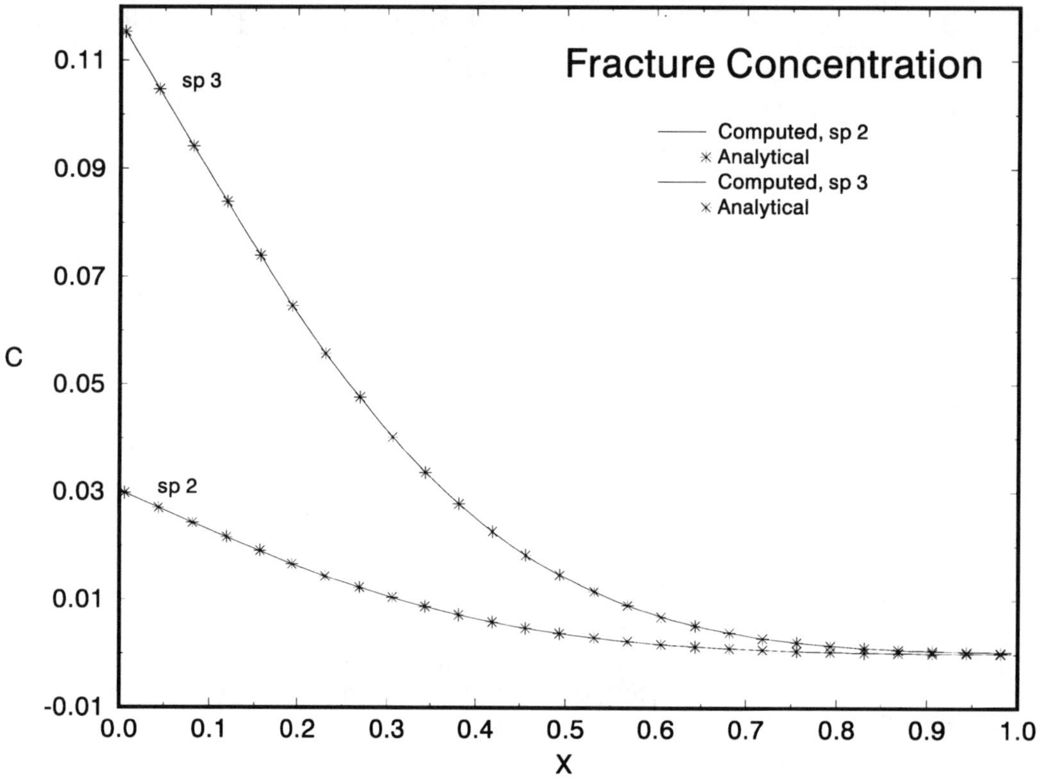

Figure 6.24.1. Sampled Verification test of the SECO_TRANSPORT 2D code using Manufactured Solutions: Multiple Species Verification. Computed concentrations *C* in the fracture system, for species 2 and 3 of a 3-species chain, compared with an analytical solution. (From Salari and Blaine, 1996, Figure 6. See Salari and Blaine, 1996, for parameters and details.)

In the case of the SECO_TRANSPORT code, the users wanted comparison with the Sudicky-Frind (1982) analytical solution. The problems with this solution were discussed previously (in Chapter 3), where it was noted that the infinite series solution has more mathematical uncertainty associated with it than does the Code being Verified. It is interesting (and from my experience, widespread) that users (and regulators, journal referees, managers, etc.) are *more impressed* with such agreements with widely accepted benchmarks in their particular discipline than they are with manufactured analytical solutions. The only real virtues seem to be familiarity and the illusion of confidence. For example, the widely accepted Sudicky-Frind solution, even when corrected and carefully evaluated, does not exercise most of the features of the code! It is based on a constant property medium and steady-state spatially-constant velocity field, yet many prefer it as a benchmark to a manufactured solution with spatially variable properties and time-and-space varying velocity field. It is often prudent to humor such irrational behavior, but only by allowing for Confirmation exercises, not by corrupting the concept of rigorous Verification. Also, managers must be warned that never-ending processes require never-ending budgets and deadline extensions.

Likewise, it is a good idea to include in Confirmation exercises some calculations on a realistic-looking problem of interest to the user. (Again, this is an open-ended process.) For the SECO_TRANSPORT code, this involved calculation of sample cases from the WIPP project (WIPP PA Dept., 1992; Helton et al, 1995, 1996.). Also, additional exercises of the TVD algorithm on high Peclet number cases were given (Salari and Blaine, 1996).

In closing, it is best to bear in mind that even with a rigorously Verified code, user errors will occur, and that *the code will be blamed*! This appears to be a fixed feature of human nature, and requires the code builder (or Sponsor, in the QA terminology of Chapter 11) to continually provide education and training to the users.

6.25 TURBULENT SEPARATED FLOW: THE ERROR ESTIMATOR OF CELIK AND KARATEKIN

Celik and Karatekin (1995) have applied a generalized Richardson Extrapolation to the calculation of turbulent flow past a backward facing step. The commercial code used (Phoenics), like many, utilizes "hybrid" first-order/second-order differencing for advection terms. This method is robust, but not very accurate, as noted in Chapter 1. See also Leonard and Drummond (1995) for warnings on the use of hybrid differencing. Celik and Karatekin used the standard k-ε two-equation turbulence model with wall functions for the turbulent inner layer, and nonuniform (power law) grid distributions. The hybrid method *ultimately* should be second order, but over practical ranges of grid sizes, it will appear to have some intermediate non-integer convergence rate with $1 < p < 2$. In spite of this ambiguity, by careful work Celik and Karatekin were able to convincingly demonstrate grid convergence, even with the grid refinement extending into the viscous sublayer, by using 5 grid refinements (with equal grid refinement factors in each coordinate direction), i.e. a total of 6 grids.

They used both the GCI (Chapter 5) and their alternative error estimator (discussed below). As anticipated (Roache, 1994) the GCI was excessively conservative with the safety factor F_s = 3 for this 6-grid sequence. (The same applies to other carefully performed grid convergence studies presented in Johnson and Hughes (1995); see Section 6.23 above.) Recall that (as acknowledged by Celik and Karatekin) the GCI was not proposed as an error *estimator* at all, but as an error *band*, equal to F_s times the error estimator, with $F_s > 1$ used to provide conservatism.

Celik and Karatekin (1995) also proposed an alternate error estimator with a built-in conservatism. Although similar to the GCI, and similarly based on a generalized Richardson Extrapolation, there are several differences. First, whereas the ε in the GCI approach (absolute or relative) is based on the difference between the fine and coarse grid solutions (and is normalized by the fine grid solution for the relative GCI), Celik and Karatekin define their relative error e_{ext} by reference to the extrapolated solution F_{ext}.

$$e_{ext} = \frac{f_{ext} - f_h}{f_{ext}} \tag{6.25.1}$$

This would be equivalent to the ε defined in Equation 5.4.3 (Chapter 5) but with f_1 (the fine grid solution) replaced by f_{ext}. Clearly, these two will be ordered approximations of one another. Also, the authors use the equivalent of the safety factor $F_s = 1$, which by itself would not be conservative.

A somewhat confusing point is that the paper involves two distinct Richardson-type Extrapolations: one used above to define the error estimate e_{ext}, and another used to obtain the best estimate of the converged solution. (Note the distinctions made in Chapter 5, that we may not want to actually use Richardson Extrapolation to obtain a final solution, for various reasons, but we can still estimate or band the error of the fine grid solution using the theory of Richardson Extrapolation.) In the above error estimator, e_{ext} is obtained using the value $p = 1$, which is (generally) conservative for the hybrid algorithm. However, the final solution (or best estimate of the converged solution) was obtained by using a value of p suggested by experiments, but limited. Their observed rate (for various quantities, over many grid combinations) was higher than $p = 1$, as expected; a representative value was observed $p = 1.3$. However, the point-by-point values of velocity components sometimes (locally) demonstrated non-monotone convergence behavior (incorporated into the Richardson Extrapolation by judicious use of the absolute value function); sometimes the observed p was < 1, and sometimes observed p was > 2. These excursions in observed p, if used in the Richardson Extrapolation, produced clearly erroneous values. The cure proposed was to simply limit the allowable p to the theoretical range of the hybrid method, i.e. $1 \leq p \leq 2$.

While this would seem to be a reasonable engineering-type approach, its limitations are clear. First of all, this is clearly non-conservative, since the actual convergence rate could be less (and was). If the same rationale were applied to anything but a hybrid or other mixed-order method, the range $1 \leq p \leq 2$ would collapse to a single number, e.g. $p = 2$ for a uniformly 2nd-order method. It is seen that the limit then simply ignores the experimental determination of (observed) p and substitutes for it the value obtained from formal analysis, as one would need to do if only two grid solutions were calculated. Thus, the limit replaces experimental *determination* or *Verification* of p with a *presumption* of p, and defeats the purpose of using more than two grid solutions in a high quality convergence study.

Another possible confusion exists in the definition of the grid refinement factor. Celik and Karatekin use the ratio $a(i)$ which is defined relative to (usually) the finest grid in the sequence. For a simple two-grid sequence, this gives $a = r$, as used in the GCI definitions in Chapter 5. For more grids in the sequence of a grid convergence test, we have

$$r_i = \frac{a_{i+1}}{a_i} \tag{6.25.2}$$

Although not incorrect, this definition of a grid refinement factor is inconsistent with previous and common usage. Clearly, if a grid is refined from 100 to 200 grids points or elements, the refinement factor "r" between these two grids should be 2, regardless of any *other* grids in the sequence of the grid refinement study. But in their definition, the grid refinement factor "a," besides being an inverse (i.e., a coarsening factor), is normalized at the finest grid in the study sequence. If the finest grid were (say) 100 in the sequence 25–50–100, then their refinement factor "*a*" for the 25 point grid would be 4, not 2. If they added another fine grid of 200 points to the study sequence, then their grid refinement factor "*a*" for the same 25 point grid would change to 8. This is an inappropriate terminology, even if it did not disagree with previous common usage. A less confusing term would be "grid index" for *a*.

Finally, the Celik and Keratekin error estimator e_{ext} can be expressed in terms of the closely related fine-grid GCI for $p = 1$ and $F_s = 1$, which we can denote by $\mathrm{GCI}_{1,1}$. That is, using Equation (5.6.1) with $p = 1$ and $F_s = 1$, we have

$$\mathrm{GCI}_{1,1} = \frac{\varepsilon}{r-1} \tag{6.25.3}$$

Then (from Appendix A of Celik and Keratekin, 1995),

$$e_{ext} = \left[\frac{r-1}{r - f_2 / f_1} \right] \mathrm{GCI}_{1,1} \tag{6.25.4}$$

In the limit of a converged solution, $f_2 \rightarrow f_1$ and $e_{ext} \rightarrow \mathrm{GCI}_{1,1}$ as expected.

For the example shown in Figure 7 of Celik and Keratekin (1995) their error estimator e_{ext} is less conservative (or less *overly* conservative) than the GCI calculated with either $p = 2$ or $p = 1$. However, as seen above, the correct comparison should be to $\mathrm{GCI}_{1,1}$, to which e_{ext} is an ordered approximation. Unfortunately, e_{ext} will not be *reliably* less overly conservative than $\mathrm{GCI}_{1,1}$ as evident from Equation 6.25.4 above. If the convergence for the quantity f is approached from above in the grid convergence sequence, i.e. if $f_2 / f_1 < 1$, then $e_{\mathrm{ext}} < \mathrm{GCI}_{1,1}$ but if $f_2 / f_1 > 1$, then $e_{ext} > \mathrm{GCI}_{1,1}$. Thus, in Figure 4 of Celik and Keratekin (1995) for velocity components above the base, e_{ext} is less conservative, but for velocity components below the base, e_{ext} is more conservative.

Although limited in the application in Celik and Keratekin (1995) to the hybrid method, the basic concept of their approach could be generalized to higher order methods. That is, instead of obtaining conservatism in the GCI (or closely related e_{ext}) by using $F_s > 1$, we could use $F_s = 1$ but $p <$ observed p. For example, one could use $F_s = 1$ and $p = 3$ in Equation 5.6.1 for a verified 4th-order accurate method. Whether this conservatism is any easier to interpret than using $F_s > 1$ is problematical, but could be worth investigation. However, the neat interpretation of the error band in relation to the ε obtained for a grid doubling with a 2nd-order method would be lost.

6.26 LEVEL OF ACCURACY ESTIMATES FROM GRID CONVERGENCE STUDIES

An important and somewhat unsettling observation about the practice of accuracy estimation or banding by grid convergence studies was made by Oberkampf et al (1995); see also Blottner

(1990). In their experience using Richardson Extrapolation on aerodynamics problems, both for local values (surface pressures) and for solution functionals (force and moment coefficients), they find that one must compute solutions with accuracy in the range of 1% to 0.1% in order to demonstrate that level of accuracy, i.e. to Verify the asymptotic convergence rate upon which the Richardson Extrapolation is based. This is not a problem if one's objective is 1% to 0.1% accuracy, but more often engineering analysts would be satisfied with something like 5–10% accuracy, and they understandably are reluctant to pay for the increased resolution necessary to achieve higher accuracy. Haworth (personal communication) had the same experience in modeling of internal combustion engines (Haworth et al, 1990, 1993), and Zingg (personal communication) in transonic airfoils (Zingg, 1991, 1992). The rule of thumb (1% to 0.1% accuracy) is probably problem dependent, but is probably reasonable for all but the easiest problems (e.g., Laplace equations).

So attaining the asymptotic range is necessary to give confidence to the error estimation, but this may already provide more accuracy than one would like to pay for. If error estimation is done outside the asymptotic range, the resulting lack of dependability motivates the large factor of safety ($F_s = 3$) in the original GCI error banding (Equation 5.6.1). In my opinion, reliable error estimation *outside* the asymptotic range is not likely for difficult problems (those with significant scale ranges, unlike Laplace solutions).

To put the dilemma succinctly, you may only want to pay for 10% accuracy, but 10% accuracy is difficult to reliably Verify!

This harsh fact of life is softened by the realization that attaining an accuracy level an order of magnitude more than required will *not* require an order of magnitude greater grid refinement, provided that higher than 1st-order methods are used. Unfortunately, it could possibly require an order of magnitude more *cost* for multi-dimensional time-dependent problems. This is where the use of error estimation for "nearby" problems is highly recommended, bringing the cost of reliability of error estimation into the same range as the accuracy requirements.

6.27 OTHER EXAMPLES OF CAREFUL USE OF RICHARDSON EXTRAPOLATION

Following Oberkampf et al (1995), we recommend four papers as examples of "careful use and estimation of error using Richardson's method" in Calculation Verification. These are Roache (1982), Shirazi and Truman (1989), Blottner (1990), and Walker and Oberkampf (1992). Also recommended is Nguyen and Maclaine-Cross (1988) for application to heat exchanger pressure drop coefficients and Caruso et al (1985).

6.28 PARAMETER CONVERGENCES OF A COMPRESSIBLE FLOW CODE NEAR THE INCOMPRESSIBLE LIMIT

Exercising a PDE code over a range of parameters is an activity that does not fit into the semantic categories of Chapter 2 very well, but it is nevertheless a worthwhile exercise in "confidence building", and uses similar techniques. In Roache and Salari (1990), we presented a cursory exercise of a CFD code built for compressible flow near the incompressible limit. There is no question of "Justification" here (see Chapter 2, Section 2.10.3) since the accuracy of the incompressible limit equations is well accepted. However, it is not at all obvious how well a compressible flow code will perform at low Mach numbers. With special care in the formulation of the equations, it is possible to build a code that handles very low Mach numbers, or even $M = 0$. But *most* compressible flow codes, given input parameters that give $M = 0$ (with non-zero velocities) will fail dramatically, e.g. a divide by zero. Many will fail at $M = 0.1$ or 0.2. Also, it is of interest to reveal the effects of low M and other parameters on the convergence rate, and to "justify" not the limit of $M \rightarrow 0$ in general, but to quantify the boundary of the incompressible approximation, which is problem dependent.

The code used in the study was based on the Approximate Factorization (AF) algorithms of Briley and MacDonald (1977,1980) and Beam and Warming (1976) using the "delta formulation." It uses boundary fitted non-orthogonal coordinates, ideal gas relations, constant specific heats, and the Sutherland viscosity-temperature relation. Significantly, the code treats cross-derivative terms without unnecessary approximations; however, it does use the common non-ordered approximation of $\partial p / \partial n = 0$ at walls. Although code versions included 3-D and turbulence models, and 2nd-order time and space accuracy had been Verified, the presently reported study considered only 2-D, laminar, steady flow.

Three steady-state problems were considered:

- a NACA 0012 airfoil at $\alpha = 0$
- a 2-D model of a submarine torpedo shuttleway
- the driven cavity problem.

Sample airfoil results are shown in Figures 6.28.1–2. The *grid* convergence results are given in the original study as well as Salari and Roache (1990). (In Roache and Salari, 1990, the *observed* spatial rate of convergence was $p > 2$, apparently an artifact of the artificial dissipation at low M.) The grid used in Figures 6.28.1–2 was 185×71. The easily attained value $M = 0.1$ gives effectively "incompressible" results. The code "works" at $M = 0.01$, but evidence of the incipient breakdown due to the singularity at $M = 0$ is beginning to show itself by the small scale pressure oscillations near the trailing edge in Figure 6.28.1.

Figure 6.28.2 presents a comparison between the compressible AF code at $M = 0.1$ and the true incompressible code of Mehta (1977) based on an entirely different mathematical formulation using the stream function-vorticity equations. The excellent agreement between the two codes was, of course, to be expected from the mathematics, but nevertheless constituted a gratifying "Confirmation" exercise (see terminology in Chapter 2). In fact, the original agreement was not so good, with a 15% discrepancy in pressure near the pressure minimum. The PDE solutions were good, but the incompressible code had to be modified (by I. P. Itty) in the post-processing quadrature for pressure, by using one-sided 2nd-order (3-point) differences for surface normal derivatives, in order to achieve the agreement shown in Figure 6.28.2.

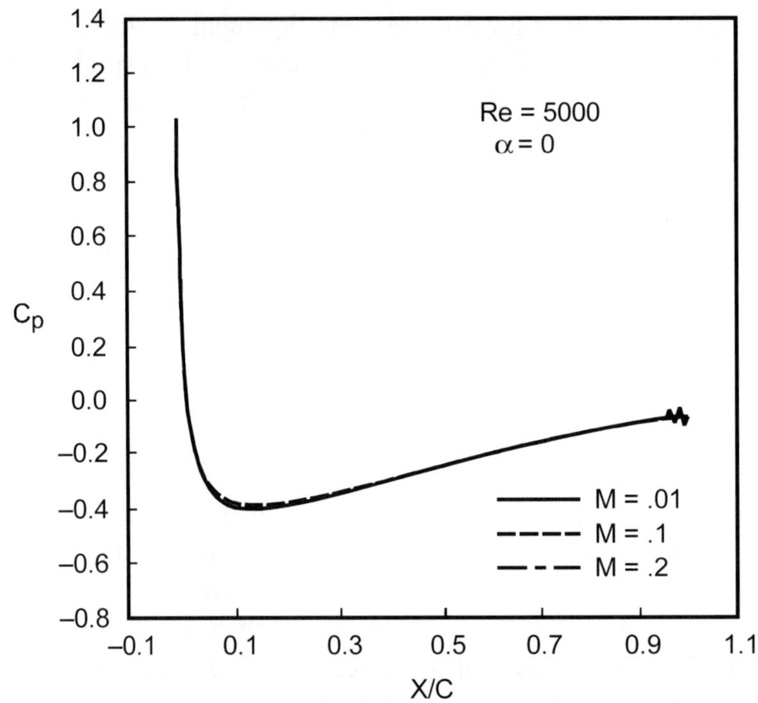

Figure 6.28.1. Pressure coefficients on NACA 0012 airfoil at three Mach numbers. Laminar flow, *Re* = 5000, α = 0. (From Roache and Salari, 1990, Figure 2.)

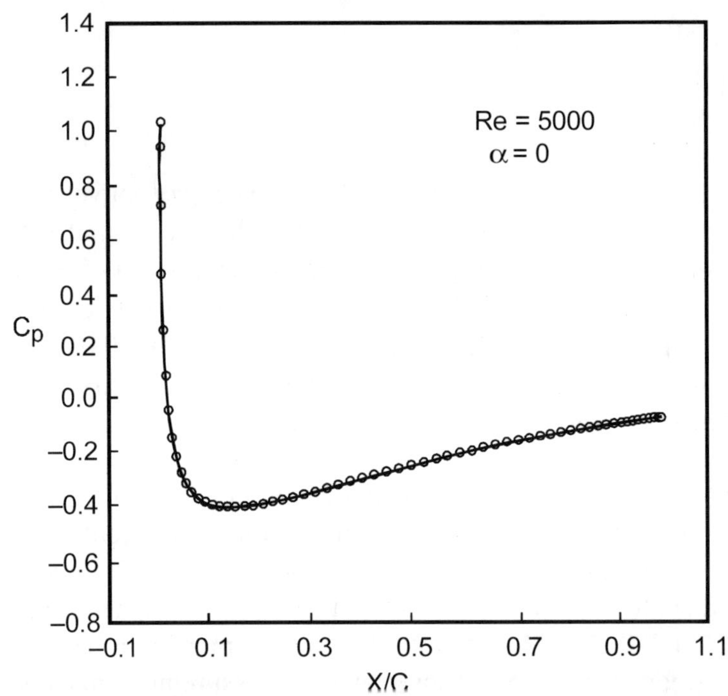

Figure 6.28.2. Pressure coefficients on NACA 0012 airfoil calculated by a compressible flow code at Mach number = 0.1 and by an incompressible code based on stream function-vorticity variables. Laminar flow, *Re* = 5000, α = 0. (From Roache and Salari, 1990, Figure 3.)

Mach Number	Total Drag C_D	Pressure Drag C_{DP}	Friction Drag C_{DF}	Stagnation Pressure C_P	Maximum $\nabla\cdot\mathbf{V}$
0.34*					0.89*
0.20	0.05282	0.01850	0.03432	1.0448	0.145
0.10	0.05268	0.01778	0.03491	1.0384	0.041
0.01	0.05331	0.01589	0.03742	1.0302	0.004
Incompressible	0.05341	0.01864	0.03477	1.0282	–

Table 6.28.1. Aerodynamic coefficients and the maximum dilatation as *M* is reduced. NACA 0012 airfoil, laminar flow, *Re* = 5000, α = 0. The "Incompress." results were obtained with a separate code based on a ψ-ζ formulation. (*The value of dilatation at $M = 0.34$ was obtained for a high *Re* turbulent flow case.) (From Roache and Salari, 1990, Table 1.)

Table 6.28.1 shows the behavior of the aerodynamic coefficients and the maximum dilatation $\nabla\cdot\mathbf{V}$ as M is reduced, again compared to the results from the incompressible stream function-vorticity (ψ-ζ) code. The incipient breakdown at $M = 0.01$ (in pressure oscillations near the trailing edge) is evident here; although the total drag looks good, in fact this is due to a lucky cancellation, as the pressure drag has undershot the incompressible value and the friction drag has overshot it. The dilatation $\nabla\cdot\mathbf{V} = 0$ identically for incompressible flow. It decreases rapidly with M, and plots of dilatation (see Figure 4 of Roache and Salari, 1990) bear a striking resemblance to experimental interferograms.

The airfoil solution at $M = 0.01$ was obtained with a time step that produced a maximum advective Courant Number = 15, and full (advective + acoustic) Courant Number = O(1500), an impressive stability performance for the AF algorithm. (Note that the breakdown is beginning, but the results are still somewhat usable at this Mach number.)

The fairly complex flow in the torpedo shuttleway (see Figures 5–11 of Roache and Salari, 1990) was obtained again using $M = 0.1$ in a 250×150 grid; no evidence of oscillations occurred, even though one region of the flow had velocities 4–5 orders of magnitude less than the free stream value at $M = 0.1$.

For the driven cavity problem, the parameter value $M = 0.1$ also proved to be adequate for achieving effectively incompressible steady solutions. Comparisons were made with the benchmark solutions of Ghia et al (1982) obtained with (another) incompressible code based on ψ-ζ variables. That study established that a uniform grid of 129×129 points was adequate, which we confirmed by also running a 65×65 grid case. A more accurate 129×129 stretched (rather than uniform) grid was used to produce Figure 6.28.3, which shows the u-velocity profile through the center of the cavity for $Re = 100$. The agreement of the $M = 0.1$ and 0.05 solutions with the incompressible benchmark solutions is excellent. The disagreement at $M = 0.01$ is pronounced, due to deterioration as the $M = 0$ singularity is approached. Unfortunately, the inaccuracy is only revealed by this convergence study (in this case, convergence as $M \rightarrow 0$) and not by any obvious qualitative breakdown in the solution. We would prefer a more precipitous

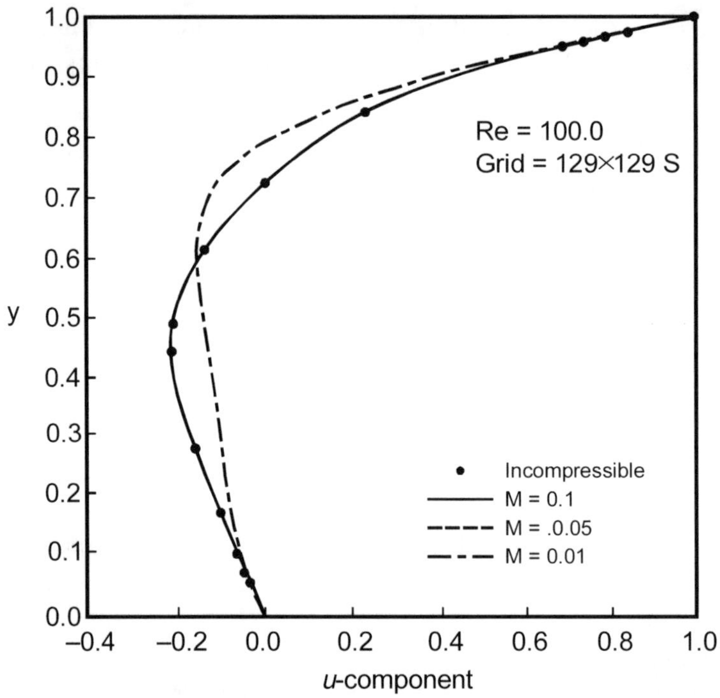

Figure 6.28.3 *u*-velocity profile through the center of the cavity for *Re* = 100.

breakdown, such as the beginning pressure oscillations near the airfoil trailing edge as in Figure 6.28.1. (There is in fact *some* evidence in a noisy quadrature for stream function near the bottom of the cavity, and poor resolution of the corner eddies.)

The agreement with the downward v-velocity component from the incompressible benchmark solution is not as good, differing by < 2.5%. (See Figure 14 of Roache and Salari, 1990.) The speculation (in Roache and Salari) that the most likely reason is the inherent error at low M in the Approximate Factorization (AF) procedure itself (see Beam and Warming, 1976) is incorrect; as pointed out by a reviewer, this error is negligible for good steady-state iterative convergence with the delta-form of AF.

Additional quantitative results are shown in Table 6.28.2, which displays convergence as $M \rightarrow 0$ of minimum density ρ, minimum stream function ψ (or primary vortex strength), and maximum dilatation $\nabla \cdot \mathbf{V}$. The minimum stream function is also given from the incompressible ψ-ζ benchmark solution of Ghia et al (1982). Note that the ψ evaluation for the compressible code involves additional numerical error in quadratures, as well as compounded errors in velocities and densities.

The study also investigated the effects of ratio of specific heats γ and the 4th-order explicit damping coefficient DC. The previously cited results were obtained with $\gamma = 1.4$ (air). At $M = 0$, the continuum results will not depend on the thermodynamic parameter γ. Consideration of the gas dynamic relationships suggest that $\gamma \rightarrow 1$ is consistent with $M \rightarrow 0$ and suggested that a low γ solution would be a better incompressible approximation. Indeed, results for $\gamma = 1.1$ and 1.005

Mach Number	γ	minimum ρ	minimum ψ	max $\nabla\cdot\mathbf{V}$
0.10	1.4	0.89028	−0.10208	0.226
0.05	1.4	0.97244	−0.10296	0.028
0.01	1.4	0.99885	−0.08531	0.00099
Incompressible	–	–	−0.10342	–
0.05	1.1	0.97977	−0.10306	0.0222
0.05	1.005	0.98218	−0.10309	0.0203

Table 6.28.2. Driven cavity solutions for minimum density ρ, primary vortex strength (minimum ψ), and maximum dilatation $\nabla\cdot$V. The "Incompress." results were obtained with a separate code based on a ψ-ζ formulations. γ is the ratio of specific heats. (From Roache and Salari, 1990, Table 3.)

at $M = 0.1$ showed a somewhat decreased range of the dilatation, as shown in Table 6.28.2. (The use of $\gamma = 1$ causes indeterminacies in the code, as $M = 0$ does.)

A more practical consideration is the effect of the artificial 4th-order explicit damping coefficient *DC*. All the previously described solutions were obtained with the code default value of $DC = 0.01$. We increased *DC* by a factor of 2, and decreased it by a factor of 50, with no significant effect on the solutions. Only when *DC* was further increased to 1.4 (140 times the nominal value) *and* the grid spacing near the boundary was doubled did a noticeable effect show up. (See Figure 16 of Roache and Salari, 1990.)

This cursory study considered the effects of Mach Number, ratio of specific heats, and artificial 4th-order damping coefficient on the behavior of an Approximate Factorization code for compressible flow in the low Mach number range for steady flows. The parameter study is far from complete, but might suggest an interesting and more complete study. Highly transient solutions are expected to be more difficult.

6.29 JUSTIFICATION OF THE DUPUIT APPROXIMATION

The previous Section 6.28 presented an example of a numerical study of parameter convergence, an activity that does not fit the definition of Verification but uses similar techniques. Herein, we consider an example of another such activity, that of "Justification" of modeling equations. As discussed in Chapter 2, Section 2.10.3, the distinction between Verification and Validation may not be obvious when one considers a code based on some sort of simplified equation set.

Agreement between results of the simplified and complete equations is not strictly included in the term "Verification," since the Verification of the simplified equation code has already been completed prior to the full equation comparison. One could say that the agreement has demonstrated that the simplified code is "solving the right equations" in one sense, i.e., it justifies the use of simplified equations. Yet to claim "Validation" would be over-reaching, since we have not demonstrated the adequacy of the more complete model by comparison with experiment. We have "solved the right equations" only in an intermediate sense of demonstrating that the simplified equations adequately represent the more complete equations, but not in the ultimate sense of "solving the right *physical* equations." Appealing to the other distinction between Verification and Validation based on mathematics vs. science, it is clear that such a comparison exercise should be categorized with Verification rather than Validation, but that the categorization is somewhat inadequate because the simplified code can (should) have been fully "Verified" in the more usual sense before the comparison exercise was started. If the only choice is "Verification" or "Validation," this "gray area" is best categorized as "Verification," with the term "Validation" reserved for experimental comparisons. Even better is to introduce another term, "Justification."

As an example of Justification, consider the Dupuit approximation, a small-angle approximation for groundwater flows that neglects any effects associated with vertical velocity. This approximation is virtually ubiquitous in groundwater flow codes, e.g. MODFLOW (McDonald and Harbaugh, 1988). The Dupuit approximation is so commonly used that a modeler can assume it with little chance of criticism in review, yet Bear (1972) gives the following requirements for its use (from Knupp et al, 1996).

- $(K_{hor} / K_{ver})s << 1$ (K = conductivity, s = slope of the water table)
- no recharge (so that the phreatic surface is a streamline)
- no seepage faces
- absence of no-flow boundaries on sides of domain
- steady state
- no local "pits or mounds" in the water table
- no well terms that would deform the water table

Knupp et al (1996) compared the results from MODFLOW, which uses the Dupuit approximation, with the results from SECO_FLOW_3D, which has an option to use the full nonlinear phreatic boundary conditions (Knupp, 1996). Each code was Verified to produce consistent (ordered) solutions to its own equation set. (Interestingly, the development of an exact Benchmark analytical solution was easier for the full equations, and use of this full solution confuses Verification of a simplified code. The fact that it was easier to Verify provided additional motivation for the development of the full equation code. See Roache et al, 1996.)

An example from the Verification exercises of the complete equation code is shown in Figure 6.29.1. The largest error, at the right boundary, was 0.005%. In Figure 6.29.2, the comparison of the full equation analytical solution with the result from the MODFLOW code shows (primarily) the effect of the Dupuit approximation. At worst, it is 8% in error. (See Knupp et al, 1996, for details of the test case parameters, resolutions, and fine modeling points required for the MODFLOW simulation.)

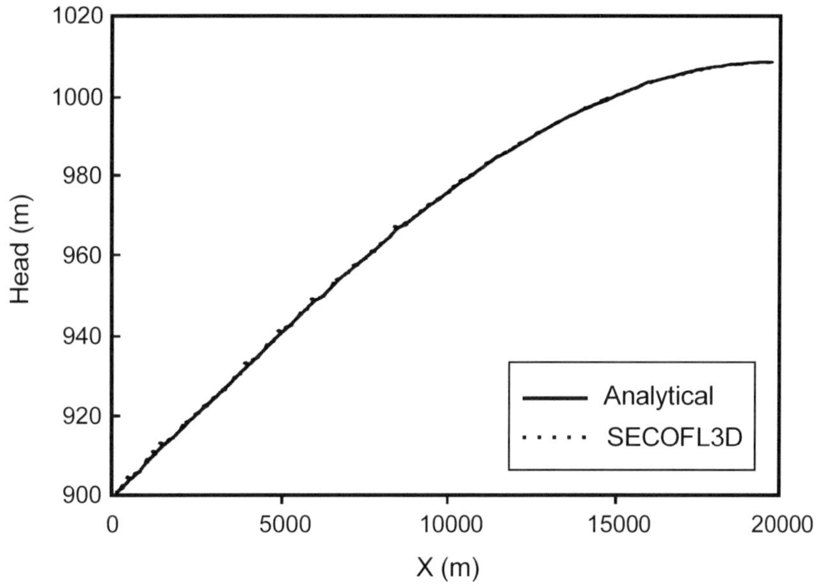

Figure 6.29.1. Water table solutions from an analytical solution of the full equations and the SECO_FLOW_3D code using the full nonlinear phreatic boundary condition. (From Figure 3-1 of Knupp et al, 1996.)

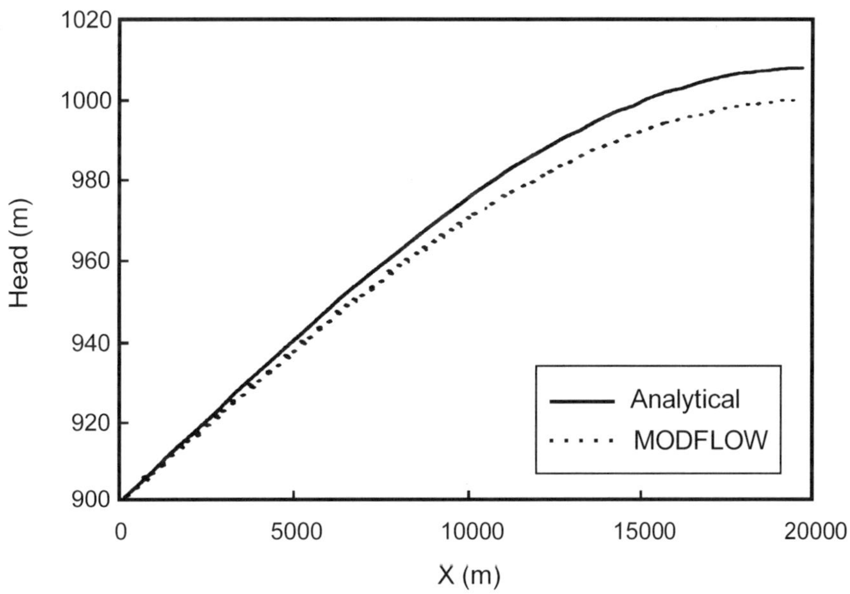

Figure 6.29.2. Water table solutions from an analytical solution of the full equations and the MODFLOW code using the Dupuit approximation. (From Figure 3-3 of Knupp et al, 1996.)

The Justification of the Dupuit approximation will obviously be problem dependent (see also Serrano, 1995). Knupp et al (1996) found that it was important to use the full phreatic condition when

- heterogeneity increased,
- recharge was higher, and
- well pumping occurred.

For some parameter combinations, the error in head was 24%. Furthermore, the true phreatic boundary condition eliminated some instances of non-unique solutions obtained with the steady-state option in MODFLOW. On the other hand, the phreatic condition code is more demanding of accurate recharge estimates (Roache et al, 1996).

Errors in fluxes are more severe than errors in head. For a suite of test cases using SECO_FLOW_3D with and without the Dupuit approximation, it caused 5–10% errors in heads and water table elevations, but up to 50% error in fluxes (Roache et al, 1996), which would correspondingly affect contaminant transport calculations.

6.30 CONCLUDING COMMENT ON PARAMETER UNCERTAINTY VS. NUMERICAL UNCERTAINTY

I offer a final comment on the philosophy expressed by many flow modelers. In the field of groundwater modeling, there is a tremendous uncertainty in the physical parameters involved. This uncertainty has often been used to excuse sloppy numerical work. For example, in one of the contributions to the international HYDROCOIN project (OECD, 1988), a grid doubling in the horizontal plane caused a 40% reduction in peak value of hydraulic head; yet most of the contributors used only a single grid and a single time step; also, the various calculations produced "a very wide spread in travel times." The same excuses have been used by aerodynamicists using turbulence models, or other approximations. The basic concept is to excuse laziness in the numerical work by noting the uncertainty in the conceptual model. Often, one hears that "we only want the answer to 10% accuracy" in the case of aerodynamics, or "order of magnitude" for groundwater flow modelers.

The first and obvious response is that there should be no *preference* for doing inaccurate numerical work, and methods are available to assess the numerical accuracy. Of course, the cost trade-offs must be decided by engineering judgment, especially in a design environment, and I would not second-guess those with deadlines to meet.

The second response, especially to the groundwater flow modelers, is to note that poor (1st-order or hybrid) numerical methods for advection terms (not so much in groundwater flow itself, but in the subsequent contaminant transport calculation) can in fact introduce an *order of magnitude error*. In the 1950's, the late Senator Dirksen of Illinois commented on military budget resolution discussions by noting "a billion dollars here, a billion dollars there, and pretty soon you are into some real money." To paraphrase the Senator for the groundwater flow and transport modelers, "an order of magnitude here, an order of magnitude there, and pretty soon you are into some real errors."

Third and finally, if a groundwater transport modeler will be satisfied with order-of-magnitude numerical accuracy, I would argue that difficult computer simulations may not be necessary. Rather than go through the charade of obtaining solutions to variable coefficient partial differential equations, the modeler can obtain the needed estimates from closed form solutions (or very simple numerical solutions of Laplace equations) by assuming constant properties, using averaged value of parameters and engineering intuition. This has the virtue of exposing the level of uncertainty, rather than concealing the uncertainty with the of gloss of mathematics.

If we are going to use partial differential equations in our conceptual model, then we should solve them honestly.

REFERENCES FOR CHAPTER 6

Beam, R. and Warming, R. F. (1976), "An Implicit Finite Difference Algorithm for Hyperbolic Systems in Conservation Law Form," *Journal of Computational Physics*, Vol. 22, 1976, pp. 87–110.

Bear, J. (1972), *Dynamics of Fluids in Porous Media*, American Elsevier, New York, 1972.

Blottner, F. G. (1990), "Accurate Navier-Stokes Results for the Hypersonic Flow over a Spherical Nosetip," *Journal of Spacecraft and Rockets,* Vol. 27, No. 2, March – April 1990, pp. 113–122.

Book, D. L., Boris, J. P., and Hain, K. (1975), "Flux-Corrected Transport II: Generalizations of the Method," *Journal of Computational Physics*, Vol. 18, 1975, pp. 248–283. See also Boris, J. P., Naval Research Laboratory, NRL–MR–2357, November 1971.

Briley, W. R., and MacDonald, H. (1977), "Solution of the Multi-Dimensional Compressible Navier-Stokes Equations by a Generalized Implicit Method," *Journal of Computational Physics*, Vol. 24, 1977, pp. 372–397.

Briley, W. R., and MacDonald, H. (1980), "On the Structure and Use of Linearized Block Implicit Schemes," *Journal of Computational Physics*, Vol. 34, No. 19, pp. 54–73.

Caruso, S. C., Ferziger, J. H., and Oliger, J. (1985), "Adaptive Grid Techniques for Elliptic Flow Problems," Report No. TF–23, Thermosciences Division, Stanford University, Stanford, California.

Celik, I. and Karatekin, O. (1995), "Numerical Experiments on Application of Richardson Extrapolation to a Turbulent Flow with Separation," in Johnson and Hughes (1995), pp. 37–43. See also *ASME Journal of Fluids Engineering*, Vol. 119, No. 3, September 1997, pp. 584–590.

Chang, S. and Haworth, D. C. (1995), "Kinetic-Energy-Balance Based Solution-Adaptive Mesh Refinement," in Johnson and Hughes (1995), pp. 7–12.

Comini, G. Cortella, G. and Manzan, M. (1995), "A Streamfunction-Vorticity-Based Finite-Element Formulation for Laminar-Convection Problems," *Numerical Heat Transfer, Part B*, Vol. 28, pp. 1–22.

de Vahl Davis, G., Jones, I. P., and Roache, P. J. (1979), "Natural Convection in an Enclosed Cavity: A Comparison Problem," *Computers and Fluids*, Vol. 7, 1979, pp. 315–316.

de Vahl Davis, G. and Jones, I. P. (1983), "Natural Convection in a Square Cavity: A Comparison Exercise," *International Journal for Numerical Methods in Fluids,* Vol. 3, No. 3, pp. 227–248.

de Vahl Davis, G. (1983), "Natural Convection of Air in a Square Cavity: A Bench Mark Numerical Solution," *International Journal for Numerical Methods in Fluids,* Vol. 3, No. 3, pp. 249–264.

Engquist, B. and Sjogreen, B. (1996), "Numerical Convergence Rate in the Presence of Shocks," to appear. (See Yee and Sweby, 1996.)

Ghia, U., Ghia, K. N., and Shin, C. T. (1982), "High-*Re* Solutions for Incompressible Flows Using the Navier-Stokes Equations and a Multigrid Method," *Journal of Computational Physics*, Vol. 48, 1982, pp. 387–411.

Haworth, D. C., El Tahry, S. H., Huebler, M. S., and Chang, S. (1990), "Multidimensional Port-and-Cylinder Flow Calculations for Two- and Four-Valve-Per-Cylinder Engines: Influence of Intake Configuration on Flow Structure," SAE Paper No. 900257.

Haworth, D. C., El Tahry, S. H., and Huebler, M. S. (1993), "A Global Approach to Error Estimation and Physical Diagnostics in Multidimensional Computational Fluid Dynamics," *International Journal for Numerical Methods in Fluids*, Vol. 17, 1993, pp. 75–97.

Helton, J. C., Anderson, D. R., Roache, P. J., et al (1995), "Effect of Alternative Conceptual Models in a Preliminary Performance Assessment for the Waste Isolation Pilot Plant," *Nuclear Engineering and Design*, Vol. 154, 1995, pp. 251–344.

Helton, J. C., Anderson, D. R., Roache, P. J., et al (1996), "Uncertainty and Sensitivity Analysis Results Obtained in the 1992 Performance Assessment for the Waste Isolation Pilot Plant," *Reliability Engineering and System Safety*, Vol. 51, 1996, pp. 251–53–100.

Holst, T. L. (1987), "Viscous Transonic Airfoil Workshop Compendium of Results," AIAA Paper 87–1460, 1987. See also *Journal of Aircraft*, Volume 25, No. 12, December 1988, pp. 1073–1087.

Hortmann, M., Peric, M. and Scheurer, G. (1990), "Multigrid Benchmark Solutions for Laminar Natural Convection Flows in Square Cavities," ASME FED Vol. 93, *Benchmark Test Cases for Computational Fluid Dynamics*, I. Celik and C. J. Freitas, Eds., Book No. H00598, June 1990, pp. 1–6.

Hugo, R. J. (1995), "Quantifying the Spatio-Temporal Effects of Optically-Active Turbulent Flowfields on a Coherent Optical Wave," Ph.D. Dissertation, Department of Aerospace and Mechanical Engineering, University of Notre Dame, 1995.

Hyman, J., Shashkov, M. and Steinberg, S. (1997), "The Numerical Solution of Diffusion Problems in Strongly Heterogeneous Non-Isotropic Materials," *Journal of Computational Physics*, Vol. 132, p. 130–148.

Jameson, A. and Martinelli, L. (1996), "Mesh Refinement and Modeling Errors in Flow Simulation," AIAA Paper 96–2050, 27th AIAA Fluid Dynamics Conference, New Orleans, Louisiana, June 17–20, 1996.

Johnson, R. W. and Hughes, E. D. (1995), Eds., *Quantification of Uncertainty in Computational Fluids Dynamics–1995,* Joint JSME–ASME Fluid Mechanics Meeting, ASME FED Volume 213, Hilton Head, South Carolina, August 14–18, 1995.

Jumper, E. J., Hugo, R. J., and Cicchiello, J. M. (1994), "Turbulent-Fluid-Induced Optical Wavefront Dynamics: Near-and Far-Field Implications," AIAA Paper 94–2547, AIAA 18th Aerospace Ground Test Conference, Colorado Springs, Colorado, June 20–23, 1994.

Jumper, E. J. and Hugo, R. J. (1995), "Quantification of Aero-Optical Phase Distortion Using the Small-Aperture Beam Technique," *AIAA Journal*, Vol. 33, No. 11, November 1995, pp. 2151–2157.

Kipp, K. L., Jr. (1987), "HST3D: A Computer Code for Simulation of Heat and Solute Transport in Three-Dimensional Ground-Water Flow Systems," U. S. Geological Survey, Water Investigations Report 86–4095, USGS, Denver, Colorado.

Knupp, P.M. and Steinberg, S. L. (1993), *Fundamentals of Grid Generation*, CRC Press, Boca Raton, Florida.

Knupp, P. M. and Luczak, R. (1995), "Truncation Error in Grid Generation: A Case Study," *Numerical Methods for Partial Differential Equations*, Vol. 11, 1995, pp. 561–571.

Knupp, P. (1996), "A Moving Grid Algorithm for 3-D Regional Groundwater Flow Calculations with Water Table and Seepage Face," *Advances in Water Resources*, Vol. 19, No. 2, 1996, pp. 83–95.

Knupp, P., Rucker, D. and Corbet, T. F. (1996), "Groundwater Modeling with the Phreatic Boundary Condition: Is It Necessary?," Ecodynamics Research Associates, Albuquerque, New Mexico, December 1996. Also, to appear.

Kreiss, H.-O. and Lundqvist, E. (1968), "On Difference Approximations with Wrong Boundary Values," *Mathematics of Computation*, Volume 22, 1968, pp. 1–12.

Kuruvila, G. and Anderson, J. D. (1985), "A Study of the Effects of Numerical Dissipation on the Calculation of Supersonic Separated Flows," AIAA Paper 85–0301, January 1985.

Leonard, B. P. and Drummond, J. E. (1995), "Why You Should Not Use 'Hybrid', 'Power Law' or Related Exponential Schemes for Convective Modelling–There Are Much Better Alternatives," *International Journal for Numerical Methods in Fluids*, Vol. 20, 1995, pp. 421–442.

Leonard, B. P. and MacVean, M. K. (1995), "Grid Refinement Study of a Benchmark Problem Using the Flux-Integral Method," in Johnson and Hughes (1995), pp. 45–50.

Lester, D. H., Jansen, G., and Burkholder, H. C. (1981), "Migration of Radionuclide Chains Through an Adsorbing Medium," *AIChE Symposium Series*, Vol. 71, No. 152.

Lotz, R. D., Thompson, B. E., Konings, C. A., and Davoudzadeh, F. (1995), "Grid Dependence and Numerical Uncertainty Analysis of a Transonic Supercritical Airfoil Calculations," in Johnson and Hughes (1995), pp. 13–18.

Lotz, R. D., Thompson, B. E., Konings, C. A., and Davoudzadeh, F. (1997), "Numerical Uncertainties in Transonic Flow Calculations for Aerofoils with Blunt Trailing Edges," *International Journal for Numerical Methods in Fluids,* Vol. 24, pp. 355–373.

McDonald, M. G. and Harbaugh, W. (1988), "A Modular Three-Dimensional Finite-Difference Ground-Water Flow Model," Techniques of Water-Resources Investigations of the United States Geological Survey, Book 6, Modeling Techniques, Chapter A1. Scientific Software Group, P.O. Box 23401, Washington, D.C. 20026.

Mehta, U. B. (1977), "Dynamic Stall of an Oscillating Airfoil," AGARD Conference Proceedings CP–227, *Symposium on Unsteady Aerodynamics*, Ottowa, Canada, Paper 23.

Nguyen, T. V. and Maclaine-Cross, I. L. (1988), "Incremental Pressure Drop Number in Parallel-Plate Heat Exchangers," *ASME Journal of Fluids Engineering.*, Vol. 110, pp. 93–96, March 1988.

Oberkampf, W. L., Blottner, F. G., and Aeschliman, D. P. (1995), "Methodology for Computational Fluid Dynamics Code Verification/Validation," AIAA Paper 95–2226, 26th AIAA Fluid Dynamics Conference, San Diego, California, 19–22 June 1995.

OECD (1988), The International HYDROCOIN Project: Groundwater Hydrology Modelling Strategies for Performance Assessment of Nuclear Waste Disposal. Level 1: Code Verification, Nuclear Energy Agency, Organisation for Economic Co-Operation and Development (OECD), Paris.

Patel, R. (1973), "An Experimental Study of a Plane Mixing Layer," *AIAA Journal*, Vol. 11, No. 1, January 1973.

Pelletier, D. and Ignat, L. (1995), "On the Accuracy of the Grid Convergence Index and the Zhu-Zienkiewicz Error Estimator," in Johnson and Hughes (1995), pp. 31–36.

Pulliam, T. H. (1986), "Efficient Solution Methods for the Navier-Stokes Equations," *Lecture Notes for the Von Karman Institute for Fluid Dynamics, Lecture Series: Numerical Techniques for Viscous Flow Computation in Turbomachinery Bladings*, Von Karman Institute for Fluid Dynamics, Brussels, Belgium, January 20–24, 1986.

Reeves, M., Ward, D. S., Johns, N. D., and Cranwell, R. M. (1986), "Theory and Implementation for SWIFT II The Sandia Waste-Isolation Flow and Transport Model for Fractured Media," SAND83–1159, Sandia National Laboratories, Albuquerque, New Mexico, 1986.

Richardson, L. F. (1910), "The Approximate Arithmetical Solution by Finite Differences of Physical Problems Involving Differential Equations, with an Application to the Stresses in a Masonry Dam," *Transactions of the Royal Society of London, Series A*, Vol. 210, pp. 307–357.

Roache, Patrick J. (1972), *Computational Fluid Dynamics*, First Edition, Hermosa Publishers, P. O. Box 9110, Albuquerque, New Mexico 87119-9110, USA. http://www.hermosa-pub.com/hermosa. Revised Printing 1976. See also Roache (1998).

Roache, P. J. (1975), "The LAD, NOS, and Split NOS Methods for the Steady-State Navier-Stokes Equations," *Computers and Fluids*, Vol. 3, 1975, pp. 179–195.

Roache, P. J. (1982), "Scaling of High-Reynolds-Number Weakly Separated Channel Flows," in *Numerical and Physical Aspects of Aerodynamic Flows,* T. Cebeci, ed., Chapter 6, Springer Verlag, New York, pp. 87–98.

Roache, P. J. and Steinberg, S. (1984), "Symbolic Manipulation and Computational Fluid Dynamics," *AIAA Journal*, Vol. 22, No. 10, October 1984, pp. 1390–1394.

Roache, P. J. and Steinberg, S. (1985), "A New Approach to Grid Generation Using a Variational Formulation," AIAA Paper 85–1527, AIAA 7th Computational Fluid Dynamics Conference, Cincinnati, Ohio, July 15–17, 1985.

Roache, P. J. and Dietrich, D. E. (1988), "Evaluation of the Filtered Leapfrog-Trapezoidal Time Integration Method," *Numerical Heat Transfer*, Vol. 14, 1988, pp. 149–164.

Roache, P. J. and Salari, K. (1990), "Weakly Compressible Navier-Stokes Solutions with an Implicit Approximate Factorization Code," AIAA Paper 90–0235, 28th AIAA Aerospace Sciences Meeting, Reno, Nevada, January 8–11, 1990.

Roache, P. J., Knupp, P. M., Steinberg, S., Blaine, R. L. (1990), "Experience with Benchmark Test Cases for Groundwater Flow," ASME FED Vol. 93, *Benchmark Test Cases for Computational Fluid Dynamics*, I. Celik and C. J. Freitas, eds., Book No. H00598, June 1990, pp. 49–56.

Roache, P. J. and Knupp, P. M. (1993), "Completed Richardson Extrapolation," *Communications in Numerical Methods in Engineering,* Vol. 9, pp. 365–374.

Roache, P. J. (1993), "The SECO Suite of Codes for Site Performance Assessment," *Proceedings of the 1993 International High-Level Radioactive Waste Management Conference*, Las Vegas, Nevada.

Roache P. J. (1994), "Perspective: A Method for Uniform Reporting of Grid Refinement Studies," *ASME Journal of Fluids Engineering,* Sept. 1994, Vol. 116, pp. 405–413. See also *Quantification of Uncertainty in Computational Fluid Dynamics*, I. Celik et al, eds., ASME Fluids Engineering Division, pp. 109–120.

Roache, P. J. (1995), *Elliptic Marching Methods for Domain Decomposition*, CRC Press, Boca Raton, Florida.

Roache, P. J., Blaine, R. L., and Baker, B. L. (1995), "SECOFL2D User's Manual," SAND90–7096, Sandia National Laboratories, Albuquerque, New Mexico, October 1995.

Roache, P. J., Knupp, P., Salari, K., et al (1996), "High Accuracy Robust Groundwater Flow and Transport Codes for Environmental Engineering: Phase II," Final Report, F08637–94–C6049 for Armstrong Laboratory, Tyndall AFB, Florida, Ecodynamics Research Associates, Albuquerque, New Mexico.

Roache, P. J. (1998), *Fundamentals of Computational Fluid Dynamics*, Hermosa Publishers, P. O. Box 9110, Albuquerque, New Mexico 87119-9110, USA. http://www.hermosa-pub.com/hermosa. To appear.

Ruge, J. W. (1995), "An Evaluation of the Grid Convergence for 1-D Two-Phase Flow in Porous Media", in Johnson and Hughes (1995), pp. 51–56.

Salari, K., and Roache, P. J. (1990), "The Influence of Sweep on Dynamic Stall Produced by a Rapidly Pitching Wing," AIAA Paper 90–0581, Reno, Nevada, January 8–11, 1990.

Salari, K., Knupp, P., Roache, P. J., and Steinberg, S. (1992), "TVD Applied to Radionuclide Transport in Fractured Porous Media," *Numerical Methods in Water Resources*, Denver, Colorado.

Salari, K. and Steinberg, S. (1994), "Flux-Corrected Transport in a Moving Grid," *Journal of Computational Physics*, Vol. 111, No. 1, March 1994, pp. 24–32.

Salari, K., Blaine, R. L., Economy, K., and Roache, P. J. (1995), "Grid Resolution Studies of Radionuclide Transport in Fractured Porous Media," in Johnson and Hughes (1995), pp. 25–30.

Salari, K. and Blaine, R. L. (1996), "The SECOTP2D User's Manual," SAND96, Sandia National Laboratories, Albuquerque, New Mexico.

Serrano, S. E. (1995), "Analytical Solutions of the Nonlinear Groundwater Flow Equation in Unconfined Aquifers and the Effect of Heterogeneity," *Water Resources Research*, Vol. 31, No. 11, November 1995, pp. 2733–2742.

Shashkov, M. and Steinberg, S. (1995), "Support-Operator Finite-Difference Algorithms for General Elliptic Problems," *Journal of Computational Physics*, Vol. 118, p. 131.

Shashkov, M. and Steinberg, S. (1996), "Solving Diffusion Equations with Rough Coefficients in Rough Grids," *Journal of Computational Physics*, Vol. 129, p. 383–405.

Shashkov, M. (1995), *Conservative Finite-Difference Methods on General Grids*, CRC Press, Boca Raton, Florida.

Shirazi, S. A. and Truman, C. R. (1989), "Evaluation of Algebraic Turbulence Models for PNS Predictions of Supersonic Flow Past a Sphere-Cone," *AIAA Journal,* Vol. 27, No. 5, pp. 560–568.

Steffen, C. J., Reddy, D. R., and Zaman, K. B. M. Q. (1995), "Analysis of Flowfield from a Rectangular Nozzle with Delta Tabs," AIAA Paper 95–2146, 26th AIAA Fluid Dynamics Conference, San Diego, California, June 19–22, 1995.

Steinberg, S. L. and Roache, P. J. (1985), "Symbolic Manipulation and Computational Fluid Dynamics," *Journal of Computational Physics*, Vol. 57, pp. 251–284.

Steinberg, S. and Roache, P. J. (1986a), "A Toolkit of Symbolic Manipulation Programs for Variational Grid Generation," AIAA Paper 86–0241, AIAA 24th Aerospace Sciences Meeting, Reno, Nevada, January 6–9, 1986.

Steinberg, S. and Roache, P. J. (1986b), "Using MACSYMA to Write FORTRAN Subroutines," *Journal of Symbolic Computation*, Vol. 2, pp. 213–216.

Steinberg, S. and Roache, P. J. (1990), "Symmetric Operators in General Coordinates," Ecodynamics Research Associates, Albuquerque, New Mexico.

Steinbrenner, J. P. and Chawner, J. R. (1992), "The GRIDGEN Version 8 Multiple Block Grid Generation Software," MDA Engineering Report 92–01, 1992.

Sudicky, E. A. and Frind, E. O. (1982), "Contaminant Transport in Fractured Porous Media: Analytical Solutions for a System of Parallel Fractures," *Water Resources Research*, Vol. 18, No. 6, 1982, pp. 1634–1642.

Tang, D. H., Frind, E. O., and Sudicky, E. A. (1981), "Contaminant Transport in Fractured Porous Media: Analytical Solution for a Single Fracture," *Water Resources Research*, Vol. 17, No. 3, 1981, pp. 555–564.

Thompson, J. F., Thames, F. C., and Mastin, C. W. (1974), "Automatic Numerical Generation of Body-Fitted Curvilinear Coordinate System for Field Containing Any Number of Arbitrary Two-Dimensional Bodies," *Journal of Computational Physics*, Vol. 15, 1974, pp. 299–319.

Thompson, J. F., Warsi, Z. U. A., and Mastin, C. W. (1985), *Numerical Grid Generation Foundations and Applications,* North-Holland, New York.

Walker, M. A. and Oberkampf, W. L. (1992), "Joint Computational/Experimental Aerodynamics Research on a Hypersonic Vehicle: Part 2, Computational Results," *AIAA Journal*, Vol. 30, No. 8, August 1992, pp. 2010–2116.

Westerink J. J. and Roache P. J. (1995), "Issues in Convergence Studies in Geophysical Flow Computations," in Johnson and Hughes (1995), pp. 1–6.

Wilcox, D. C. (1993), *Turbulence Modeling for CFD,* DCW Industries, Inc., La Cañada, California.

Winslow, A. (1967), "Numerical Solution of the Quasilinear Poisson Equations in a Nonuniform Triangle Mesh," *Journal of Computational Physics*, Vol. 2, 1967, pp. 149–172.

WIPP PA Dept. (1992), *Annual Performance Assessment for the Waste Isolation Pilot Plant, December 1992, Vol. 1: Preliminary Comparison with 40 CFR 191, Subpart B Vol. 2: Technical Basis,* Performance Assessment Department, Sandia National Laboratories, Albuquerque, New Mexico.

Yee, H. C. and Sweby, P. K. (1996), "Some Aspects of Numerical Uncertainties in Time-Marching to Steady-State Numerical Solutions," AIAA Paper 96–2052, 27th AIAA Fluid Dynamics Conference, New Orleans, Louisiana, June 17–20, 1996.

Zingg, D. W. (1991), "Viscous Airfoil Computations Using Richardson Extrapolation," AIAA Paper 91–1559–CP, AIAA 30th Aerospace Sciences Meeting, Reno, Nevada, January 6–9, 1992.

Zingg, D. W. (1992), "Grid Studies for Thin-Layer Navier-Stokes Computations of Airfoil Flowfields," *AIAA Journal*, Vol. 30, No. 10, October 1993, pp. 2561–2564. See also AIAA Paper 92–0184.

7

SINGLE GRID ERROR ESTIMATORS

This Chapter, taken primarily from Roache (1997), "Quantification of Uncertainty in CFD," covers methods of error estimation and error banding that do not involve systematic grid convergence testing. These methods are in Categories B, C, and D of the following taxonomy (previously given in Chapter 4) for sources of additional information for error estimation.

B. Additional Solution(s) of the Governing Equations on the Same Grid
   B.1 Higher Order Accuracy Solution(s)
   B.2 Lower Order Accuracy Solution(s)
C. Auxiliary PDE Solutions on the Same Grid
D. Auxiliary Algebraic Evaluations on the Same Grid; Surrogate Estimators
   D.1 Non-Conservation of Conservation Variables
   D.2 Non-Conservation of Higher Moments
   D.3 Zhu-Zienkiewicz Type Estimators
   D.4 Convergence of Higher Order Quadratures

As noted in Chapter 4, these methods (unlike Category A methods, grid convergence studies) cannot be used for Verification of Codes, but can be used with Verified Codes for the Verification of individual Calculations. They have the advantage of requiring no additional grid generation, but involve significant additional algorithm and code development, beyond that required for the basic solution code.

7.1 ERROR ESTIMATION FROM HIGHER OR LOWER ORDER ACCURACY SOLUTIONS ON THE SAME GRID (CATEGORY B)

7.1.1 Higher Order Accuracy Solution(s) (Category B.1)

Category B1 methods estimate the accuracy of a base solution by comparison with "higher order accuracy solution(s)." (Again, note the somewhat abusive but common terminology.) Richardson (1910) again scooped modern error estimation papers by inventing Category B.1, error estimation from higher order solutions on the same grid, noting that the difference between a 2nd-order accurate solution and a 4th-order accurate solution is itself an ordered error estimator.

The higher order accuracy solution might be obtained via a new solution of higher-order discretizations (FDM, FVM, FEM, etc.) or by deferred corrections, compact differences (again by direct or deferred corrections, etc.).

For the technique of error estimation from higher order solutions on the same grid, much the same advantages and limitations apply as with the grid convergence technique, i.e., it applies to all point values and functionals (lift, drag, etc.); x and t errors may be estimated independently or coupled; the error estimate includes nonlinear coupling. These methods are not commonly used because they require additional code capability, unlike grid convergence tests (Category A). On the other hand, these Category B methods do not require additional grid generation.

The development costs of the additional code capability may be reduced by noting several points. If we were intending to use the higher order solution itself, many restrictions and requirements could apply, such as full iterative convergence, strict conservation, etc. However, if our only use of the higher order solution is to estimate the error of the base solution, these considerations are not so important. The point is that error estimation of (say) a 2nd-order solution using 4th-order methods is less demanding numerically than obtaining a 4th-order solution for direct use. Also, though not generally recognized, directional splitting works (Roache, 1997). (The following development has not been completed for cross derivative terms, which in any case may require some careful formulation for higher order stencils and should be rigorously verified.)

We use a notation similar to that above for multiple grid solutions, with f denoting the exact solution, f_2 the 2nd-order accurate solution, and f_4 the 4th-order accurate solution. C_{2x} are the coefficients of the Taylor's theorem expansion for the 2nd-order solution in the x-direction, etc., and R_{2x} are the remaining terms in the complete 2nd-order expansion in x, etc. Then, in 2-D,

$$f = f_2 + C_{2x}\Delta x^2 + R_{2x} + C_{2y}\Delta y^2 + R_{2y} \tag{7.1.1.1}$$

$$f = f_4 + C_{4x}\Delta x^4 + R_{4x} + C_{4y}\Delta y^4 + R_{4y} \tag{7.1.1.2}$$

Defining the error of the 2nd-order solution E_2 to be

$$E_2 = f - f_2 \tag{7.1.1.3}$$

and substituting for f from Equation (7.1.1.2) we obtain

$$E_2 = f_4 - f_2 + \mathrm{O}(\Delta^4) \tag{7.1.1.4}$$

This is the basic (Richardson, 1910) result that the difference between the 2nd- and 4th-order solutions on the same grid is itself a 4th-order error estimate for the 2nd-order solution (a result that it hardly deserves to be named a theorem). Likewise obvious, if we define the error of the 4th-order solution E_4 to be

$$E_4 = f - f_4 \tag{7.1.1.5}$$

and substitute for f from Equation (7.1.1.2) we obtain

$$E_4 = f_2 - f_4 + \mathrm{O}(\Delta^2) \tag{7.1.1.6}$$

for a 2nd-order error estimator for the 4th-order solution.

The somewhat less obvious result is for the directional splitting of the higher order solutions, which can be much easier to implement than a fully directional higher-order solution. Let f_{4x} denote the solution obtained with 4th-order discretization in x and 2nd-order in y, and, f_{4y} denote the solution obtained with 4th-order discretization in y and 2nd-order in x. The C_{2x}, etc., coefficients are unchanged from the previous definitions. Then

$$f = f_{4x} + C_{4x}\Delta x^4 + R_{4x} + C_{2y}\Delta y^2 + R_{2y} \tag{7.1.1.7}$$

$$f = f_{4y} + C_{4y}\Delta y^4 + R_{4y} + C_{2x}\Delta x^2 + R_{2x} \tag{7.1.1.8}$$

We now estimate the difference between f_{4x} and f_2, using Equations (7.1.1.7) and (7.1.1.1) respectively. (If we dropped the higher order remainder terms at this point, we would show the resulting estimator to be 2nd-order accurate, but by retaining these terms presently, we will show that the estimator is 4th-order accurate.)

$$\begin{aligned} f_{4x} - f_2 &= f - C_{4x}\Delta x^4 - R_{4x} - C_{2y}\Delta y^2 - R_{2y} \\ &\quad -\left\{ f - C_{2x}\Delta x^2 - R_{2x} - C_{2y}\Delta y^2 - R_{2y} \right\} \\ &= C_{2x}\Delta x^2 + R_{2x} - C_{4x}\Delta x^4 - R_{4x} \end{aligned} \tag{7.1.1.9}$$

Similarly, we estimate the difference between f_{4y} and f_2 using Equations (7.1.1.8) and (7.1.1.1) respectively.

$$\begin{aligned} f_{4y} - f_2 &= f - C_{4y}\Delta y^4 - R_{4y} - C_{2x}\Delta x^2 - R_{2x} \\ &\quad -\left\{ f - C_{2x}\Delta x^2 - R_{2x} - C_{2y}\Delta y^2 - R_{2y} \right\} \\ &= C_{2y}\Delta y^2 + R_{2y} - C_{4y}\Delta y^4 - R_{4y} \end{aligned} \tag{7.1.1.10}$$

Adding (7.1.1.9) and (7.1.1.10) gives

$$\begin{aligned} f_{4x} + f_{4y} - 2f_2 &= C_{2x}\Delta x^2 + R_{2x} + C_{2y}\Delta y^2 + R_{2y} \\ &\quad - C_{4x}\Delta x^4 - R_{4x} - C_{4y}\Delta y^4 - R_{4y} \end{aligned} \tag{7.1.1.11}$$

Comparing to Equation (7.1.1.1), the first 4 terms on the RHS of (7.1.1.11) are identically equal to $(f - f_2)$ without approximation. (That is, the remainder terms R_{2x} and R_{2y} are included.) The last 4 terms on the RHS of (7.1.1.11) are $O(\Delta^4)$. With the definition (7.1.1.3), we have the error of the 2nd-order solution E_2 estimated as

$$E_2 = f_{4x} + f_{4y} - 2f_2 + O\left(\Delta^4\right) \tag{7.1.1.12}$$

Thus the estimate for the 2nd-order solution can be obtained to 4th-order by directionally split 4th-order solutions.

7.1.2 Lower Order Accuracy Solution(s) (Category B.2)

Certainly a lower-order solution on the same grid could be used to estimate the error of the higher order solution (Category B.2), in the same way that grid coarsening can be used rather than grid refinement, but I know of no such applications.

7.2 AUXILIARY PDE SOLUTIONS ON THE SAME GRID (CATEGORY C)

The technique of Auxiliary PDE Solutions on the Same Grid does not simply involve a local evaluation of something. The key aspect here is that errors are transported - advected, diffused, etc. [Babuska et al (1994,1997) refer to this phenomenon as "error pollution."]

Most of the papers using this approach have appeared in the FEM literature, but the concepts are equally applicable to FDM and FVM. An early development for FDM limited to cartesian grids was presented by Schonauer et al (1981). Here, we follow Van Straalen et al (1995) and introduce the concepts with the simple non-conservation form of the steady-state 2-D linear advection-diffusion equation with a source term,

$$V\nabla\Phi - \nabla(\Gamma\nabla\Phi) = q \qquad (7.2.1)$$

where V is the advection velocity, Φ is the concentration of the transported scalar, Γ is the diffusivity, and q is a source of Φ (per unit mass). We write a simplified form of (7.1.13) using $\Im$ to represent the entire partial differential equation operator.

$$\Im(\Phi) = q \qquad (7.2.2)$$

Let ϕ_{approx} be some *continuum* approximate solution of Equation (7.2.1) obtained numerically. It will not be sufficient to consider only point-wise (or node-wise) numerical solutions; rather, ϕ_{approx} must be defined in the operator domain of $\Im$, i.e., for this example problem, it must be a 2-D continuous function with a least second degree derivatives. (How this continuum approximate solution will be generated from nodal FDM or FVM solutions, which we denote by ϕ_{nodal} for the sake of emphasis, will be discussed momentarily.) Then the error function is defined as $\varepsilon \equiv \Phi - \phi_{approx}$ and from the properties of linear operators we can define the so-called "*Exact Operator Residual*" as

$$R(\phi_{approx}) \equiv \Im(\phi_{approx}) - q \qquad (7.2.3)$$

The terminology of "*Exact Operator Residual*," though common, is somewhat misleading, in my opinion, because it is "exact" only for a given continuum ϕ_{approx} but is not uniquely defined for any ϕ_{nodal}. That is, if the ϕ_{nodal} we obtain from a CFD code is all we have, we cannot define an operator residual until we extend ϕ_{nodal} to ϕ_{approx} and this procedure is somewhat arbitrary. For example, generation of ϕ_{approx} by piecewise linear interpolation (i.e., basis functions) will not be adequate for the method as described so far, because derivatives of this ϕ_{approx} will generally not exist at cell interfaces, and such ϕ_{approx} will not be in the operator domain of $\Im$.

Once we have defined a ϕ_{approx} from ϕ_{nodal} and defined $R(\phi_{approx})$, we note that the error ε is given *exactly* by the solution of the error equation,

$$\Im(\varepsilon) = -R(\phi_{approx}) \qquad (7.2.4)$$

The numerical solution of this equation provides the "Auxiliary PDE Solutions on the Same Grid" that constitutes Category C of error estimators.

The auxiliary error equation (7.2.4) utilizes the exact operator residual R as a source term just as the original PDE (7.2.2) contains the source term q. This provides an instructive conceptual model for the error propagation process; the exact operator residual R is a "source" for the global error distribution, which is transported by advection and diffusion. Clearly, the solution of the error equation is non-local.

Following Van Straalen et al (1995), we also distinguish the Exact Operator Residual (7.2.3) from the *Approximate Operator Residual* r_L which also appears in the numerical analysis literature, defined as

$$r_L = L(\Phi) - q \tag{7.2.5}$$

where L is the discretized operator that one has solved, i.e. that approximates $\Im$. Study of the residual r_L leads to a discussion of *order*, since r_L is typically expressed as a Taylor-series expansion about the exact solution. For example, see Ferziger (1988).

The numerical solution of Equation (7.2.4) provides the approximate global solution for the exact error ε. There are two decisions to be made on the approximations: extending ϕ_{nodal} to ϕ_{approx} and then a discretization of $\Im(\varepsilon)$, which is (virtually) the same decision one had to make for the original problem. (The differences are that boundary conditions for the error equation will be homogeneous, and any algorithmic aspect affected by the magnitude of the solution, e.g. nonlinear flux limiters, will be triggered differently.) Thus, many variations on this methodology are possible. By considering an integral form of the error equation (7.2.4), Van Straalen et al (1995) eliminate the requirement for smoothness in ϕ_{approx} and use piecewise linear functions. This is a methodology from FEM theory, including the ubiquitous integration by parts, adapted to FVM. Note, however, the following universal point in all these methods. It is essential that the discretizations used to define ϕ_{approx} and the operator in the error equation be distinct from those used on the original PDE, otherwise the method becomes a meaningless circular definition. In Van Straalen et al (1995), the error estimation works only because the original PDE is discretized using 1st-order FVM for advection, and advection is the dominant error source, and the error equation is solved using 2nd-order methods.

The procedure for error estimation in Category C (Auxiliary PDE Solutions on the Same Grid) then involves selection of an alternate functional form for the inter-cell (inter-element) solution and evaluation of the (differential) residual. It is possible that the derivatives may not exist at element boundaries. Whether one selects a C^2 smooth function to begin with, or one blends (interpolates) the differentials obtained at element (or cell) centers to produce smooth functions over the entire regions, or puts off such questions to the quadrature rules, etc. is not really very significant. The residuals are then transported in an error equation for Category C methods; alternately, they could be used as local error estimators (as in Category D methods discussed below) which is indeed a significant distinction. The major point (for both Category C and D methods) is that the residual evaluation must involve a discretization rule different from that used in obtaining the solution. In this aspect, these methods are after all not so distinct from category B methods, which use higher (or lower) order stencils to evaluate the error. In category C, a higher order stencil is used to obtain a higher-order solution, but if the stencil were used only locally to obtain a local residual, it would be like Category D. The approach in which a rigorously iterated solution is replaced with a less completely converged solution is intermediate, the "globalness" of the error estimate now being dependent on the amount and type of iteration.

For examples of the error transport equation approach in FDM and FVM, see Schonauer et al (1981), Ferziger (1993), Van Straalen et al (1995). The FEM literature is extensive, e.g. see Mills (1987), Babuska et al (1994,1997), Strouboulis and Oden (1990), Ewing (1990), Padra and Larreteguy (1995). In all cases, the error transport equation will be some linearized version of the governing equations. It will be somewhat cheaper to solve than the original fluid dynamics equations (e.g., full Navier-Stokes) but, as expected, will be a less accurate error estimator near boundary layer separations, etc. This approach is truly global, but not as reliable Category A or B methods.

7.3 AUXILIARY ALGEBRAIC EVALUATIONS ON THE SAME GRID: SURROGATE ESTIMATORS (CATEGORY D)

Category D methods, involving auxiliary algebraic evaluations on the same grid, are relatively cheap, need no additional grid generation, and (sometimes) use no significant dynamic memory. However, the "error" evaluated usually has no direct relation to any error measure of engineering or scientific interest; hence my use of the term "Surrogate Estimators." An alternate and preferable terminology would be "single grid error *indicator*" or "*index*" rather than error *estimator*.

7.3.1 Non-Conservation of Conservation Variables (Category D.1)

Category D.1, non-conservation of conservation variables, applies only to codes which do not use fully conservative algorithms for "conservation variables." For example, mass is not identically conserved in most pre-1980 boundary layer codes, in many FEM codes, in codes based on collocation methods (including the very accurate wavelet methods, e.g. Vasilyev et al, 1995; Vasilyev and Paolucci, 1996) as well as other CFD codes. The error estimate then involves numerical evaluation by quadrature for the erroneous loss or gain of mass in the computational domain. If the coded algorithm is consistent, this error will $\rightarrow 0$ but only in the limit of $\Delta \rightarrow 0$ for a non-trivial problem. (Even for a nominally mass-conserving full Navier-Stokes code, the satisfaction of mass conservation will usually depend on the degree of strict iterative convergence achieved, but this gives no indication of the discretization error of the solution, which is our interest here.) Momentum, vorticity, or internal energy are other possibilities. Note that the evaluation of this error depends on the accuracy of the quadrature, which probably should be consistent with the algorithm for solving the PDE's (see also discussion of D.4 below) but this does not appear to be a strict requirement. Conservative codes and algorithms are generally preferred, but non-conservative codes are seen to offer a readily evaluated and understood (although still surrogate) error measure.

In an earlier section, I noted that, if there is any suspicion that the grid resolution is not in the asymptotic range, three grid solutions are necessary to verify (or determine) the rate of convergence and thereby estimate the error, when the exact solution is not known. The qualifier used, "when the exact solution is not known," may seem redundant when the context is a realistic problem, but there is an important and useful distinction to be made in regard to Category D.1 and D.2 methods that are based on conservation errors. Although the conservation imbalance may not be of direct engineering or scientific relevance, it has the advantage that the exact solution of this quantity *is* known, namely zero. Thus, in these surrogate error measures based on conservation errors, one requires only a single grid solution to calculate the surrogate error, and only two grid solutions to extract the observed convergence rate p. The same is true if other functionals have known values, e.g. the drag coefficient = 0 for symmetric airfoils in inviscid flow, and can be used as an exact solution value to monitor convergence (Jameson and Martinelli, 1996).

7.3.2 Non-Conservation of Higher Moments (Category D.2)

Category D.2 involves evaluation of conservation errors for higher-order moments. In a typical turbulent (Reynolds-Averaged) Navier-Stokes code, fully conservative discretization may be used for the "primitive" variables of mass, momentum, and internal energy. However, turbulent kinetic energy is typically not identically conserved. An evaluation of its global and local conservation by quadrature, including carefully evaluated boundary inflow terms and dissipation (Haworth et al, 1993; Chang and Haworth, 1995, 1997), then gives a surrogate indication of general discretization errors. In Chang and Haworth (1995, 1997) it is also used to reliably guide local grid refinement for solution adaptive grid generation.

The fully detailed balance equations for linear momentum, angular momentum, mean flow kinetic energy, and turbulence kinetic energy have been presented by Haworth et al (1990, 1993). For clarity, following Chang and Haworth (1995, 1997), we restrict attention in this example to steady, laminar, incompressible flows, but this is not a limitation of the methodology. In laminar flow, the turbulence kinetic energy is of course zero, and we drop the distinction between mean and instantaneous quantities. The total kinetic energy in an arbitrary volume V is K,

$$K \equiv \int_V \frac{1}{2} \rho U_j U_j d\upsilon \qquad (7.3.2.1)$$

The kinetic energy budget for an arbitrary volume V with bounding surface S can be written symbolically as

$$\dot{K} = FLUX + PRES + SHEAR + DISS \qquad (7.3.2.2)$$

The terms on the right-hand-side represent the rate at which kinetic energy is advected out of V through S, ($FLUX$), the rate at which pressure forces on S extract kinetic energy from V, ($PRES$), the rate at which viscous stresses over S extract kinetic energy from V, ($SHEAR$), and the rate at which viscous stresses convert kinetic energy to sensible energy (heat) over the interior of the volume *V*, or viscous dissipation, (*DISS*).

In the FVM methods used by Haworth et al, mass and momentum are conserved at the cell level, but kinetic energy is not conserved. The left-hand-side $\dot{K}$ represents imbalance in kinetic energy resulting from discretization of the momentum equation. For any convergent discretization, $\dot{K}$ approaches zero in the limit as the grid spacing approaches zero, and departures from zero are suitable surrogate error measures.

The full evaluation of $\dot{K}$ for turbulent flow requires a fairly complex code development in itself, and the correlation with useful engineering measures of accuracy must be established by suites of computations for any new class of problems. However, this approach has some significant advantages. As noted above in Section 7.3.1, in these surrogate error measure based on conservation errors, one requires only a single grid solution to calculate the error, and only two grid solutions to extract the observed convergence rate *p*. Also, Haworth (1993) has noted that, in his experience with turbulent combustion problems, when the grid refinement is sufficient to clearly exhibit asymptotic behavior, the numerical accuracy is already higher than he would like to pay for. Other CFD practitioners have noted the same experience in other turbulent flow problems; see previous discussion in Chapter 6, Section 6.26, especially Blottner (1990) and Oberkampf et al (1995). While systematic grid convergence tests are reliable and self-contained (i.e., they do not depend on establishing correlations with some other accuracy study) they often do not become reliable (i.e., exhibit clear asymptotic performance) until the accuracy is somewhat excessive for engineering purposes. In Haworth's experience, the kinetic energy imbalance provides a more economical indicator.

7.3.3 Zhu-Zienkiewicz Type Estimators (Category D.3)

Category D.3, the Zhu-Zienkiewicz type Estimators, were developed and intended primarily to guide grid adaptation. Zhu-Zienkiewicz (ZZ) estimators are not a single method, but a rather extensive family of methods. See Zhu and Zienkiewicz (1990), Zienkiewicz and Zhu (1987, 1992), Wu et al (1990), Hetu and Pelletier (1992), Pelletier and Ilinca (1994,1997), Ilinca et al (1995), Pelletier et al (1995), Pelletier and Ignat (1995), and other FEM methods that have a similar flavor such as those of Strouboulis and Oden (1990), Oden et al (1993), Babuska et al (1994,1997), Ewing et al (1990), and Hugger (1997).

Hugger (1997) gives the following general description of the "equilibrium" or "equilibration" method. "First postulate an error estimator (possibly justified by physical arguments); then prove that the estimated error is close (upper and/or lower bound in an *a priori* given norm, generally the energy norm) to the exact error under certain regularity assumptions." These methods allow the global energy norm to be well estimated and often give good evaluation of local errors (and provide the local estimate of stress accuracy, certainly important for structures problems). More relevant in the present context of the Quantification of Uncertainty, these estimators can be used as surrogate error estimators, and are often cheap compared to direct solution of FEM equations using inefficient solvers. However, when extended to FDM and FVM (or to efficient FEM solvers) their cost, when amortized only over the more efficient solver, is not insignificant (Pelletier, 1996). They still have the significant advantage compared to grid convergence studies of not requiring additional grid generation and multiple grid runs.

The ZZ family of estimators can be described in broad terms (Pelletier, 1996) by reference to the following 4 steps described for the simple non-conservation form of the steady-state linear advection-diffusion equation with a source term, this time written simply in 1-D for constant diffusivity.

$$V\frac{d\Phi}{dx} = \Gamma\frac{d^2\Phi}{dx^2} + q(x) \tag{7.3.3.1}$$

where V is the advection velocity, Φ is the concentration of the transported scalar, Γ is the diffusivity, and q is a source of Φ (per unit mass). Assume that the solution is obtained using linear elements, so that the continuum approximate solution Φ is obtained with linear interpolation between nodes.

Step 1. Perform a least-squares local projection of derivatives in the form $d\Phi^*/dx = ax + b$.

Step 2. Evaluate the projected derivatives at the nodes.

Step 3. Evaluate the error estimate on the elements.

$$\|e_i\|^2 = \int_{x(i)}^{x(i+1)} \left(\frac{dC^*}{dx} - \frac{dC}{dx}\right)^2 dx \tag{7.3.3.2}$$

Step 4. Evaluate the global error estimate by summing over the *NEL* elements.

$$\|e_{total}\|^2 = \sum_{k=1}^{NEL} \|e_k\|^2 \tag{7.3.3.3}$$

Each of these 4 steps allows a variety of implementations, leading to the extensive family of ZZ estimators.

Although developed for FEM, the Zhu-Zienkiewicz approach is easily adapted to FDM and FVM (Pelletier, 1996) using the concepts discussed above in Section 7.2 on residual evaluation. Pelletier and Ignat (1995) have shown a good correlation of their ZZ estimator with the GCI (Chapter 4) for unstructured grid turbulent flow problems. ZZ methods naturally provide error estimates for the derivative of the solution, but sometimes can be structured to provide error estimates for flux or for the dependent variable itself (Pelletier, 1996).

The Zhu-Zienkiewicz type methods share the shortcoming of all surrogate indicators for fluid dynamics; other than guidance for grid adaptation, there is little inherent engineering or scientific interest in the error measure as defined. Therefore, unless the only interest is mathematics for its own sake, it is necessary to establish a correlation of the Zhu-Zienkiewicz indicators with an error measure of interest. This can only be accomplished by expensive numerical experimentation for a given class of problems. For example, it would be highly unlikely that a correlation based on numerical experiments for internal combustion engine modeling would provide any guidance for external aerodynamics, nor even for a large range of flow parameters for geometrically similar problems. However, when one is involved in extensive suites of calculations, one may build up such correlations by experience, and arrive at a practical and relatively inexpensive single-grid error estimator. The ZZ family does not account for transport of errors; although Step 4 above gives a "global" error estimate, this evaluation is simply a summing of local error terms, and does not account for advection, diffusion, etc.

This need for establishing a correlation in error measures is also true if one direct error measure (say, obtained by a grid convergence study) is to be used as a surrogate for another. For example, in an airfoil calculation, does a 1% error estimate on C_L insure a 1% (or 5%, etc.) error estimate for C_M? The correlation providing the (fuzzy) answer will only be accurate for a restricted range of angle of attack, Reynolds number, Mach Number, and airfoil class.

One can just as well use the various weight functions designed to guide solution adaptivity for single-grid estimators, e.g. Lee and Yeh (1994a,b). These are loosely related to Zhu-Zienkiewicz methods and carry the same caveats. But, as noted previously, this task of grid adaptation as typically practiced has little connection to the Quantification of Uncertainty for a final calculation with a useful error measure, and the success of these local "error" estimators in guiding grid adaptation must not be taken as demonstrations of their efficacy for the Quantification of Uncertainty. See discussion in Chapter 4, Section 4.2.

Finally, the inherent limitation of the energy norm itself, the basis of Zhu-Zienkiewicz and similar error indicators, is easily demonstrated by the following counter-example (MacKinnon, 1996). Consider the following simple 1-D constant coefficient advection-diffusion-source equation, contrived to produce a quadratic solution.

$$-f_{xx} + cf_x = g, \quad 0 \le x \le 1 \tag{7.3.3.4}$$

$$f(0) = 0, \quad f'(1) = 1 \tag{7.3.3.5}$$

$$g = -2 + 2cx \tag{7.3.3.6}$$

These produce the quadratic solution

$$f = x^2 \tag{7.3.3.7}$$

so that use of a 2nd-order discretization (say, centered differences with Dirichlet boundary conditions and the usual node-on-boundary configuration) produces the exact answer. In spite of the answer being exact, the commonly used error norms are *not* zero. (They do, however, converge toward zero as the grid is refined.) By contrast, an error estimate based on grid convergence (e.g., the GCI of Chapter 5) will produce the correct zero error estimate.

7.3.4 Convergence of Higher Order Quadratures (Category D.4)

Category D.4 is the simplest, involving convergence of higher order quadratures. For example, quadrature (numerical integration of a known function) for pressure and shear force on the surface leads to evaluation of a drag coefficient C_D. For non-trivial problems (i.e., problems with significant solution structure), 2nd-order quadrature will give a different answer than 4th-order accurate quadrature. As the discretization refines, these two quadratures will converge to each other. For coarse grids, the difference may be used as a surrogate error estimator.

The same philosophy has been used point-wise. The simplest approach ultimately reduces to comparing the calculated pointwise value of some variable of interest with the value interpolated between the neighbors of the point. Indeed, these will converge as the solution converges, and therefore can (and have) been used as surrogate error estimates. However, the poverty of the concept is shown by the fact that the "error estimate" is obtained without recourse to the governing PDE's. It is difficult to take seriously an error estimate for point-wise pressures from an incompressible Navier-Stokes code when the same error estimate would obtain from the same pointwise values in a hypersonic solution or in Darcy flow in a porous medium. In fact, since the governing PDE's are not invoked, these "error estimators" would be better named simply "resolution indicators."

7.4 TIME ACCURACY ESTIMATION

Estimation of the numerical error of the time discretization can be performed in the same manner as the spatial errors, but it is also possible to use simpler methods. In fact, it is relatively straightforward to estimate the temporal error as the calculation evolves, and to build a code with a solution adaptive time step to control the temporal error to a predetermined level, even when only 1st-order time differencing is used.

Consider the following inexpensive temporal error estimator (Oden et al, 1993; Roache, 1993a,b) for a fully implicit (backward) time differencing method. We write the backward time method for a general equation system in terms of an operator L (not necessarily linear) as

$$\frac{f_{i,j}^{n+1} - f_{i,j}^{n}}{\Delta t} = L^{n+1} \qquad (7.4.1)$$

Generally, the solution would appear to involve an expensive matrix solution for all f^{n+1}. The inexpensive time error estimator uses the difference between backward and forward time integration, implemented as an extrapolation. The method is very cheap to implement because it does not require another implicit matrix solution, nor even another explicit stencil evaluation. It includes the effects of time-dependent boundary conditions and source terms.

In the current time step, Equation (7.4.1) is advancing the solution for f from time level n to time level $(n + 1)$ with increment Δt using fully implicit (backward) time differencing, so that the operator L is being evaluated at $(n + 1)$. In the previous time step, the solution was advanced from $(n - 1)$, relative to the current indexing, to n with increment Δt_{OLD} and the L evaluated at n. We *could* explicitly evaluate L^n and make a separate, parallel estimate of the values f^{n+1} with an explicit step, as in

$$\frac{F_{i,j}^{n+1} - f_{i,j}^{n}}{\Delta t} = L^n \tag{7.4.2}$$

where F^{n+1} signifies the new value predicted by the explicit algorithm. (Since L^n involves values of f only at the location (i, j) but not at the neighboring locations $(i \pm 1, j)$ and $(i, j \pm 1)$ the equation is explicit, i.e., it does not involve a matrix solution of f at all values of i and j simultaneously.) The difference between the new value of f^{n+1} predicted by the implicit algorithm and F^{n+1} predicted by the explicit algorithm, both of which predictions are $O(\Delta t)$ accurate, is itself an error estimator of accuracy $O(\Delta t^2)$ for the time discretization error for that time step.

Explicit evaluation of L^n would not be expensive compared to the computer time necessary for the implicit matrix solution, but it does involve coding storage penalties and complexities (storage of old values of boundary conditions, source terms, etc.) An even more economical and elegant approach is to recognize that the L from the *previous implicit* step is identical to the L for the *present explicit* step. Thus, the L for the present explicit step can be evaluated from the knowledge of the previous change in f, requiring only the temporary storage of previous solution arrays. The explicit solution for f at $(n + 1)$, F^{n+1}, is then obtained by simple linear extrapolation of previous solutions. For constant Δt,

$$F_{i,j}^{n+1} = 2f_{i,j}^{n} - f_{i,j}^{n-1} \tag{7.4.3}$$

For the more general case of variable Δt,

$$F_{i,j}^{n+1} = f_{i,j}^{n} + \frac{\Delta t}{\Delta t_{OLD}}\left(f_{i,j}^{n} - f_{i,j}^{n-1}\right) \tag{7.4.4}$$

The pointwise difference $F^{n+1} - f^{n+1}$ is then a pointwise temporal error estimator of $O(\Delta t^2)$.

Usually, a user would be interested in the maximum over the spatial domain of the percentage error, so one would evaluate the error estimator E_{ET} for the single time step as

$$E_{ET} = 100 \cdot \max_{i,j}\left\{abs\left(F_{i,j}^{n+1} - f_{i,j}^{n+1}\right)\right\} / f_{RANGE} \tag{7.4.5}$$

E_{ET} is thus calculated as the % maximum deviation (or L^{∞} norm) of the absolute value of the difference between the new f values f^{n+1}, predicted by the fully implicit algorithm, and F^{n+1}, predicted by the explicit algorithm, normalized by f_{RANGE}, which is the total range of f^{n+1}. (In practice, ghost point evaluations with Dirichlet boundary conditions tend to exaggerate the error, so error estimates and f_{RANGE} should be calculated only over interior points.)

Note that the explicit calculation is used only as an error estimator within a time step of an implicit method, not as the solution algorithm, i.e., its effects do not accumulate. Thus, stability limitations and/or conservation issues of explicit time stepping as a solution algorithm are irrelevant.

Although the extrapolation procedure is equivalent to explicit time-stepping, the extrapolation cannot be started until there are two time levels. Also, in the event that the initial conditions are set arbitrarily by the analyst (without setting initial conditions as a steady-state solution), it is likely that the initial conditions are incompatible with the boundary conditions applied at the first time step. This means that the change in boundary values during the first time step is virtually fixed, i.e., does not depend on the time resolution. Consequently, the first time step would not provide a meaningful estimate of $\partial f / \partial t$ and the error estimator would be inaccurate. Thus for this common situation, the time error estimation by extrapolation cannot be applied until the fourth time step.

E_{ET} can readily be used (internal to the code) as the basis for a solution-adaptive time-stepping algorithm, adjusting Δt so that the error estimate E_{ET} is acceptable, either by re-calculating the previous time step (preferred) or by simply adjusting the next time.

7.5 CONCLUDING REMARKS ON SINGLE-GRID ERROR ESTIMATORS

Although single-grid error estimators are not as reliable as classical grid convergence studies, they all have a significant advantage. Because they require only a single grid, they can more easily be incorporated into codes for users other than the code developer, e.g. commercial codes. I strongly believe that some single-grid error estimator should be included in *any* commercial code, and that the error estimate output should *not* be an option, but an automatic hard-wired feature of all calculations.

Also, although my own experience and preference has always been for systematic grid convergence studies (e.g., Chapters 4, 5, 6), they too have their difficulties. As Haworth (1993) has pointed out to me, "The difficulties of grid generation and of ensuring that one lies within the radius of convergence cannot be dismissed lightly. Unstructured hex-cell grid generation remains a manually intensive process." For his work on internal combustion engines (Haworth et al, 1993; Chang and Haworth, 1995, 1997) the "minimum allowable mesh density is determined not only by accuracy considerations but also by the need to resolve geometric features" and by the "requirement that mesh integrity...be maintained as the grid deforms to accommodate valve and piston motion." Although automatic unstructured grid generation has been claimed in the FEM literature for decades, the experience of Haworth (and many others) remains that unstructured grid generation for such moving boundary problems requires weeks, even months. "Until robust auto-hex generation for arbitrary configurations is a reality, these applications are unlikely candidates for multiple solutions on multiple grids." [Of course, the time-step convergence test is cheap, and involves no problem of spatial grid generation.]

From an algorithmic viewpoint, *automatic* grid convergence testing is also possible. Some CFD developers would claim that this is already accomplished by solution-adaptive grid generation (of the h and/or p types) but, as already noted, the "error" upon which adaptive gridding is based is usually not a kind of measure of accuracy of interest to engineering or science code users. Automatic grid convergence testing is straightforward conceptually, but unfortunately the data structures of scientific programming languages make such codes difficult to develop and maintain. The SciNapse system under development by SciComp, Inc. (see Akers et al, 1997) will address automated grid convergence testing.

REFERENCES FOR CHAPTER 7

Akers, R. L., Kant, E., Randall, C. J., Steinberg, S. and Young, R. L. (1997), "SciNapse: A Problem-Solving Environment for Partial Differential Equations," *IEEE Computational Science and Engineering*, Vol. 4, No. 3, July–September 1997, pp. 32–42.

Babuska, I. , Strouboulis, T. and Upadhyay, C. S. (1994), "A Model Study of the Quality of A Posteriori Error Estimators for Linear Elliptic Problems. Error Estimation in the Interior of Patchwise Uniform Grid of Triangles," *Computer Methods in Applied Mechanics and Engineering*, Vol. 114, 1994, pp. 307–378.

Babuska, I. , Strouboulis, T. Gangaraj, S. K. and Upadhyay, C. S. (1997), "Pollution Error in the *h*-Version of the Finite Element Method and the Local Quality of the Recovered Derivatives," *Computer Methods in Applied Mechanics and Engineering*, Vol. 140, 1997, pp. 1–37.

Blottner, F. G. (1990), "Accurate Navier-Stokes Results for the Hypersonic Flow over a Spherical Nosetip," *Journal of Spacecraft and Rockets,* Vol. 27, No. 2, March–April 1990, pp. 113–122.

Chang, S. and Haworth, D. C. (1995), "Kinetic-Energy-Balance Based Solution-Adaptive Mesh Refinement," Joint JSME-ASME Fluid Mechanics Meeting, August 14–18, 1995, *Quantification of Uncertainty in Computational Fluids Dynamics–1995*, ASME FED Vol. 213, Hilton Head, South Carolina, R. W. Johnson and E. D. Hughes, Eds., pp. 7–12.

Chang, S. and Haworth, D. C. (1997), "Adaptive Grid Refinement Using Cell-Level and Global Imbalances," *International Journal for Numerical Methods in Fluids*, Vol. 24, pp. 375–392.

Ewing, R. E. (1990), "A Posteriori Error Estimation," *Computer Methods in Applied Mechanics and Engineering*, Vol. 82, pp. 59–72.

Ferziger, J. H. (1988), "A Note on Numerical Accuracy," *International Journal for Numerical Methods in Fluids*, Vol. 8, pp. 995–996.

Ferziger, J. H. (1993), "Estimation and Reduction of Numerical Error," *FED Vol. 158, Symposium on Quantification of Uncertainty in Computational Fluid Dynamics,* ASME Fluids Engineering Division, Summer Meeting, Washington, D.C., June 20–24, 1993, pp. 1–8.

Haworth, D. C., El Tahry, S. H., Huebler, M. S., and Chang, S. (1990), "Multidimensional Port-and-Cylinder Flow Calculations for Two- and Four-Valve-Per-Cylinder Engines: Influence of Intake Configuration on Flow Structure," SAE Paper No. 900257.

Haworth, D. C., El Tahry, S. H., and Huebler, M. S. (1993), "A Global Approach to Error Estimation and Physical Diagnostics in Multidimensional Computational Fluid Dynamics," *International Journal for Numerical Methods in Fluids*, Vol. 17, 1993, pp. 75–97.

Haworth, D. C. (1993), personal communication.

Hetu, J.-F. and Pelletier, D. H. (1992), "Adaptive Remeshing for Viscous Incompressible Flows," *AIAA Journal*, Vol. 30, 1992, pp. 1986–1992.

Hugger, J. (1997), "The Equilibrium Method for Local A Posteriori, Pointwise Estimation of the Finite Element Error," *Computer Methods in Applied Mechanics and Engineering*, Vol. 145, Nos. 3–4, 1997, pp. 263–283.

Ilinca, A., Camarero, R., Trepanier, J. Y., and Reggio, M. (1995), "Error Estimator and Adaptive Moving Grids for Finite Volumes Schemes," *AIAA Journal*, Vol. 33, No. 11, November 1995, pp. 2058–2065.

Jameson, A. and Martinelli, L. (1996), "Mesh Refinement and Modeling Errors in Flow Simulation," AIAA Paper 96–2050, 27th AIAA Fluid Dynamics Conference, New Orleans, Louisiana, June 17–20, 1996.

Lee, D. and Yeh, C. L. (1994a), "A Hybrid Adaptive Gridding Procedure for Three-Dimensional Flow Problems," *Computers and Fluids*, Vol. 23, 1994, pp. 39–53.

Lee, D. and Yeh, C. L. (1994b), "Computation of Turbulent Recirculating Flows Using a Hybrid Adaptive Grid," *Numerical Heat Transfer, Part A*, Vol. 26, 1994, pp. 415–430.

MacKinnon, R. (1996), personal communication.

Mills, R. J. (1987), *A Posteriori Error Estimates in Computational Fluid Mechanics*, Ph.D. Dissertation, University of New South Wales, School of Mechanical and Industrial Engineering, Sydney, Australia, March 1987.

Oberkampf, W. L., Blottner, F. G., and Aeschliman, D. P. (1995), "Methodology for Computational Fluid Dynamics Code Verification/Validation," AIAA Paper 95–2226, 26th AIAA Fluid Dynamics Conference, San Diego, California, 19–22 June 1995.

Oden, J. T., Kennon, S. R., Tworzydlo, W. W., Bass, J. M., and Berry, C. (1993), "Progress on Adaptive *h-p* Finite Element Methods for the Incompressible Navier-Stokes Equations," *Computational Mechanics*, Vol. 11, 1993, pp. 421–432.

Padra, C. and Larreteguy, A. (1995), "A-Posteriori Error Estimator for the Control-Volume Finite-Element Method as Applied to Convection-Diffusion Problems," *Numerical Heat Transfer, Part B*, Vol. 27, pp. 63–80.

Pelletier, D. and Ilinca, F. (1994), "Adaptive Remeshing for the *k*-ε Model of Turbulence," AIAA Paper 94–0818, 32nd AIAA Aerospace Sciences Meeting and Exhibit, Reno, Nevada, January 10–13, 1994.

Pelletier, D., Ignat, L. and Ilinca, F. (1995), "An Adaptive Finite Element Method for Conjugate Heat Transfer," AIAA Paper 95–0637, AIAA 33rd Aerospace Sciences Meeting and Exhibit, Reno, Nevada, January 9–12, 1995.

Pelletier, D. and Ignat, L. (1995), "On the Accuracy of the Grid Convergence Index and the Zhu-Zienkiewicz Error Estimator," Joint JSME–ASME Fluid Mechanics Meeting, August 14–18, 1995, *Quantification of Uncertainty in Computational Fluids Dynamics–1995*, ASME FED Vol. 213, Hilton Head, South Carolina, R. W. Johnson and E. D. Hughes, Eds., pp. 31–36.

Pelletier, D. (1996), personal communication.

Pelletier, D. and Ilinca, F. (1997), "Adaptive Remeshing for the *k*-ε Model of Turbulence," *AIAA Journal*, Vol. 35, No. 4, April 1997, pp. 640–646.

Richardson, L. F. (1910), "The Approximate Arithmetical Solution by Finite Differences of Physical Problems Involving Differential Equations, with an Application to the Stresses in a Masonry Dam," *Transactions of the Royal Society of London, Series A*, Vol. 210, pp. 307–357.

Roache, P. J. (1992), "Computational Fluid Dynamics Algorithms Developed for WIPP Site Simulations," in Russell, T. F. et al, Eds., *Proceedings of the IX International Conference on Computational Methods in Water Resources,* Denver, Colorado, Elsevier Applied Science, New York, June 9–12, 1992, pp. 375–382.

Roache, P. J. (1993a), "The SECO Suite of Codes for Site Performance Assessment," *Proceedings of the 1993 International High-Level Radioactive Waste Management Conference,* Las Vegas, Nevada, April 26–30, 1993.

Roache, P. J. (1993b), "The SECO Code Algorithms for Groundwater Flow and Transport," *Finite Elements in Fluids: New Trends and Applications*, Part II, Proceedings of the VIII International Conference on Finite Elements in Fluids, Univ. Politecnica de Cataluna, Barcelona, Spain, K. Morgan, E. Onate, J. Periaux, J. Peraire, and O. C. Zienkiewicz, Eds., 20–23 September 1993, pp. 939–948.

Roache, P. J. (1997), "Quantification of Uncertainty in Computational Fluid Dynamics," *Annual Review of Fluid Mechanics*, Vol. 29, 1997, pp. 123–160.

Schonauer, W. (1981), private communication; see also, Schonauer, W., Raith, K., and Glotz, G. (1981), "The Principle of the Difference of Difference Quotients as a Key to the Self-Adaptive Solution of Nonlinear Partial Differential Equations," *Computer Methods in Applied Mechanics and Engineering,* Vol. 28, No. 3, 1981, pp. 327–359.

Strouboulis, T. and Oden, J. T. (1990), "A Posteriori Error Estimation of Finite Element Approximations in Fluid Mechanics," *Computer Methods in Applied Mechanics and Engineering*, Vol. 78, pp. 201–242.

Van Straalen, B. P., Simpson, R. B., and Stubley, G. D. (1995), "*A Posteriori* Error Estimation for Finite Volume Simulations of Fluid Flow Transport," *Proceedings of the Third Annual Conference of the CFD Society of Canada*, Vol. I, Banff, Alberta, 25–27 June 1995, P. A. Thibault and D. M. Bergeron, Eds.

O. V. Vasilyev, O. V., Paolucci, S. and Sen, M. (1995), "A Multilevel Wavelet Collocation Method for Solving Partial Differential Equations in a Finite Domain," *Journal of Computational Physics*, Vol. 120, 1995, pp. 33–47.

Vasilyev, O. V. and Paolucci, S. (1996), "A Dynamically Adaptive Multilevel Wavelet Collocation Method for Solving Partial Differential Equations in a Finite Domain," *Journal of Computational Physics*, Vol. 125, 1996, pp. 498–512.

Wu, J., Zhu, J. Z., Szmelter, J. and Zienkiewicz, O. C. (1990), "Error Estimation and Adaptivity in Navier-Stokes Incompressible Flows," *Computational Mechanics*, Vol. 6, 1990, pp. 259–270.

Zhu, J. Z. and Zienkiewicz, O. C. (1990), "Superconvergence Recovery Technique and *A Posteriori* Error Estimates," *International Journal for Numerical Methods in Engineering*, Vol. 30, 1990, pp. 1321–1339.

Zienkiewicz, O. C. and Zhu, J. Z. (1987), "A Simple Error Estimator and Adaptive Procedure for Practical Engineering Analysis," *International Journal for Numerical Methods in Engineering*, Vol. 24, 1987, pp. 337–357.

Zienkiewicz, O. C. and Zhu, J. Z. (1992), "The Superconvergent Patch Recovery and *A Posteriori* Error Estimates. Part I: The Recovery Technique," *International Journal for Numerical Methods in Engineering*, Vol. 33, 1992, pp. 1331–1364.

8

HARD STORIES

This Chapter covers features of Verification which can be troublesome. Examples are given of difficult problems that require special care, and can involve ambiguities, judgment calls, and subtleties.

8.1 FACTORS INFLUENCING CONVERGENCE RATES

The following discussion of a variety of factors that can influence the observed convergence rates in a discrete solution is taken from Westerink and Roache (1995) and Roache (1997). For the sake of continuity and completeness in this section, the following paragraph repeats the distinctions already given in Section 6.3.1 between formal, actual and observed convergence rates.

Formal convergence rates are always (by definition) assumed to be indicated by the leading order space and/or time truncation error terms. However, even in the asymptotic range (where discrete space and time steps tend to zero), formal convergence rates may never be achieved, leading to the definition of the *actual asymptotic* convergence rate. Even neglecting the well-recognized problems with computer round-off error, this *actual asymptotic* convergence rate may be different for a particular problem from the *formal* convergence rate for reasons to be discussed shortly. Finally, the convergence rate actually *observed* in numerical experiments may be different from *either* the *formal* or *actual asymptotic* convergence rates, simply due to grid resolution not being adequate to achieve the asymptotic range. It is not unusual for the fine grid Richardson Error Estimator to be less than the coarse error estimator for the same grid, indicating that the *observed* convergence rate is less than the *formal* rate.

8.1.1 Asymptotic and Non-Asymptotic Convergence Behavior due to Higher Order Truncation Term Competition

In establishing the formal convergence rate, it is assumed that the leading order truncation error dominates all other terms in the truncation series and hence the exponent of the discrete space or time step is taken as the order of the method. However, subsequent terms in the truncation series may be competitive with the leading order truncation term for several reasons. For nonlinear flows that cascade energy to the highest resolvable wavenumber range, the derivatives of the response function continue to steepen as we refine the grid. This leads to competition of the leading and subsequent order truncation terms as we refine the grid. *Even in the asymptotic limit*, the actual convergence rate will not necessarily match the formal convergence rate given by the leading order truncation term.

Another common cause of higher truncation term interaction occurs with "relatively" coarse discretizations. The resulting interaction of these error terms leads to non-asymptotic convergence behavior and typically causes the observed convergence rate to be less than the formal and/or actual convergence rates. We note that this situation often arises in unstructured graded finite element meshes which attempt to provide high grid resolution in regions where response functions (i.e., the solutions) vary rapidly (with corresponding high derivatives) and low grid resolution in regions where the response varies slowly (with corresponding low derivatives). Although the overall local error in the low grid resolution regions is modest due to the low derivatives of the response functions, the leading order spatial truncation error may be of the same magnitude as the next spatial truncation error, leading to differences between the observed and formal convergence rates.

Also, note that in general 2-D and 3-D problems, spatial cross derivative truncation terms arise that can profoundly affect the character and magnitude of the solution. These terms have been studied in the context of upwinding schemes for convection dominated transport equations but are in general not well studied or understood for complicated flow scenarios (e.g. see Cantekin and Westerink, 1990).

A minor point on fluid dynamics terminology is appropriate here. The term "convection-dominated flow" is often a misnomer. Consider the 1-D steady convection-diffusion (or advection-diffusion) model equation,

$$-u_{xx} + Re\, u_x = 0 \tag{8.1.1}$$

Such problems are usually described as "convection dominated" when the parameter range is such that *Re*, the Reynolds number (or more generally, the Peclet number) is large, $Re >> 1$, since this parameter multiplies the convection derivative. However, for this *steady* equation, convection *never* dominates diffusion. In fact, both convection and diffusion are everywhere in perfect balance, the convection term $Re\ u_x$ being exactly equal to the diffusion term u_{--}. (If we tried to define the convection term as being u_x rather than $Re\ u_x$ we would find that the diffusion term u_{-} is dominant, by the factor *Re*.) The more appropriate term is simply "high Reynolds number flow."

8.1.2 The Effect of Space-Time Truncation Term Cancellation and Superconvergence

It is a common feature of discrete solutions to PDE's to encounter conditions in which various space-time truncation error terms cancel. For example in the Forward-Time-Centered-Space solution to the 1-D constant coefficient diffusion equation, if

$$\Delta t = \frac{\Delta x^2}{6\mu} \qquad (8.1.2.1)$$

where μ is the diffusion coefficient, then the leading order time and space truncation terms cancel each other, and the method becomes 2nd-order accurate in time and 4th-order accurate in space. Also, in the 1st-order upwind solution to a pure 1-D convection equation, the choice

$$\text{Courant Number } c = \frac{V\Delta t}{\Delta x} = 1 \qquad (8.1.2.2)$$

leads to the cancellation of all space and time truncation errors and yields an exact solution at all nodes. Also, generalized wave-continuity equation (GWCE) solutions to the shallow water equations are subject to space-time truncation error cancellation as c ranges between 0.5 and 1.5 (Luettich and Westerink, 1994). This latter case corresponds to the phase error distribution changing from a dominant phase lead error to a phase lag error, passing through the zero phase lag axis.

These three examples point to the fact that space and time truncation errors certainly interact, modify or even eliminate each other. This cancellation of truncation terms can decrease *or increase* the effective convergence rate p depending on the problem and how the space/time-step ratio is maintained in the convergence study. Finally we note that mixed space time truncation errors (e.g., the Lax-Wendroff family) even further complicate the convergence behavior (Roache, 1998). This category includes characteristics-based methods, e.g. Roache (1992).

8.1.3 The Effect of Physical Parameter Resolution on Grid Convergence

As the level of grid resolution is increased, the level of physical parameter resolution can also change. The most common instance is geometry definition. As an example, consider a convergence study of tidal computations within the Western North Atlantic ocean using a shallow water equation based model (Westerink et al, 1994; Luettich and Westerink, 1994). The scale of definition of the bathymetry (i.e., the ocean floor) depends on the grid resolution. As the grid is refined, we can either interpolate the coarsely resolved bathymetry onto the finer grid or increase the level of resolution of bathymetry by re-interpolating actual available bathymetry data onto the finer grid. The results are not identical, of course. These types of grid studies indicate that under-resolution in the *grid* leads to over-predictions in the response, while under-resolution in the *bathymetry* leads to under-predictions in response. Thus, under-resolution in the grid and the inherent under-resolution in bathymetry lead to a (partial) cancellation of errors, which corresponds to a cancellation of truncation errors from the grid and bathymetric terms. Again, this type of truncation error cancellation can lead to observed convergence rates being higher than the formal convergence rates.

Similar questions will arise whenever geometry (or other parameter) resolution changes with grid refinement. For complex mechanical systems such as under-the-hood heat transfer in automobiles, coarse grid simulations will lump together some components into some idealized shape. As the grid is refined, the modeler must decide on whether to redefine the components with smaller scale features, or to simply increase the resolution of the idealized shape. This is a subtle modeling question.

A related but more fundamental question (repeated here from Chapter 5, Section 5.10.9) arises when geostatistical methods are used to generate particular realizations of grid-block property variations with specified statistical parameters. Only the statistical results are of interest, not the solutions of the individual realizations. The question is then, should the grid refinement studies be performed separately from the geostatistical realizations? That is, should the solution of the partial differential equations be converged on finer grids with the assumed continuum property variation fixed at a geostatistically generated coarse-grid distribution, or should the geostatistical generation also change as the grid is refined? This is not a trivial question, and although definition of a fixed continuum problem for the grid refinement studies is conceptually easier, it is clear that substantial computer savings could accrue to the combined convergence approach. In either case, the grid increments should be less than (be able to partially resolve) the correlation length of the property variation.

The question may arise as to where in the taxonomy should parameter uncertainty be included. Specifically, should parameter uncertainty be part of Verification or Validation? The answer is clear after a little thought, i.e. this is not a "gray area" at all. The answer is, in Validation. The determinant is the working definition of Validation, deciding whether or not one is "solving the right equations," which equations obviously *must* include the parameters of the equations. The exercise may look "gray" in that the calculational burden required to add the error bar of parameter uncertainty to model calculations yields an "error bar" similar to that of a grid convergence test. But it really is a "model" test rather than a "code" or "calculation" test, and therefore must be part of Validation.

Purely mathematical studies on the *sensitivity* of solutions to parameters are neither Verification nor Validation, but just results. (As already noted, these studies do require

Verification of both Code and Calculation in order to be trustworthy.) Indeed, this type of study is one of the most powerful uses of simulation, performing "virtual experiments" that would be difficult or impossible to accomplish physically. Of many possible examples, an excellent one is the study of Oldenburg and Pruess (1995) on groundwater-brine systems showing the surprising sensitivity of gross flow patterns to molecular diffusivity. Another purely mathematical modeling exercise showed the effect of ratio of specific heats on the numerical behavior of compressible flow simulations as the incompressible limit is approached (Roache and Salari, 1990; see also Chapter 6, Section 6.28).

8.1.4 Summary of Formal vs. Actual Asymptotic vs. Observed Convergence Rates

We have noted that *observed* convergence rates for complicated geophysical and other flow problems are often less and sometimes higher than the *formal* rates associated with a simple analysis of the leading truncation error terms, and have presented some of the possible origins of this problem. Some of these issues can be taken into account by carefully designing numerical experiments that define convergence. Whether grid size dependence is influencing the observed convergence rate for a problem can only be checked by going to a sufficiently fine level of resolution. This will produce the *actual asymptotic* convergence rate (i.e., the rate observed for asymptotically fine resolution without round-off errors) which may or may not equal the *formal* rate. Space-time truncation error dependence can be checked by performing a thorough study of the space-time error interaction found through Taylor and Fourier series analysis. It may be desirable to isolate the effects of the spatial errors by carefully keeping time steps small enough such that time errors do not interact with the space error, and by refining the time step with the space step in an appropriate fashion. (For some problems this would consist of linear variation of the time step with the spatial grid size, for others quadratic.)

However, it actually may be advantageous to work with time steps that incur space-time truncation error cancellation since they lead to overall lower errors for a given discretization even though convergence rates may not reach formal or actual asymptotic rates (depending how one sets up the convergence testing). For example, for solutions of the GWCE based shallow water equation, the best phase propagation properties are achieved with Courant numbers which lead to space-time truncation error cancellation (Luettich and Westerink, 1994).

Finally, parameter convergence can be separated out by freezing the parameter definition while refining the grid (Luettich and Westerink, 1994). We can then use the finest discretization and sequentially refine the level of resolution with which the parameters are defined.

Despite one's care in separating out these issues, *observed* convergence rates may still be less than *formal* or *actual asymptotic* convergence rates. In fact, it may not even be beneficial to try to separate them out when making an assessment of the overall error in an operational model. Therefore we should account for these uncertainties when applying Richardson-based error estimators (or others). It is certainly justified to account for this type of problem in the GCI by including the safety factor $F_s > 1$ as recommended in Chapter 5, using the GCI as an error *band*, equal to F_s times the error estimator.

8.1.5 Other Considerations in Defining Convergence Rates

Geophysical flows are very rich in their scale content and the temporal and spatial complexity of the associated flow features. Wave lengths in a geophysical flow range from the basin scale to the dissipation scale, which cannot be resolved except in the lowest Reynolds number computations. Furthermore, the features of a geophysical flow can be quite intricate. For example, a tidal flow includes local response minima for harmonic response constituents that include very high gradients in space in both amplitude and phase. (In fact, phase lines come together at the center of the amphidrome; Westerink et al, 1989.) The wavenumber range and complexity of the flow will make the task of assessing convergence and errors of such flow computations quite difficult. This observation also applies to DNS calculations of aerodynamic turbulence.

8.1.6 Defining Wavenumber Dependent Convergence

Due to the scale richness of a geophysical flow, it appears very reasonable to look at convergence on a wavenumber- or frequency-selective basis. This is routinely done in tidal studies where tidal frequencies are separated out and examined individually. As grid refinement increases, more wavenumbers and therefore more frequencies appear. We note that the effect of new frequencies will be reflected in other constituents as the convergence study proceeds. This process works well, although we must keep in mind that the accuracy of a specific constituent does not necessarily reflect the accuracy of all wavenumbers in the spectrum. Note that, for Verification of a *Code*, rather than Verification of a realistic Calculation, it is advisable to develop a model problem which uses the same governing PDE's but does not exhibit the strongly resolution-dependent flow features, as in Dietrich et al (1990).

8.1.7 Artificial Flow Features

As we examine any converging geophysical flow, we must carefully scrutinize the response functions as to their physical relevance. Specifically, not all flow features are converging functions which incur a certain relative percentage of error. In fact, artificial flow features can appear and disappear at a given wave number. Common artificial features are the $2\Delta x$ waves which appear due to phase lag errors, folded dispersion relationships, and nonlinear cascading of energy into the higher wave number range where there may not be sufficient physical dissipation to eliminate this energy. The $2\Delta x$ waves are relatively simple to identify as non-physical flow features, since this energy moves to the highest resolvable wavenumbers as we refine the grid. However, there are other artificial flow features that can appear in the long-wave physically relevant portion of the response spectrum.

In particular, boundary placement and boundary condition implementation can influence the generation of artificial flow features. For example, in hurricane storm surge computations, placement of an "open ocean" boundary adjacent to a resonant basin such as the Gulf of Mexico can excite an artificial 18 hour period mode with significant amplitude (Blain et al, 1994). The generation of this artificial mode is very strongly dependent on where the open ocean boundary is placed and on how the boundary conditions are implemented. This again points to the importance of performing convergence studies on domain size (i.e. boundary placement), as well as boundary condition implementation and non-ordered approximations on the boundary in general. Analogous situations are common in aerodynamics and mechanical engineering flows with outflow boundary conditions (Roache, 1972b, 1994). The key feature here is that these errors (perhaps better classified as modeling errors than numerical errors) *are not ordered* in grid increment and therefore will not be disclosed in grid convergence studies. However, as noted previously, these errors may prove in numerical experiments to be ordered in $1/l$ where l is the distance from the region of interest to the "open" boundary, and this error may be estimated (or banded, as in the GCI) by the same methods used for the grid convergence studies. (See Roache, 1994, and Chapter 6, Section 6.10.)

8.1.8 L_2 and L_∞ Norms as Error Indicators; CAFE Curves

Different fields of endeavor have different requirements for the variables and functionals of interest. To generalize too much, engineering fields (mechanical, aerospace, chemical) often require functionals of the solution such as lift coefficient, heat transfer rates, mixing rate, etc. It is less common that practical interest is in accuracy of the entire field calculation; one example that comes to mind is aero-optical propagation (Truman and Lee, 1990). However, meteorological and ocean modelers are usually interested in accuracy of all the field variables. In these cases, it is meaningful to emphasize global accuracy of the entire computational field, and difficulty arises in attempting to characterize global accuracy with just one or a few numbers.

Typically, L_2 or L_∞ error norms are used as error indicators. However, questions arise whether these error norms are always reliable, and more significantly how they can be interpreted. For example, in a tidal convergence study, L_∞ norms are poor indicators of error due to the fact that shifting amphidromes, associated with very high gradients in response amplitude and phase, could cause extremely high localized errors, which do not converge smoothly or at the same rate as an L_2 norm. Furthermore, there is the general problem of interpreting how representative an L_2 or L_∞ norm is to the response as a whole.

The concept of Cumulative Area Fraction Error (CAFE) curves introduced by Luettich and Westerink (1994) presents domain errors in a more complete and meaningful way. These curves plot the fraction of the total domain that exceeds a particular error level (y-axis) against that error level (x-axis). Under- and over-prediction are indicated separately. Assuming that under- and over-prediction are distributed approximately evenly throughout the domain, median under- and over-prediction errors correspond approximately to a cumulative area fraction of 0.25.

As an example of the cumulative area fraction error curves, consider some results of a grid intercomparison for tidal flow computations as reported by Westerink et al (1995). Three grids were examined: Grid SS4, a regular 24255 node grid; Grid T1, a graded 11712 node grid with increasing resolution in shallower waters; and Grid T2, a graded 28889 node grid that is based on grid T1 with added resolution over steep bathymetric gradients. Results from all three grids are compared to those from a 95999 node regular grid. Figure 8.1.8.1 plots the three CAFE curves and clearly indicates that the grid T2 outperforms the other two grids over the entire

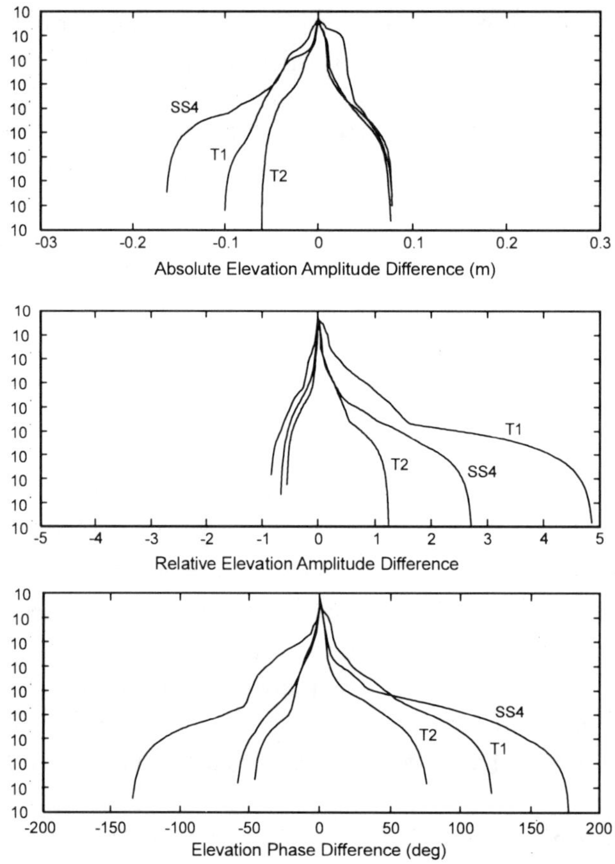

Figure 8.1.8.1. Example of Cumulative Area Fraction Error (CAFE) curves of Westerink et al (1995). (From Figure 2 of Westerink and Roache, 1995.)

distribution of error levels. Note that the error curves are continuous and that every error level in the domain is incorporated. Furthermore, extreme L_∞ errors correspond to the end points of the error curves. For a perfect solution the under- and over-prediction curves would collapse to a line on the zero error level. Finally, note that these CAFE curves provide a direct relationship between any error level and the fraction of the domain which exceeds that error level.

8.1.9 Summary of Other Considerations in Defining Convergence Rates

We have noted that the flow spectrum must be carefully considered in any convergence study. Certainly looking at convergence behavior in a frequency or wavenumber specific way appears very reasonable. However, the overall spectrum should be considered as a whole and related to the level of convergence of a specific frequency. Furthermore, the response spectrum must be carefully scrutinized for artificial flow features such as $2\Delta x$ waves and resonant waves artificially set up through boundary placement or boundary condition implementation. The Cumulative Area Fraction Error (CAFE) distribution concept gives a complete and meaningful picture of the error distribution; the CAFE curves can be particularly useful in grid convergence studies and grid intercomparisons.

For other examples of grid convergence studies in geophysical ocean and lake modeling, see Johns et al (1983a,b), Le Provost and Vincent (1986), Bennett and Campbell (1987), Dietrich and Roache (1991), Dietrich et al (1990), Lardner and Song (1992), Piacsek and Allard (1993), Luettich and Westerink (1994), Westerink et al (1994, 1995).

8.2 BEHAVIOR OF QUASI-HIGHER-ORDER METHODS

Leonard and MacVean (1995) have performed a grid convergence study that is uniquely valuable for covering different algorithms in a consistent framework, allowing comparison of error estimation across methods. The benchmark chosen is a commonly used 2-D unsteady meteorological problem consisting of an initial elliptical blob of cold air aloft that descends under gravity towards a ground plane. As it spreads laterally, a strong (nonlinear) density current is formed, with a sharp leading-edge front and a number of internal fronts associated with Kelvin-Helmholtz rotors. Although there is no exact solution available, numerous fine-grid Benchmark calculations have been reported previously; see Straka et al (1993). Figure 8.2.1 shows the initial conditions and Benchmark (fine grid) solutions for perturbation potential temperature θ.

The authors considered 7 different methods, 3 of which are flux-limited schemes. Only data on the 4 non-limited schemes were presented to allow evaluation of the error estimators. These 4 are as follows.

i. LF, a 2nd-order leapfrog method (e.g., see Roache, 1972b)
ii. UTOPIA, Leonard's nominally third-order flux-integral method (Leonard 1991)
iii. Q5, a quasi-5th-order scheme in which higher-order cross-differences are neglected and 2nd-order methods are used for diffusion (Leonard and MacVean, 1995)
iv. U5, a higher-order reference method (Leonard and MacVean, 1995)

The authors perform a thorough grid convergence study over 5 grids with 4 doublings, with mesh increments (in meters) of $\Delta = 400, 200, 100, 50, 25$. Sample results for the 200 *m* mesh are shown in Figure 8.2.2.

Importantly, the convergence rates p (in the present notation) were determined experimentally. The authors' experience was that determining p values at every grid point was unworkable, leading to a wide range of different p values at different grid points "due in large part to slight phase shifts near strong gradient regions." They then used the global L_1 error norm,

$$L_1 = \frac{1}{N}\sum\left|\theta_{computed} - \theta_{reference}\right| \tag{8.2.1}$$

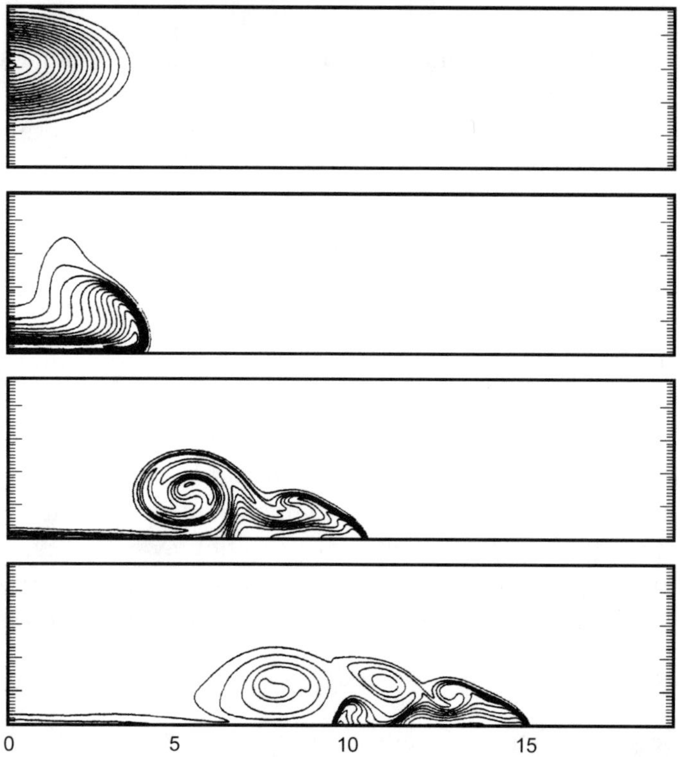

Figure 8.2.1. Initial conditions and Benchmark solutions for perturbation potential temperature θ for the descending blob of cold air. Solution times are 0, 300 *s*, 600 *s*, 900 *s*. (From Figure 1 of Leonard and MacVean, 1995.)

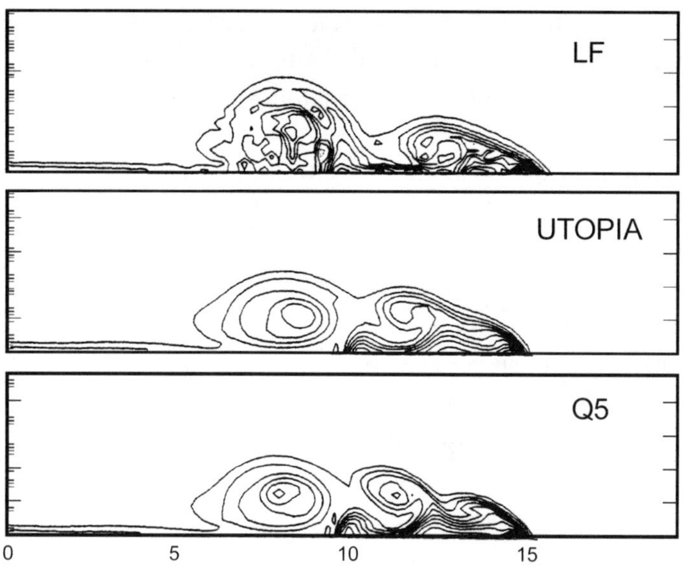

Figure 8.2.2. Results after 900 *s* on the 200 *m* mesh for the leapfrog, UTOPIA, and Quasi-5th-order method. (From Figure 6 of Leonard and MacVean, 1995.)

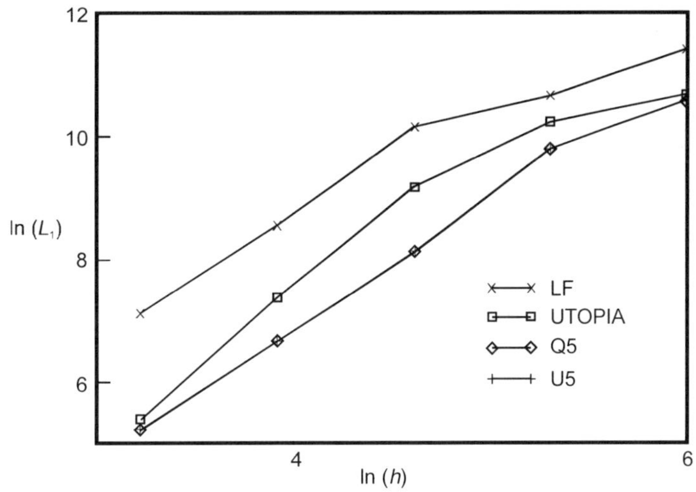

Figure 8.2.3. Error norm (L_1) vs. mesh size for leapfrog (LF), UTOPIA, Quasi-5th-order (Q5) and reference method (U5). (From Figure 4 of Leonard and MacVean, 1995.)

which normalizes the error over all N cells, calculated for all but the coarsest grid. $\theta_{reference}$ was calculated by Richardson Extrapolation using experimentally determined values of p (solved by graphical iteration). The L_1 error norm vs. mesh size is shown in Figure 8.2.3.

The "rather less expensive" Q5 results are graphically indistinguishable from those of the U5 reference method in Figure 8.2.3, but the error estimates will be distinguishable.

The relative error RelErr for each numerical solution was calculated from the reference solution. The relative error on grid h_1 was also estimated as E_1 calculated by the (generalized) Richardson Extrapolation (Roache, 1994; Equation 5.4.2, Chapter 5) using the experimental values of p for each method and grid pair. Also, the GCI value (Chapter 5) using the original "factor of safety" = 3 was used, i.e.

$$\text{GCI} = 3E_1 \tag{8.2.2}$$

(Note the change in notation from Leonard and MacVean, 1995, whose "ε_1" corresponds to E_1 in the present notation.) The results are given in Table 8.2.1.

The following discussion of these results is taken verbatim (except for the change in notation noted above) from Leonard and MacVean (1995). "One sees immediately that simple uncertainty diagnostics are difficult to devise. In particular, Roache's GCI appears to be (usually) much too conservative. Our E_1 (= GCI / 3) appears to be somewhat more useful as a relative error estimator but not particularly accurate in absolute terms."

	LF	UTOPIA	Q5	U5
$\Delta = 25$ m				
RelErr	0.30	−0.04	0.03	0.03
E_1	6.33	−0.05	0.03	0.03
GCI ($F_s = 3$)	19.00	−0.15	0.10	0.10
$\Delta = 50$ m				
RelErr	20.30	−0.41	0.14	0.14
E_1	12.15	+1.40	0.92	0.95
GCI ($F_s = 3$)	36.44	+4.21	2.76	2.85
$\Delta = 100$ m				
RelErr	67.71	+6.58	3.10	3.17
E_1	51.79	+2.66	4.90	4.83
GCI ($F_s = 3$)	155.36	+7.99	14.69	14.50
$\Delta = 200$ m				
RelErr	101.07	+11.44	24.40	24.22
E_1	17.38	−13.20	7.48	3.87
GCI ($F_s = 3$)	52.15	−39.60	22.43	11.62

Table 8.2.1. Uncertainty diagnostics for $|\theta_{max}|$, in %. (From Table 1 of Leonard and MacVean, 1995.)

My own evaluation differs somewhat. First, I repeat that the GCI was not intended to provide an error estimate, but a conservative error band, equal to 3× the error estimate. Given that clarification, we still see that the results of Table 8.2.1 are still somewhat favorable to my philosophy of conservatism. There are 16 cases (4 methods, 4 fine grids) considered. For 10 of these 16 cases, the GCI is very conservative and E_1 is either a good estimate or is itself very conservative, the incidences being strongly skewed to the finest grids, as expected (Roache, 1994; see also Chapter 5). In 3 of the remaining 6 cases, the GCI is still conservative (less than the factor of 3) but E_1 is *not*. As anticipated (Roache, 1994) and as experienced for the simple 1-D Burgers equation (Roache, 1995; see also Chapter 5) the generalized Richardson Estimator is *not* conservative. More surprising, even the GCI itself, with the full (original) "factor of safety" = 3, is not conservative in the remaining 3 out of 16 cases. As expected, these failures occur on the coarsest grid of 200 *m*. (Note that this is the coarsest *fine* grid in the coarse-fine pairs; i.e., the results in Table 8.2.1 for the 200 *m* grid were obtained from the pair of grid solutions on a coarse grid of 400m and a fine grid of 200 *m*.) Considering the noise evident in the top contour plot of Figure 8.2.2, one might argue that inspection would suggest that the LF method was not in the asymptotic range, so a CFD practitioner would know better than to expect the original GCI to be conservative. Based on folklore horror stories of the general competence level of CFD commercial code users, I doubt this. With more certainty, I suggest that many sophisticated users would be fooled by the plausible results in Figure 8.2.2 for the Q5 method, yet even for this method the GCI with $F_s = 1$ is not conservative in Table 8.2.1. Furthermore, note that the conservatism of the calculations has been enhanced (compared to standard minimal practice) because the authors have experimentally determined values of p. The "nominally 3rd-order" UTOPIA method shows a convergence rate of $p = 2.83$. This is good performance for the method, since the "nominal" value was determined by analysis of the constant velocity problem, whereas the present benchmark advection field is far from uniform. But the point is that, if only the "nominal" result of the formal analysis $p = 3$ had been used, the UTOPIA result would be less conservative. In summary, in my evaluation, the thorough study of Leonard and MacVean makes a rather strong case for maintaining conservatism in the GCI error band, unless (as I originally suggested (Roache, 1994; see also Chapter 5) three or more grid solutions are calculated to establish p.

Further evidence of the importance of experimentally determining p comes from the results of Leonard and MacVean (1995) for 3 flux-limited schemes:

i. the MUSCL scheme of Van Leer (1979 applied one-dimensionally, i.e. with the transverse gradient terms;
ii. LQ5, the earlier Q5 scheme with the first multidimensional limiter of Leonard et al (1995); and
iii. 1UP, a first-order upstream flux integral method for advection with second-order physical diffusion.

The convergence behaviors are shown in Figure 8.2.4. One would expect Q5 to have at best 2nd-order convergence, and less when the flux limiter is active (at the fronts), and the MUSCL scheme applied one-dimensionally to be somewhat less than 2nd-order. In fact, both exhibit a higher-than-theoretical value of $p = 2.4$, "rather intriguing but, as yet, unexplained." If $p = 2$ had been used in a two-grid application of the GCI, the conservatism would have been increased even beyond the safety factor of $F_s = 3$. On the other hand, the 1UP method of nominally 1st-order convergence might be expected to differ from $p = 1$, if at all, with a slight improvement (in the non-rigorously asymptotic range) because of the 2nd-order accuracy in the diffusion terms.

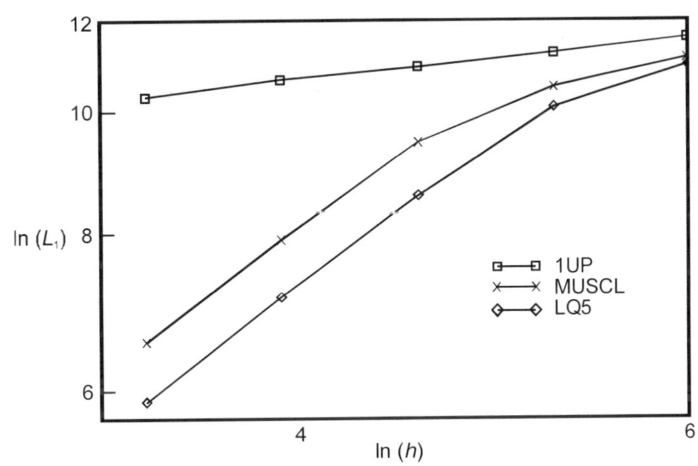

Figure 8.2.4. Error norm (L1) vs. mesh size for first-order (1UP), MUSCL, and limited quasi-5th-order (LQ5) methods. (From Figure 7 of Leonard and MacVean, 1995.)

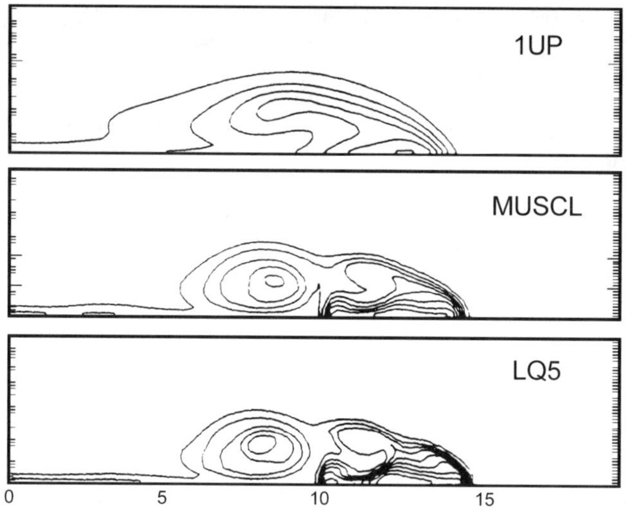

Figure 8.2.5. Results after 900 *s* on the 200 *m* mesh for first-order (1UP), MUSCL, and limited quasi-5th-order (LQ5) methods. (From Figure 9 of Leonard and MacVean, 1995.)

In fact, the observed value is a miserable $p = 0.35$. (The well-known smearing of solution detail by 1st-order methods is shown in Figure 8.2.5.) The GCI denominator $(r^p - 1)$ would have been calculated (with $p = 2$) as 1, whereas the value with observed $p = 0.35$ would be 6.25. The conservative $F_s = 3$ is overwhelmed by this non-conservative factor 1/6.25, so that *even the GCI is non-conservative* by more than a factor of 2. The simple E_1 estimator would be non-conservative by the factor of 6.25. This behavior would be disastrous for the Quantification of Uncertainty; imagine publishing an "uncertainty diagnostic" for (say) a lift coefficient of 4%, then being shown that your calculation was actually in error by 25%.

To reiterate my position (Chapter 5), based upon careful multiple grid resolution studies like those of Leonard and MacVean (1995) and others (see papers in Johnson and Hughes, 1995), I recommend that the GCI be calculated with $F_s = 1.25$ *only* for grid convergence studies based on three or more grids. Recall that two grid solutions are necessary to estimate the error if p is already known, three grid solutions are necessary to determine the observed p, and *four* grid solutions are necessary to determine that the observed p is constant, i.e. that the sequence of solutions is within the asymptotic range. (Of course, if three grids are used and the observed $p \sim$ the theoretical value, e.g. observed $p \sim 2$ for a theoretically 2nd-order method, then one would have high confidence in using $p \sim 2$ without performing the fourth grid solution.) For the minimal two-grid solutions on a new problem (i.e., a problem for which p cannot be confidently inferred from a nearby problem), I recommend that the GCI should still be calculated with the usually (but not always) conservative value of $F_s = 3$. See also discussion in Chapter 5.

8.3 SOME GOOD NEWS FOR TURBULENCE MODELING

As noted in Chapter 3, the error for mixed order methods should, strictly speaking, be banded using the lowest order of the method, i.e. 3rd-order differencing for advection terms used with 2nd-order differencing for other terms should be estimated using $p = 2$. Although this will be true in all cases *strictly* speaking, a *practical* and most advantageous exception exists for turbulence modeling.

In k-ε and k-ω turbulence models, it is a matter of computational experience that the numerical accuracy of the transport equations for these variables does not have a major effect on flow average quantities of usual engineering interest, notably skin friction and pressure distribution. If 1st-order or hybrid differencing is used for the advection terms in the turbulence equations, the convergence rate for these equations will of course be $p = 1$, but the experimentally observed rate of convergence for flow average quantities, over realistic grid resolutions, is $p = 2$. (Salari, 1997; see e.g., Bergstrom and Gebart, 1997.) Bergstrom and Gebart (1997) experimentally showed $p \sim 2$ ($p = 2.08$) for flow average quantities over the finest three grids in a four-grid calculation, demonstrating that a practical asymptotic range had been attained, and was converging at 2nd-order accuracy. This is a significant Verification and confidence builder for both the algorithm and the commercial code used, CFDS-Flow3D (AEA Technology, 1994). (The code is now known as CFX.) The use of 1st-order differencing in the turbulence equations greatly stabilizes them (Ekaterinaris and Menter, 1994, Ekaterinaris, 1995) and is recommended, *provided only* that the user Verifies the experimental observed $p = 2$ for each new (not "nearby") calculation.

8.4 THE MYTH OF THE CONVERGED "SOLUTION"

CFD practitioners, especially industrial engineers using general-purpose commercial CFD codes, would like to be able to obtain, at least in principle, a "converged solution" (converged both iteratively and in time-space discretization) and then be able to confidently post-process that solution for any engineering measure of interest. Unfortunately, this cannot be possible in any general sense. The problem is again the *a priori* unknown correlation between possible error measures of engineering and scientific interest. Of course, expert judgment (based on experience with a nearby class of problems) can be valuable, but the problem is not soluble in any general sense.

For example, in our experience with dynamic stall and oscillating airfoil and wing calculations (Salari and Roache, 1990), lift was well converged at the finest discretization achieved ($141 \times 55 \times 55$ in 3-D, and 461×71 in 2-D), drag was more problematical, and moment definitely was not converged. If one had performed the grid convergence study (or other error estimation exercise) examining only the lift, one could only infer from engineering judgment (experience on related problems) whether or not resolution was adequate for drag and moment. Viegas et al (1988) noted the same variance of convergence with this additional distinction. Their calculations of lift and *pressure* drag were perhaps adequately converged (with a ~1% persistent oscillation) but moment (as above) and *friction* drag were not. If some new error measure became of interest, say position of transition, or location of trailing edge separation, or second harmonic component of the unsteady pressure coefficient (of interest for helicopter dynamic stall analysis; McCroskey, 1981), the convergence study would have to be repeated for this error measure, strictly speaking.

Other examples are plentiful.

Rosenfeld (1994) used grid resolution studies of bluff-body wakes with resolution up to 513 $\times$ 513 points, and examined convergence in both the physical and Fourier domains. He showed that very fine meshes were required to obtain convergence in the physical domain, but that Fourier components converge on coarser grids than either the phase angle or the solution in the physical domain. A major factor in this behavior is the resolution required for the initial transients, which do effect the phase angle of the solutions, but are of no importance (physically or numerically) to the fully developed periodic solution. "These findings indicate that the mesh resolution has a more pronounced effect on the phase velocity of the vortices than on phenomena related to magnitude," so that coarser grids may be acceptable for prediction of force coefficients.

Blottner (1990) noted that hypersonic sphere-cones calculations require much higher grid resolution for heat-flux calculations than for surface pressure calculations with the same level of uncertainty.

Sengupta et al (1995) showed that calculations of the branching between symmetric and non-symmetric flow patterns require special attention to grid resolution.

Rumsey et al (1996) calculated unsteady laminar flow over a circular cylinder, and unsteady turbulent flow over an 18% thick circular arc airfoil. They explored the sensitivity of grid convergence to many physical and numerical parameters, including turbulence model (algebraic vs. one-equation), implicit vs. explicit time differencing, and various time-step sub-iteration schemes. They note that for some conditions, "even results on the finest 369 $\times$ 129 grid are not completely code converged," yet [as is well known] the "oscillation frequency is relatively insensitive to grid density for both turbulence models."

Certainly, it is always possible to devise some error measure that is exquisitely sensitive to discretization error, e.g., some high-order statistical correlations in turbulent flow calculations, so that this measure is far from converged even when other, more benign measures are well converged. A practical example is a boundary layer stability calculation. For most engineering applications, skin friction and wall heat transfer are of principal interest, and these are sensitive to the first normal derivative of velocity at the wall. However, it is known from classical stability theory (e.g., Lin, 1967) that laminar boundary layer stability is dependent on the diffusion of vorticity across the critical layer (the y-position at which the mean flow speed equals the disturbance wave speed). The principal component of boundary layer vorticity magnitude being $\partial u / \partial y$, this means that the appropriate measure of accuracy is of the term $\partial^3 u / \partial y^3$, which may be expected to be more difficult to "converge" than simply the velocity component u. Conversely, accurate convergence of pressure may be unnecessary.

Thus, the concept of a "converged solution," ascertained to be so *independent of the intended error measure*, is a myth.

8.5 ESOTERIC CODING MISTAKES

As noted in Chapter 3, general PDE codes (more general than the simple Poisson equation in nonorthogonal coordinates) would be difficult to include in a theorem on code Verification because of esoteric errors. The difficult aspects of the codes are not *algebraic* complexity; in Steinberg and Roache (1985) we convincingly verified 1800 lines of dense Fortran. The more difficult and vexing problems come from totally distinct types of complexity, arising from option combinations and switches like conditional differencing. Esoteric errors can arise because of nonlinear flux limiters like FCT, TVD, hybrid or type-dependent differencing, etc. This does not invalidate my claim that the Method of Manufactured Solutions (Chapter 3) Verifies the Code. Obviously, the method cannot Verify a feature that is not turned on! This merely says the obvious, that in order to test a code with many features, one may require many test runs to turn on these features.

A type of error unlikely to be included in a accuracy Verification theorem arose in a code for calculating supersonic base pressure (Mueller et al, 1970; Mueller and Roache, 1973) using the Chapman-Korst flow model. Several Fortran variables were defined for groups of terms involving Mach number M, such as MS = M**2, M1 = M – 1.0, M2 = SQRT (MS – 1.0), etc. One of these variables was not declared REAL. The original code was initially exercised and considered to be bug-free for Mach numbers of 1.5, 2, 3, 4, 5, and 6, with excellent agreement with experiment. It was of course correct for integer M only, and quite good for M = 1.5, leading to a false conclusion on accuracy. For M = 1.3, it failed unambiguously, leading to a correction.

In the first example in Chapter 1, Section 1.3 (from Roache, 1982), a factor of 2 error in cross derivatives in a grid generation code did not show up during extensive testing. The test suite was composed of homogeneous elliptic grid generating equations that produce almost conformal transformations (with discrepancies due only to boundary distributions) for which the cross derivative terms vanish in the limit of $\Delta \to 0$. Although revealed by the general code accuracy Verification procedure, as described in Chapter 3, one can easily envision being misled by particular problems.

8.6 A FALSE VERIFICATION TEST OF A PARTICLE TRACKER

The 2-D and 3-D SECO_TRACKER codes (Roache, 1993a,b) were built to provide temporal high-order particle tracking. They use a modified version of a 5th-order Runge-Kutta-Fehlberg ODE integrator called RKF45 (Shampine et al, 1976) for integration in time combined with linear velocity interpolation in space and time to produce particle tracking with Verified accuracy of $O[(\Delta x)^2, (\Delta t)^5]$. This may appear to be inconsistent with the typical 2nd-order flow field accuracy, but is in fact justifiable. The key concept is that the flow codes are Eulerian, whereas the particle tracking is intrinsically Lagrangian. For a steady Eulerian velocity field with linear variation in space, there is *zero* time truncation error in the velocity field, yet the (Lagrangian) solution for particle position involves exponential functions in time. The higher time variability of particle position, compared to Eulerian velocity fields, justifies the use of higher accuracy tracking algorithms. See also Baptista (1987).

In Roache et al (1990), we described the following "hard story" in the debugging experience with SECO_TRACKER.

8.6.1 Spatial Convergence of Tracker Codes

The first benchmark test case for the Tracker codes was designed to verify spatial convergence. The benchmark solution is a 1-D trajectory. The 1-D velocity field varies in space and time as

$$|V| = 0.5\frac{\cos(time)}{\cos(x)} \tag{8.6.1.1}$$

The 1-D solution is

$$\text{x\_part} = \text{asin}\left[0.5\sin(time)\right] \tag{8.6.1.2}$$

which solution has sufficient structure to exercise all terms in the Taylor series expansion of the discretization error. The solution is rotated with the direction cosine $dcos1 \leq 1$, and $dcos2 = \sqrt{1 - dcos1^2}$. Using $dcos1 = 0.5$ gives the trajectory angle $\alpha_1 = 60°$, so the trajectory is not through the grid corners (to assure that interpolation is exercised). The test results are shown in Table 8.6.1.1.

The term *relerr_p* is the relative error in final position, and *relerr_coef* is calculated as $relerr\_p \cdot (il - 1)^2$. For a method which is 2nd-order accurate in space, this coefficient should become asymptotically constant as *il* is increased. The time discretization error is $O(\Delta t^5)$ and is negligible compared to the spatial discretization error for this problem.

As seen in Table 8.6.1.1, this coefficient is indeed roughly constant. (The slight decrease in *relerr_coef* as the grid is refined would theoretically indicate faster than 2nd-order convergence, but this distinction is not significant, and is evidence of less than asymptotic behavior, round-off error, and time discretization error.)

$il \times jl$	*relerr_p* (%)	*relerr_coef*	error ratio
5×5	2.88 E+00	4.61 E–01	–
9×9	5.74 E–01	3.68 E–01	5.0
17×17	6.97 E–02	1.78 E–01	8.2
33×33	5.06 E–02	5.18 E–01	1.4

TABLE 8.6.1.1 2-D Tracker Results For Test #1, Uniform Grid, MAC = 0.
(From Table 8 of Roache et al, 1990.)

$il \times jl$	*relerr_p* (%)	*relerr_coef*
5×5	1.6238 E+00	2.5981 E–01
9×9	4.0202 E–01	2.5729 E–01
17×17	5.3066 E–02	1.3585 E–01
33×33	5.1116 E–02	5.2343 E–01

Table 8.6.1.2. 2-D Tracker Results For Test #1, Uniform Grid, MAC = 1.
(From Table 9 of Roache et al, 1990.)

$il \times jl$	*relerr_p* (%)	*relerr_coef*
5×5	4.4230 E+00	7.0768 E–01
9×9	7.6609 E–01	4.9030 E–01
17×17	8.4578 E–02	2.1652 E–01
33×33	4.7120 E–02	4.8250 E–01

Table 8.6.1.3. 2-D Tracker Results For Test #1, Stretched Grid, MAC = 0.
(From Table 10 of Roache et al, 1990.)

The successive ratios of relative error in particle position are displayed in the last column of Table 8.6.1.1. The ratio of successive errors as the spatial step is halved, from $(il - 1) \equiv N = 4$, 8, 16, is better than the theoretical value for the 2nd-order method, $2^2 = 4$. This test of 2nd-order spatial discretization error is not as sensitive to computer round-off error as is the 5th-order time discretization test. The better-than-theoretical performance will not generally hold for arbitrary velocity fields. The failure in Table 8.6.1.1 at $N = 32$ is indicative of accumulated round-off error using the single precision version of the code on an old Vax computer with about 7–8 significant figures of accuracy for floating point calculations. (The single precision version was developed for later conversion to the Cray-XMP, for which any use of double precision variables prevents vectorizing.)

The above test results are obtained for collocated grids (MAC = 0); similar results are obtained for a staggered (Marker-and-Cell or Arakawa-C grid; MAC = 1) in Table 8.6.1.2, and for a stretched grid in Table 8.6.1.3.

8.6.2 Temporal Convergence of Tracker Codes: A False Negative Test

The second benchmark test case for the particle tracker codes was designed to verify temporal convergence. A solid-body rotation velocity field with rotational speed = 2π radians/unit time should return the particle to its original position at final time = 1.

For an initial position at *x_part* = (0.75, 0.0) the following errors in the final *x*-position at N time steps were obtained, as shown in Table 8.6.2.1.

N	error in x_part	error ratio
10	4.9E–02	–
20	1.7E–03	28.8
40	5.5E–05	30.9
80	4.0E–05	1.4

Table 8.6.2.1. 2-D Tracker Results For Test #2, Uniform Grid, MAC = 0.
(From Table 11 of Roache et al, 1990.)

The ratio of successive errors as the time step is halved, from $(il - 1) \equiv N = 10, 20, 40$, very closely fits the theoretical value for the 5-th order method, $2^5 = 32$. The failure from $N = 40$ to 80 is again indicative of accumulated round-off error using the single precision version of RKF45 on an old Vax computer with about 7–8 significant figures of accuracy for floating point calculations.

Although this benchmark test case #2 does verify the 5th-order time discretization error, there is no spatial interpolation error, since the velocity field for solid body rotation is linear in space.

This is an important distinction. In actual application, a *serious* coding mistake was not detected by this seemingly convincing benchmark test case, i.e. we obtained a "false negative" error test. The code logic which located the cell indexes containing the particle position was in error. However, since the solid body rotation gives $u = -cy$ and $v = +cx$, i.e. a linear variation in velocity components, the linear interpolation to find $u(x, y, t)$ and $v(x, y, t)$ is algebraically exact, no matter what cell is used as the basis for the interpolation. (The wrong cell merely results in linear *extrapolation*, which is still algebraically exact.) Only when the previous benchmark test case #1, involving the (non-trivial) rotated 1-D trajectory, was exercised was the coding error discovered.

Similar results were obtained from the 3-D tracker codes. The benchmark test case was the 3-D analog of the first 2-D problem above. The same 1-D solution was rotated with direction cosines $dcos1, 2 \leq 1$, and $dcos3 = (1 - dcos1^2 - dcos2^2)^{1/2}$. Using $dcos1 = 0.5$ and $dcos2 = 0.6$ gives $\alpha_1 = 60°$ and $\alpha_2 = 53.13°$, so the trajectory is not through the grid corners. Results are shown in Tables 8.6.2.2 and 8.6.2.3.

Particle trajectory calculations can be quirky. Some FEM codes provide discontinuous flow velocity solutions that can result in peculiar and totally erroneous qualitative behavior of particle tracking. If the discontinuous velocity solutions are used without smoothing, they can result in trajectory calculations with particle "trapping" at an element interface; with smoothing, they can result in "a peculiar 'kick-back' on the trajectory close to a fracture zone". See Figure 3.23 of OECD (1988).

il = *jl* = *kl*	*relerr_p* (%)	*relerr_coef* =
5	2.8987 E+00	4.6379 E–01
9	5.2918 E–01	3.3868 E–01
17	6.5718 E–02	1.6824 E–01
33	5.0346 E–02	5.1555 E–01

Table 8.6.2.2. 3-D Tracker Results For Test #1, Uniform Grid, MAC = 0.
(From Table 12 of Roache et al, 1990.)

N	*relerr_p*	*N* error ratio
4	2.8786	
8	0.5292	5.4
16	0.0667	7.9
32	0.0503	1.3

Table 8.6.2.3. 3-D Tracker Results For Test #2, Uniform Grid, MAC = 0.
(From Table 13 of Roache et al, 1990.)

8.7 Inadequacy of Single Grid Calculations for Parameter Trends

Considerable effort has been invested in developing methods which always give smooth, reasonable looking solutions at coarse grid resolution, with the idea that even under-resolved solutions could provide some useful engineering information, e.g. trends of some flow functional with parameters. Even the hypercritical Oreskes et al (1994) and pessimistic Konikow (1992) (see discussion in Chapter 2), while claiming that Validation and Verification are impossible, state that computational simulations can provide reliable information on trends.

In fact, nothing can be taken for granted in this respect. In a classic paper for the Finite Element community entitled "Don't Suppress the Wiggles—They Are Trying to Tell You Something," Gresho and Lee (1981) warned against the design of algorithms which provide smooth solutions at coarse resolutions. The most common such method is simple upstream differencing for advection, which tends to obliterate small scale flow features, but there are more subtle methods with more subtle qualitative errors. Brown and Minion (1995) gave examples of misleading behavior of a Godunov projection method for incompressible Navier-Stokes equations. The method always produced "smooth, apparently physical solutions" even when under-resolved. However, closer examination indicates the presence of "spurious nonphysical vortices that are artifacts of the under-resolution." That is, rather than *obliterate* small-scale features as simple upstream differencing does, this method *introduces* false small-scale features. The authors note that "these artifacts are not unique to Godunov methods" but "are observed with other difference approximations as well." Their lesson is that "it is not sufficient that the computed solutions appear smooth and well-resolved," and they note again the observation of Gresho and Lee (1981) and many others, that the *obvious* failure of simpler methods is less dangerous. "Since the centered methods fail rather badly in the under-resolved case, it is somewhat easier to know when one is properly resolving the computed solutions for those methods." On the other hand, one prefers smooth solutions at all levels of grid resolution to guide solution adaptive grid generation, so we should not eliminate these methods entirely.

Even the use of simple 2nd-order centered differencing does not guarantee correct parameter trends for under-resolved grids. In a very early pioneering paper, Burggraf (1966) computed the now-common Benchmark problem of steady incompressible Navier-Stokes flow in a driven cavity using 2nd-order centered differences. At the low Reynolds numbers, his solutions were smooth and realistic looking, clearly showing not only the principal vortex but also corner eddies, even for coarse resolutions of 20×20 and 40×40 cells. Burggraf noted the trend of movement of the position of these corner eddies with change in Reynolds number. This is surely a simple trend, and one that might be expected to be reliably predicted with grids so slightly under-resolved that the qualitatively correct flow features were present. In fact, Burggraf was not mislead, noting that the *opposite trend* of eddy position with Reynolds number was obtained with the 20×20 grid compared to the 40×40 grid.

Trends and sensitivity calculations (derivatives of flow solution variables or functionals with respect to parameters that determine the flow) are complicated by discontinuities such as shock waves and even the end points of (continuous) rarefaction waves. See Appel and Gunzburger (1997) for a taxonomy of sensitivity calculation approaches and a clear exposition of difficulties with discontinuous solutions, including numerical algorithmic issues and grid refinement.

These examples do not indicate a hopeless situation. They simply indicate the need for grid resolution studies, no matter what the numerical method used, and no matter if the first grid solution looks plausible. The danger is in the commonly expressed opinion that qualitative accuracy of parameter trends can be taken for granted, which is nothing more than *wishful thinking*.

8.8 HARD-WIRED DATA VS. USER INPUT DATA

Data on physical properties is a source of error in simulations, and can cloud the distinctions between Verification of Codes vs. Verification of Calculations, between Verification and Validation, and between conceptual model errors and code errors.

For example, consider thermodynamic properties for individual chemical species, as discussed by Oberkampf et al (1995). Experimental errors in input values should be considered in Validation, clearly. However, the variation of properties with temperature will be accounted by table look-up and interpolation, which introduces additional error and could arguably be considered a numerical error of the simulation, or a numerical error of the experimental data reduction, depending on who supplies the interpolation algorithm. (Oberkampf et al note that for "table look-up/interpolation or curve fits...surprisingly, for many cases in the past, the approaches resulted in large errors.") Likewise, if the data and interpolation algorithm are hard-wired in the code, the issue is Verification and Validation of the code, but if it supplied by the user, it is a conceptual modeling question of Validation. Code QA (Quality Assurance) issues (see Chapter 11) are also clouded if the code has options for *both* hard-wired data (which may have gone through Certification/QA) *and* for user over-ride with other input data. This code flexibility is obviously a desirable feature for the engineering analyst, yet it causes genuine problems in engineering management: the code Certification/QA process cannot cover undefined user input, and the very meaning of Certification/QA becomes questionable. For an analyst user to claim use of a "Certified" or "Quality Assured" code may provide misleading confidence in the simulation if the user has over-ridden the hard-wired property data that was an inherent part of the Certification/QA process.

8.9 DEGRADED RATE OF CONVERGENCE DUE TO USER MODELING ERRORS

Because of its importance, we again describe Example 3 from Chapter 3, a groundwater contaminant transport calculation. The point in Chapter 3 was the remarkable sensitivity of systematic grid convergence testing; the point herein is that the code itself was proven to be 2nd-order accurate for this type of problem, yet a subtle modeling error by the *user* during the grid convergence testing degraded the rate of convergence to 1st order.

In the groundwater contaminant transport calculations (advection-diffusion + decay, retardation, and matrix diffusion), use of a plausible single-grid-block representation for a point source as the grid is refined introduces error in a finite volume (or block-centered finite difference) formulation. In this cell configuration, the cell faces align with the boundaries of the computational domain, and doubling the number of cells requires the *location* of the single cell representing the source to *shift by* $\Delta/2$. It is to be expected that the solution accuracy in the neighborhood of the source would be affected, but surprisingly, the accuracy of time-integrated *discharge* across boundaries far from the source was also degraded to 1st-order accuracy. See Chapter 6, Section 6.12 for details.

8.10 LESSONS FROM NONLINEAR DYNAMICS

H. Yee and colleagues (Yee et al, 1991; Lafon and Yee, 1992; Yee and Sweby, 1996) have given a large number of examples of aerodynamics solutions involving "nonlinear dynamics" (see additional literature citations in Yee and Sweby, 1996). Spurious (grossly erroneous) numerical solutions for steady state problems with strongly nonlinear source terms can be obtained for stable implicit algorithms applied beyond the linear stability limit. The nonlinear dynamics solutions, and the strange and chaotic convergence behavior, are interesting and are valuable to explain iterative convergence behavior relative to initial conditions through the representation of *basins of attraction*. These are maps of initial condition parameters which lead to iterative convergence; an example is shown in Figure 8.10.1. The system being solved is for a nonequilibrium flow field relaxation of the 1-D Euler equations for a (N_2, N) mixture, representative of shock tube experiments or hypersonic wind tunnels. The model is expressed as a single ODE,

$$\frac{dz}{dx} = S(\rho, T, z) \tag{8.10.1}$$

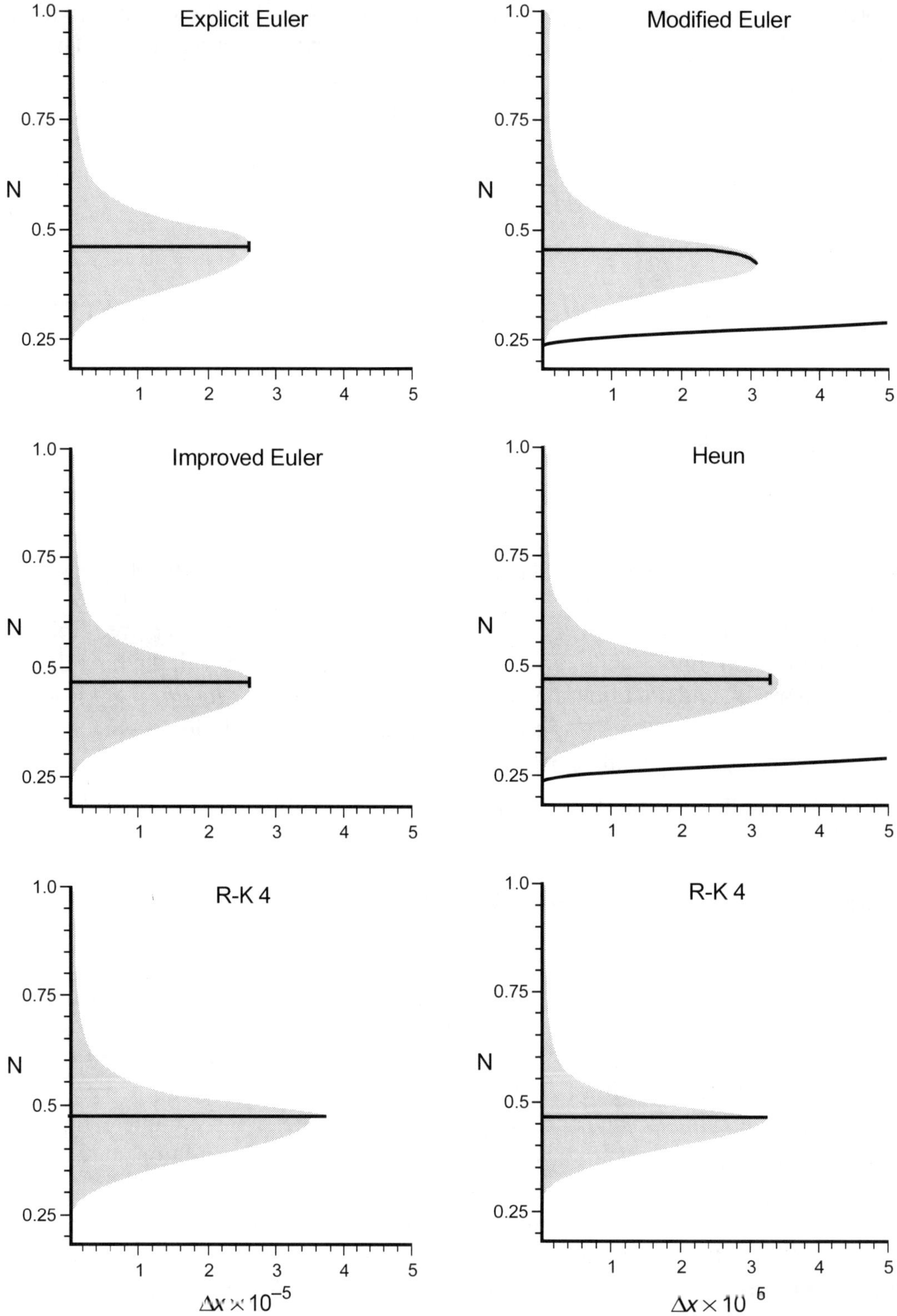

Figure 8.10.1. Basins of attraction for Equation 8.10.1 for various numerical methods.
(From Figure 4.1 of Yee and Sweby, 1996.)

where z is the mass fraction of the N_2 species, ρ is the density of the mixture, and T is the temperature. (There are two algebraic equations for ρ and T.) See Yee and Sweby (1996) for details on the system of equations.

Figure 8.10.1 shows the experimentally determined basins of attraction for this system using various numerical methods. In each of these plots, the shaded region denotes the basin of attraction, in which combinations of initial z values and step size Δx converge to the asymptotes, which are depicted by the solid black line. The unshaded region indicates combinations of initial z and Δx values that do not converge to a physical solution of the problem. "As can be seen, the basins of attraction narrow considerably for the larger values of Δx. (Note that the axis scale is 10^{-5}!)"

Unfortunately, these calculations are extremely computationally intensive, and cannot be expected to be performed as a matter of course. In Yee et al (1991), the authors warn against obtaining spurious solutions, especially with implicit methods applied well beyond the stability limit of explicit methods. The points of their work are the danger of being misled by such calculations, the elucidation of the convergence behavior through description of the basins of attractions, and the need for further research to guarantee *a priori* accuracy. Their studies on basins of attraction shed some light on the "phenomenon of near (but lack of) convergence in large stiff systems," or what other authors have described as "stalling" of iterative convergence (e.g., see Roache, 1972b), by associating the behavior with the bifurcation of spirals to limit cycles. The authors claim that "isolation of the sources of numerical uncertainties is of fundamental importance," but I do not agree that it is a necessary prelude to Verification of Calculations. Results from further such research are of course inherently interesting, and perhaps "can be a viable complement to the standard guidelines for numerical analysis and CFD practices," but we already have the tools necessary to Verify Calculations.

The major goal motivating their work (Yee, 1992) is the development of numerical methods that guarantee *a priori* accuracy, without the need for grid convergence testing or other error estimators. I consider this goal to be unrealistic for any non-trivial problem (perhaps for anything but the Laplace equation), and that the real lesson of their examples is simply the need for *a posteriori* error estimation. I reject the claim (Yee and Sweby, 1996) that standard practices such as grid convergence tests may be "not possible (e.g., too CPU intensive)," as previously discussed (Chapter 4).

In all but one of the examples in Yee et al (1991), the most cursory time-step convergence study would quickly reveal the inadequacy of the temporal resolution, with no reasonable chance of being misled into a non-physical solution. The one remaining example (involving a fourth-order Runge-Kutta integration) could be misleading with a cursory study, but would be revealed in a thorough systematic study. Many of the other extensive list of CFD topics discussed, especially in Yee and Sweby (1996), are overly broad and constitute non-issues to numerical uncertainties and Verification, e.g., solution strategies, or improving iterative convergence rates.

(My remarks apply to computational simulations of physical processes, not to studies of numerical chaos for its own mathematical interest, which I do not share.)

In my opinion, the most important and surprising point to be gained from these examples is the following. With nonlinear iterative methods (not necessarily nonlinear *problems*, but simply nonlinear *methods*, such as classical 4th-order Runge-Kutta time integration, or nonlinear flux limiters such as FCT, TVD, ENO, etc.) it is necessary to check the effect of changing Δt (or equivalently, the relaxation parameter for non-time-accurate iterative methods) on the solution *even for a steady-state problem.*

As noted, the nonlinearity need not be part of the problem definition, but can arise simply from the numerical method itself. Nonlinear flux limiters introduce nonlinearity to linear problems whenever wave fronts in the solution are not well resolved. As noted by Yee et al (1991) such methods are especially likely to cause spurious steady-state solutions when they are implicit and when the time step used greatly exceeds the allowable stable time step for an explicit method. Note that some more straight-forward methods are also nonlinear, notably the common family of Lax-Wendroff methods. Applied to the constant coefficient scalar advection equation,

$$\frac{\partial f}{\partial t} = -V\frac{\partial f}{\partial x} \tag{8.10.2}$$

all the Lax-Wendroff methods reduce to (Roache, 1972a,b)

$$f_i^{n+1} = -c\left(f_{i+1}^n - f_{i-1}^n\right) + \frac{c^2}{2}\left(f_{i+1}^n + 2f_i^n + f_{i-1}^n\right) \tag{8.10.3}$$

where the Courant number is

$$c = V\frac{\Delta t}{\Delta x} \tag{8.10.4}$$

Thus, although the modeled equation is linear, the discrete equation is nonlinear not in the dependent variable f, but in the discretization parameters Δt and Δx (through the c^2). That the steady state discrete solution is actually a function of Δt for a Lax-Wendroff Euler code was demonstrated by Roache (1972a). (See also Appendix B of Roache, 1972b.) The interpretation given was in terms of the artificial viscosity for *steady state* solutions, which is different from the *transient* artificial viscosity for some methods (notably Lax-Wendroff). Once forewarned, the practitioner can easily test for this effect by changing the Δt (or analogous steady state iteration parameter) after the solution has been obtained. Such Verification is economical because the bulk of the computational cost will have been expended in obtaining the base solution. If there is no sensitivity to Δt, this fact is verified with the work equivalent of an additional *single time step.* Also, unlike automated spatial grid convergence testing, this feature would be trivial to include in a code as an automatic feature, along with a warning message to the user if the Verification fails.

Since most of the examples of spurious solutions are trivial to check with time-step convergence testing, the sometimes high costs of spatial grid convergence testing are not an issue here. However, it is *very significant* that single-gird convergence estimators (Chapter 7) by themselves (i.e., without time-step convergence tests) presumably would *not* reveal these spurious solutions. Of course, the time-step convergence test is cheap, and involves no problem of spatial grid generation.

8.11 ADAPTIVE AND LOCAL TIME STEPPING, AND STEADY STATE

This phenomenon of spurious nonlinear dynamics solutions also suggests to me that the simplest adaptive time stepping approach will not be adequate for such nonlinear algorithms. That is, a simple adaptive time stepping algorithm of the *forward type* (not repeating the previous time step calculation, but only adjusting the next time step) could be tricked into accepting a spurious nonlinear solution. However, this can be adequately checked by storing the Δt sequence produced by the adaptive time stepping algorithm, and then repeating the calculation with the entire sequence reduced or increased by some factor, analogous to spatial grid refinement or coarsening.

In a related issue, Benek et al (1996) express concern about "unsteady acceleration schemes based upon local time stepping." These schemes basically consist of choosing a *local* Δt close to the optimum value indicated by analysis applied locally, and usually amount to setting the local Courant number ~ 1. Obviously, physical transient behavior will not be faithfully represented, since the method is inconsistent with the original PDE's. Benek et al (1996) state the following. "It is assumed that such a procedure will produce the same converged solution as an unaccelerated method for steady state flows. Is there an effect on the approximation of unsteady flow with this type of acceleration?"

The answer follows. In the limit of all $\Delta t \rightarrow 0$ but at different rates, i.e. maintaining some Δt's larger than others, the PDE consistently simulated is one with a distorted time variable, so unsteady solutions will not be correct. Clearly, a continuum steady state solution for distorted time PDE will also be a continuum steady state for the undistorted time PDE, so the method is not dangerous. However, it is also conceivable that severely distorted local time stepping could confuse the non-uniqueness issue (see Chapter 2), possibly favoring a different steady state solution than a time-consistent method. Since uniqueness must be dealt with *ad hoc* in any case, I do not consider that local time stepping schemes are a significant problem.

8.12 OTHER QUESTIONS RELATED TO THE STEADY STATE

There are other questions regarding the attainment of a "steady state," raised by Benek et al (1996) and others.

- "How long should the simulation be run before the starting transients no longer affect the solution?"
- "What are the optimal initial conditions from which to start the simulation [to quickly attain the steady state]?"
- "If the steady state approximation is used [in the choice of the governing equations] will the same time averaged quantities be predicted [as would be obtained by time-averaging the solution of the time-dependent equations]?"

The response to all these questions is that they do not involve computational issues *per se*. That is, the answers are not computational but physical.

For example, it is well known that pitot pressure probes, which give a single physically-averaged pressure reading, do not produce the average pressure in a fluctuating field, but a reading higher than the average; e.g., see Dolling (1996.) Computational simulations can provide answers, but not *a priori*, anymore than experiments can. Like the (strictly computational) question of what will constitute adequate grid resolution, it is simply asking too much of Computational Science to answer such questions *a priori*, but the questions are answerable *a posteriori* from the simulations. Certainly, it is not true *generally* that steady state equations will produce the same time-averaged quantities as would be obtained by time-averaging the solution of the time-dependent equations. If they did, turbulence modelers would be out of business, since time-averaged turbulence would simply reduce to the laminar flow equations!

REFERENCES FOR CHAPTER 8

AEA Technology (1994), *CFDS-Flow3D User Guide*, Computational Fluid Dynamics Services, Building 8.19, Harwell Laboratory, Oxfordshire OX11 ORA, United Kingdom.

AGARD (1988), *Validation of Computational Fluid Dynamics*, Lisbon, Portugal, 2–5 May 1988, NATO Advisory Group for Aeronautical Research and Development, AGARD CP 437, December 1988. Volume 1: Symposium Papers and Round Table Discussion. Volume 2: Poster Papers.

Appel, J. R. and Gunzburger, M. D. (1997), "Difficulties in Sensitivity Calculations for Flows with Discontinuities," *AIAA Journal*, Vol. 35, No. 5, May 1997, pp. 842–848.

Baptista, A. M. (1987), *Solution of Advection-Dominated Transport by Eulerian-Lagrangian Methods using the Backward Methods of Characteristics*, Ph.D. Thesis, Dept. of Civil Engineering, Massachusetts Institute of Technology, 1987.

Benek, J. A., Kraft, E. M., and Lauer, R. F. (1996), "Validation Issues for Engine/Airframe Integration," AIAA Paper 96–2031, 27th AIAA Fluid Dynamics Conference, New Orleans, Louisiana, June 17–20, 1996.

Bennett, J. R. and Campbell, J. E. (1987), "Accuracy of a Finite-Difference Method for Computing Lake Currents," *Journal of Computational Physics*, Vol. 68, 1987, pp. 262–271.

Bergstrom, J. and Gebart, R. (1997), "Estimation of Numerical Accuracy for the 3D Flow Field in a Hydraulic Turbine Draft Tube," *Computers and Fluids*, to appear.

Blain, C. A., J. J. Westerink, and R. A. Luettich, Jr. (1994), "The Influence of Domain Size on the Response Characteristics of a Hurricane Storm Surge Model," *Journal of Geophysical Research*, Vol. 99, C9, 1994, pp. 18467–18479.

Blottner, F. G. (1989), "Accurate Navier-Stokes Results for the Hypersonic Flow Over a Spherical Nosetip," *AIAA Journal of Spacecraft and Rockets*, Vol. 27, No. 2, 1990, March – April 1990, pp. 113–122. See also AIAA Paper 89–071, June 1989.

Brown, D. L. and Minion, M. L. (1995), "Performance of Under-Resolved Two-Dimensional Incompressible Flow Simulations, *Journal of Computational Physics*, Vol. 122, 1995, pp. 165–183.

Burggraf, O. R. (1966), "Analytical and Numerical Studies of the Structure of Steady Separated Flows," *Journal of Fluid Mechanics*, Vol. 24, Part 1, 1966, pp. 113–152.

Cantekin, M. E. and Westerink, J. J. (1990), "Non-Diffusive $N + 2$ Degree Petrov-Galerkin Methods for Two-Dimensional Transient Transport Computations," *International Journal for Numerical Methods in Engineering*, Vol. 11, 1990, pp. 397– 418.

Dietrich, D. E., Roache, P. J. and Marietta, M. G. (1990), "Convergence Studies with the Sandia Ocean Modeling System," *International Journal for Numerical Methods in Engineering*, Vol. 11, 1990, pp. 127–150.

Dietrich, D. E., and Roache, P. J. (1991), "An Accurate Low Dissipation Model of the Gulf of Mexico Circulation," *Proceedings International Symposium on Environmental Hydraulics*, 14–16 December 1991, University of Hong Kong, J. H. W. Lee, et al, eds., H. Balkeema, Amsterdam.

Dolling, D. S. (1996), "Considerations in the Comparison of Experimental Data with Simulations–Consistency of Math Models and Flow Physics," AIAA Paper 96–2030, 27th AIAA Fluid Dynamics Conference, New Orleans, Louisiana, June 17–20, 1996.

Ekaterinaris, J. A. and Menter, F. R. (1994), "Computation of Separated and Unsteady Flows with One and Two-Equation Turbulence Models," AIAA Paper 94–0190, January 1994.

Ekaterinaris, J. A. (1995), "Numerical Investigation of Dynamic Stall of an Oscillating Wing, *AIAA Journal*, Vol. 33, No. 10, October 1995, pp. 1803–1808.

Gresho, P. M. and Lee, R. L. (1981), "Don't Suppress the Wiggles–They're Telling You Something," *Computers and Fluids*, Vol. 9, No. 2, 1981, pp. 223–255.

Johns, B., Dube, S. K., Mohanty, U. C. and Rao, A. D. (1983a), "Simulation of Storm Surges Using a Three-Dimensional Numerical Model: An Application to the 1977 Andhra Cyclone," *Quarterly Journal of Research of the Meteorology Society*, Vol. 109, 1983, pp. 211–224.

Johns, B., Dube, S. K., Mohanty, U. C. and Rao, A. D. (1983b), "On the Effect of Bathymetry in Numerical Storm Surge Simulation Experiments," *Computers and Fluids*, Vol. 11, 1983, pp. 161–174.

Johnson, R. W. and Hughes, E. D. (1995), Eds., *Quantification of Uncertainty in Computational Fluids Dynamics–1995,* Joint JSME–ASME Fluid Mechanics Meeting, ASME FED Volume 213, Hilton Head, South Carolina, August 14–18, 1995.

Konikow, L. F. and Bredehoeft, J. D. (1992), "Groundwater Models Cannot be Validated," *Advances in Water Resources*, Vol. 15, 1992, pp. 75–83.

Lafon, A. and Yee, H. C. (1992), "On the Numerical Treatment of Nonlinear Source Terms in Reaction-Convection Equations," AIAA Paper 92–0419, 30th AIAA Aerospace Sciences Meeting and Exhibit, Reno, Nevada, 6–9 January 1992.

Lardner, R. W. and Song, Y. (1992), "A Comparison of Spatial Grids for Numerical Modelling of Flows in Near-Coastal Seas," *International Journal for Numerical Methods in Engineering*, Vol. 14, 1992, pp. 109–114.

Leonard, B. P. (1991), "The ULTIMATE Conservative Difference Scheme Applied to One-Dimensional Scalar Advection," *Computer Methods in Applied Mechanics and Engineering*, Vol. 88, 1991, pp. 17–94.

Leonard, B. P., MacVean, M. K., and Lock, A. P. (1995), "The Flux Integral Method for Multidimensional Convection and Diffusion," NASA TM 106679, ICOMP–94–13. Also, *Applied Mathematical Modeling*, Vol. 19, pp. 333–342.

Leonard, B. P. and MacVean, M. K. (1995), "Grid Refinement Study of a Benchmark Problem Using the Flux-Integral Method," in R. W. Johnson and E. D. Hughes (1995), Eds., *Quantification of Uncertainty in Computational Fluids Dynamics–1995,* Joint JSME–ASME Fluid Mechanics Meeting, ASME FED Vol. 213, Hilton Head, South Carolina, August 14–18, 1995, pp. 45–50.

Le Provost, C. and Vincent, P. (1986), "Some Tests of Precision for a Finite Element Model of Ocean Tides," *Journal of Computational Physics*, Vol. 65, 1986, pp. 273–291.

Lin, C. C. (1967), *The Theory of Hydrodynamic Stability*, Cambridge University Press, London.

Luettich, R. A. and Westerink, J. W. (1994), "Continental Shelf Scale Convergence Studies with a Baratropic Tidal Model," *Quantitative Skill Assessment for Coastal Ocean Models*, D. R. Lynch and A. M. Davies, Eds., AGU Press.

McCroskey, W. J. (1981), "The Phenomenon of Dynamic Stall," NASA TM 81264.

Mueller, T. J., Hall, C. R., and Roache, P. J. (1970), "Influence of Initial Flow Direction on the Turbulent Base Pressure in Supersonic Axisymmetric Flow," *AIAA Journal of Spacecraft and Rockets*, Vol. 7, No. 12, December 1970.

Mueller, T. J. and Roache, P. J. (1973), "Base Drag Calculations in Supersonic Turbulent Axisymmetric Flows," *AIAA Journal of Spacecraft and Rockets*, Vol. 10, No. 4, April 1973.

Oberkampf, W. L., Blottner, F. G., and Aeschliman, D. P. (1995), "Methodology for Computational Fluid Dynamics Code Verification/Validation," AIAA Paper 95–2226, 26th AIAA Fluid Dynamics Conference, San Diego, California, 19–22 June 1995.

OECD (1988), *The International HYDROCOIN Project: Groundwater Hydrology Modelling Strategies for Performance Assessment of Nuclear Waste Disposal. Level 1: Code Verification*, Nuclear Energy Agency, Organisation for Economic Co-Operation and Development (OECD), Paris.

Oldenburg, C. M. and Pruess, K. (1995), "Dispersive Transport Dynamics in a Strongly Coupled Groundwater-Brine Flow System," *Water Resources Research*, Vol. 31, No. 2, February 1995, pp. 289–302.

Oreskes, N., Shrader-Frechette, K., and Belitz, K. (1994), "Verification, Validation, and Confirmation of Numerical Models in the Earth Sciences," *Science*, Vol. 263, 4 February 1994, pp. 641–646.

Piacsek, S. and Allard, R. (1993), "The Importance of Horizontal Resolution in Coupled Ice/Ocean Models of the Arctic," *SIAM Conference on Mathematical and Computational Issues in the Geosciences*, Houston, Texas, April 19–21, 1993.

Roache, P. J. (1972a), "On Artificial Viscosity," *Journal of Computational Physics*, Vol. 10, No. 2, October 1972, pp. 169–184.

Roache, Patrick J. (1972b), *Computational Fluid Dynamics*, First Edition, Hermosa Publishers, P. O. Box 9110, Albuquerque, New Mexico 87119-9110, USA. http://www.hermosa-pub.com/hermosa. Revised Printing 1976. See also Roache (1998).

Roache, P. J. (1982), "Semidirect/Marching Methods and Elliptic Grid Generation," *Symposium on the Numerical Generation of Curvilinear Coordinate Systems and Use in the Numerical*

Solution of Partial Differential Equations, J. F. Thompson, Ed., April 1982, Nashville, Tennessee, Elsevier Science Publishing Company, pp. 729–737.

Roache, P. J. and Salari, K. (1990), "Weakly Compressible Navier-Stokes Solutions with an Implicit Approximate Factorization Code," AIAA Paper 90–0235, 28th AIAA Aerospace Sciences Meeting, Reno, Nevada, January 8–11, 1990.

Roache, P. J., Knupp, P. M., Steinberg, S., Blaine, R. L. (1990), "Experience with Benchmark Test Cases for Groundwater Flow." ASME FED Vol. 93, *Benchmark Test Cases for Computational Fluid Dynamics*, I. Celik and C. J. Freitas, Eds., Book No. H00598, June 1990, pp. 49–56.

Roache, P. J. (1992), "A Flux-Based Modified Method of Characteristics," *International Journal for Numerical Methods in Fluids,* Vol. 15, pp. 1259–1275

Roache, P. J. (1993a), "The SECO Suite of Codes for Site Performance Assessment," 1993 International High-Level Radioactive Waste Management Conference, Las Vegas, Nevada, April 26–30, 1993.

Roache, P. J. (1993b), "The SECO Code Algorithms for Groundwater Flow and Transport," *Finite Elements in Fluids: New Trends and Applications*, Part II, Proceedings of the VIII International Conference on Finite Elements in Fluids, Univ. Politécnica de Cataluña, Barcelona, Spain, K. Morgan, E. Onate, J. Periaux, J. Peraire, and O. C. Zienkiewicz, Eds., pp. 939–948, 20–23 September 1993.

Roache, P. J. (1994), "Perspective: A Method for Uniform Reporting of Grid Refinement Studies," *ASME Journal of Fluids Engineering*, Vol. 116, 1994, pp. 405–413.

Roache, P. J. (1995), "Verification of Codes and Calculations," AIAA Paper 95–2224, 26th AIAA Fluid Dynamics Conference, San Diego, California, June 19–22, 1995. Also, *AIAA Journal*, May 1998.

Roache, P. J. (1997), "Quantification of Uncertainty in Computational Fluid Dynamics," *Annual Review of Fluid Mechanics*, Vol. 29, 1997, pp. 123–160.

Roache, P. J. (1998), *Fundamentals of Computational Fluid Dynamics*, Hermosa Publishers, P. O. Box 9110, Albuquerque, New Mexico 87119-9110, USA. http://www.hermosa-pub.com/hermosa. To appear.

Rosenfeld, M. (1994), "Grid Refinement Test of Time-Periodic Flows over Bluff Bodies," *Computers and Fluids*, Vol. 23, No. 5, 1994, pp. 693–709.

Rumsey, C. L., Sanetrik, M. D., Biedron, R. T., Melson, N. D. and Parlette, E. B. (1996), "Efficiency and Accuracy of Time-Accurate Turbulent Navier-Stokes Computations," *Computers and Fluids*, Vol. 25, No. 2, 1996, pp. 217–236.

Salari, K. and Roache, P. J. (1990), "The Influence of Sweep on Dynamic Stall Produced by a Rapidly Pitching Wing," AIAA Paper 90–0581, Reno, Nevada, January 8–11, 1990.

Salari, K. (1997), personal communication.

Sengupta, R., Orkwis, P. D., and Davis, S. M. (1995), "Unsteady Solutions, Solution Transients and Local Grid Resolution Requirements for Asymmetric Vortex Formation," AIAA Paper 95–0364, AIAA 33rd Aerospace Sciences Meeting, Reno, Nevada, January 9–12, 1995.

Shampine, L. F., Watts, W. A., and Davenport, S. (1976), "Solving Non-Stiff Ordinary Differential Equations–The State of the Art," Sandia Laboratories Report SAND 75–0182, 1975, Sandia National Laboratories, Albuquerque, New Mexico. Also, *SIAM Review*, Vol. 18, No. 3, July 1976, pp. 376–411.

Steinberg, S. and Roache, P. J. (1985), "Symbolic Manipulation and Computational Fluid Dynamics," *Journal of Computational Physics*, Vol. 57, 1985, pp. 251–284.

Straka, J. M., Wilhelmson, R. B., Wicker, L. J., Anderson, J. R. and Droegemeier, K. K. (1993), "Numerical Solutions of a Non-Linear Density Current: A Benchmark Solution and Comparisons," *International Journal for Numerical Methods in Engineering*, Vol. 17, 1993, pp. 1–22.

Truman, C. R. and Lee, M. J. (1990), "Effects of Organized Turbulence Structures on the Phase Distortion in a Coherent Optical Beam Propagating through a Turbulent Shear Flow," *Physics of Fluids A*, Vol. 2, No. 5, 1990, pp. 851–857.

Van Leer, B. (1979), "Towards the Ultimate Conservative Difference Scheme. V. A Second-Order Sequel to Godunov's Method," *Journal of Computational Physics*, Vol. 32, 1979, pp. 101–136.

Viegas, J. R., Rubesin, M. W., and MacCormack, R. W (1988), "On the Validation of a Code and a Turbulence Model Appropriate to Circulation Controlled Airfoils," in AGARD (1988), Paper No. 6.

Westerink, J. J., K. D. Stolzenbach and J. J. Connor (1989), "General Spectral Computations of the Non-Linear Shallow Water Tidal Interactions Within the Bight of Abaco," *Journal of Physical Oceanography*. Vol. 19, 1989, pp. 1350–1373.

Westerink, J. J., R. A. Luettich, and J. C. Muccino (1994), "Modeling Tides in the Western North Atlantic Using Unstructured Graded Grids," *Tellus*, Vol. 46A, 1994, pp. 187–199.

Westerink, J. J., R. A. Luettich, C. A. Blain, and S. C. Hagen (1995), "The Utility of the Finite Element Method in Computing Surface Elevation and Circulation in Continental Margin

Waters," in *Finite Element Modeling of Environmental Problems*, G. F. Carey, Ed., John Wiley and Sons, Chichester, pp. 39–59.

Westerink, J. J. and Roache, P. J. (1995), "Issues in Convergence Studies in Geophysical Flow Computations," Joint JSME–ASME Fluid Mechanics Meeting, August 14–18, 1995, Hilton Head, South Carolina, Session F137. See also Proceedings of the EPA Workshop on *Next Generation Environmental Models Computational Methods*, G. Delic and Mary F. Wheeler, eds., Society for Industrial and Applied Mathematics, Philadelphia, Pennsylvania, 1997, pp. 95–107.

Yee, H. C., Sweby, P. K., and Griffiths, D. F. (1991), "Dynamical Approach Study of Spurious Steady-State Numerical Solutions of Nonlinear Differential Equations. I. The Dynamics of Time Discretization and its Implications for Algorithm Development in Computational Fluid Dynamics," *Journal of Computational Physics*, Vol. 97, 1991, pp. 249–310.

Yee, H. C. (1992), personal communication.

Yee, H. C. and Sweby, P. K. (1996), "Some Aspects of Numerical Uncertainties in Time-Marching to Steady-State Numerical Solutions," AIAA Paper 96–2052, 27th AIAA Fluid Dynamics Conference, New Orleans, Louisiana, June 17–20, 1996.

PART III
VALIDATION

Part II of this book covered "Verification", the process of determining whether or not a code is "solving the equations right", a purely mathematical question. Part III now covers "Validation", the process of determining whether or not a code is "solving the right equations" for a particular physical problem. Verification is essentially and strictly an activity in *mathematics*, the mathematics of numerical analysis, whereas Validation is essentially and strictly an activity in *science*: physics, chemistry, fluid dynamics, even the "soft" sciences of economics, sociology, etc. Common sense and the testimony of experienced practitioners strongly indicates that rigorous Verification, first of the Code, then of the particular Calculation, should *precede* any Validation exercise.

I will have somewhat less to say on Validation than Verification, because there is less methodology and more experiment involved, and therefore Validation is more specific to a particular area. Still, some useful general observations are possible.

It is worthwhile to repeat here the limitations of our attempts at semantic distinction, already discussed in the conclusion to Chapter 2. Although valuable, it is clear that we (the scientific - mathematical - engineering community) are *not* going to achieve uniform, non-overlapping terminology, even in the most nearly universally accepted terms, Verification and Validation. Nevertheless, following the advice of Chuang Tzu (see Introduction to Chapter 2), it is worthwhile to keep in mind the ideas *behind* Verification and Validation. In reading journal papers and reports, it is not a good idea to try to enforce terminology or taxonomies, but rather to try to detect the authors' terminology or taxonomies (often implicit, rather than explicitly defined) and to learn from the authors' experiences and perspectives.

9

DIFFICULTIES WITH EXPERIMENTS AND VALIDATION

This Chapter 9 covers general concepts of experimentation and Validation exercises. The following Chapter 10 will present specific examples. The theme chosen for this Chapter is the *difficulties* with experiments, for good reason. Experimental data is not as sacrosanct as many computational practitioners believe.

9.1 CREDULOUSNESS

There is a saying in the aerodynamics community. "No one believes the CFD results except the one who performed the calculation, and everyone believes the experimental results except the one who performed the experiment." (On a personal note, my early career experience was as an experimentalist, so I have never had a problem with excessive gullibility.)

Very significant progress has been made in "non-intrusive" experimental techniques (e.g., Marvin, 1995) including 3-D laser velocimetry of mean and fluctuating velocities, laser fluorescence for mean and fluctuating densities and temperatures, laser interferometry for local skin friction, and pressure-sensitive and thermographic paints for surface pressure and heating. These are vast improvements over previous clearly intrusive techniques such as pitot-static tubes, hot wire anemometers, thermocouples, etc. (Even flow visualization can be intrusive for liquid free convection flows, as the illumination provides a heat source; Jankowski, 1985.) At the other extreme, removal of rock from its field location for laboratory testing can clearly have order of magnitude effects on permeability measurements. Alternately, the preferential selection of field samples which are intact can bias the sampling towards lower permeabilities (Brinster, 1995). But even for virtually non-intrusive measurement techniques, there are still experimental errors involved. For example, wind tunnels suffer from flow angularity and blockage effects, which are further complicated because they vary with angle of attack of the models. Likewise, hypersonic facilities often are plagued with axial static pressure variation that has no counterpart in flight.

Briefly, it is naive to accept a single experiment as the final word. As Marvin (1995) notes, "Reliance on single experiments or measurement procedures during the code Validation process should be viewed with caution because of facility and instrumentation limitations. Therefore, redundant measurements and similar experiments performed in more than one facility are desirable." Also, Aeschliman and Oberkampf (1997, p. 13) make an interesting observation based on their extensive experience. "The general point is that as one progresses down the list to more difficult quantities for CFD to predict, the experimental uncertainty generally increases also."

Uncritical acceptance of experimental values can and has held up progress for years. Both false invalidation (i.e., a negative conclusion on Validation) and false Validation have occurred. Wilcox (1994) described one case where disagreement with ablating heat transfer data was originally attributed to inadequacy of a turbulence model. After several years, direct communication with the experimentalists clarified previously incomplete boundary conditions and produced excellent agreement. Conversely, a turbulent pipe flow computation (with a different turbulence model) that agreed well with experiment proved to be false Validation; later reviews showed *both* experiment and computation were seriously in error (Aeschliman and Oberkampf, 1997).

Barber (1996) provides a review of several code Validation/Certification studies in the aerodynamics literature, and cites examples wherein missing experimental details have lead to poor Benchmark calculation comparisons. The details he discusses are

1. the geometrical definition,
2. the data reduction procedures, and
3. the dominant physics.

Similarly, Marvin (1995) notes that "Accuracy assessments for experimental data are essential; otherwise there is no quantitative means for determining the Validated range of a code." Required are "error estimates for test geometry dimensions, test operating and freestream conditions, model and flowfield measured variables, and instrumentation." In the very difficult experiments associated with hypersonic airbreathing engines, "geometrical definition" errors include the axial and vertical cowl *positions*. Other sources of error are model wall temperature and tunnel operating conditions (Van Wie and Rice, 1996). In these experiments on hypersonic inlet performance, the data reduction errors reach a new level. "Given that CFD techniques play a critical role in the measurement process (through estimating certain quantities and sensitivities), it becomes difficult to separate the Validation of the measurement procedure from the Validation of the [computational] analysis procedure. In this situation, the experimental and [computational] analytical techniques are intertwined in a single process." (Van Wie and Rice, 1996).

Also, Marvin (1995) states that "Uncertainty analysis is a well established method for determining experimental data accuracy, and it should be a prerequisite for all Validation databases," for which he cites the well-known paper by Moffat (1981). For more recent systematic discussion of experimental uncertainty, see Coleman and Steele (1995), Coleman et al (1995), Coleman (1996), Steele et al (1996), and Aeschliman and Oberkampf (1997).

In the aerodynamics community, there is a growing consensus among those who have seriously considered Validation issues that results from older experiments not designed specifically for CFD Validation are usually inadequate for the task. For example, Aeschliman and Oberkampf (1997, p. 10) state the following. "No rational computational fluid dynamicist would suggest there is no need for Verification or Validation of CFD codes. A common view among CFDers, however, is that while code V&V is indeed necessary, the process—particularly the Validation step—can be accomplished through comparison to existing data, documented in reports or archival journals, obtained for some purpose other than CFD code Validation. We most strongly disagree."

Their reasons, based on wide experience, are as follows. "Almost invariably, critical details are missing from published data, particularly for archival journal publications where discussion is limited in the interest of reducing paper length. It is critically important that the boundary and/or initial conditions assumed by the code be accurately known from the experiment." Some specific examples will be given in Chapter 10. I suspect that this situation in the aerodynamics community is representative of other areas of computational science and engineering.

Similarly, Porter (1996), reporting on several workshops in the aerodynamics community on Validation/Certification, includes these consensus statements (as of 1996).

- There is no focused effort to provide the understanding of fundamental physical phenomena so they can be modeled in CFD codes.
- There is a dearth of archival quality, benchmark experimental data specifically designed for CFD code Validation and Calibration.
- Experimental data that do exist have deficiencies, inadequate documentation, and are not readily accessible to the community at large.

9.2 HISTORICAL METHODS OF VALIDATING SCIENTIFIC THEORIES

The task of Validating a conceptual model and code has been likened to the general problem of Validating scientific theories (Mehta, 1996). "In the philosophy of science, the historical methods of validating theories are rationalism, empiricism, and positive economics. Rationalism involves the logical development of the model based on indisputable axioms...Empiricism requires that every axiom, deduction, assumption, or outcome be empirically confirmed... Positive economics is concerned with the model's predictive capability and not its structure, assumptions, or derivation. An example is the calibration of a model to predict certain features and quantities." (Mehta, 1996; Naylor and Finger, 1967).

Pure rationalism is perhaps appropriate for ethereal regimes of pure science, but most scientists and all engineers would not trust major projects to involved theories developed by difficult-to-follow derivations of "indisputable axioms." On the other end of the spectrum, positive economics could be regarded as nothing more than thinly disguised blind extrapolation, e.g. stock market models. As a practical matter for code Validation in science and engineering projects, empiricism seems to be the operative philosophy. It is not so daunting a task when approached modularly, i.e. by building on past work.

Ultimately, code Validation will come down to comparison (directly or indirectly) of code predictions with physical experiments. (I do not agree with Mehta, 1996, Figure 6 and elsewhere, that Sensitivity-Uncertainty Analysis can be included with rationalism, empiricism, and positive economics as a method of Validation.) The comparison can be direct or indirect. Indirect comparison occurs when a previously Validated code is taken as a Benchmark. In this philosophy, the Benchmark code may be regarded as a repository of experimental information as well as a means of interpolating and smoothing previous experimental results. The Validation exercise still compares numerical answers to experimental data, but second-hand experimental data, one level removed from the original experiment. As a practical matter, the error bars of the original experiments will be widened by additional computational and conceptual uncertainties of the Benchmark code. Thus, indirect Validation may be convenient and even appropriate for a project, but it does not solve any of the difficulties with experiments and Validation; rather, it compounds them.

9.3 THE THEORY-LADEN EXPERIMENT

As we noted in Chapter 2, "Validation" in a real sense applies to the Conceptual Model embodied in a code, rather than the code itself. (The code correctness is the subject of Verification, etc.) In regard to the present subject of experiments and Validation, we note the following.

A Conceptual Model is needed prior to experimental data gathering.

For a broad definition of "Conceptual Model" required here, I adopt that of Johnson (1996, p. 172) described in a rather different context (historical-critical models or methods) but that still fits our needs. "I use the term *model* to mean an imaginative construal of the subject being studied, as well as a structured picture of both process and product: a model is a paradigm within which the data pertinent to a discipline makes sense."

For example, an experimental measurement of "permeability" for flow in porous media *requires* the conceptual model of Darcy Law, which defines permeability. Likewise, hot-wire anemometry requires the Newtonian law of heat transfer. Pressure measurement requires non-directionality of pressure (diagonalization of *P* tensor, which is not so obvious—see any fluid dynamics text). There are also assumed scales of unsteadiness and spatial variation involved in any experimental data taking. As a practical matter (vs. rarefied philosophizing) the "model" is so minimal for Aristotelian pure observation that it has no consequences, but for practical data gathering in science, engineering, and other disciplines, it is important to keep this fact in mind (Johnson, 1996).

"Every observation is laden with theory."

Likewise, most experiments involve computation, and many now involve *simulations*. Examples will be given in Chapter 10.

9.4 RANDOM AND SYSTEMATIC ERRORS IN EXPERIMENTS

Certainly every experiment has some uncertainty associated with the results, and certainly it makes no sense to expect or even look for computational comparisons finer than this experimental uncertainty. As Coleman (1996) states, "the [experimental] uncertainties should set the scale at which comparisons (or Validations) should be made or attempted. This holds for data-data, model-data, and model-model comparisons."

"Total error can be considered to be composed of two components: a *precision* (random) component and a *bias* (systematic) component...”...“An error is classified as precision if it contributes to the scatter of the data [in repeat experiments in the same facility]; otherwise, it is a bias error. It is assumed that corrections have been made for all systematic errors whose values are known." Unfortunately, all experimentalists know that random component errors are straightforward (though tedious) to evaluate, whereas systematic (bias) errors are difficult. It is only bias errors that can be removed by Calibration; e.g., see Van Wie and Rice (1996).

Coleman (1996) gives a partial classification of possible sources of bias errors in experiments:

- calibration errors,
- data acquisition errors,
- data reduction errors,
- test technique errors, etc.

He also considers the propagation of random and bias errors through the data reduction process, and the important but difficult consideration of *correlated* precision uncertainties.

Aeschliman et al (1995) provide and excellent overview of ground-based experimental aerodynamics uncertainty analysis, including a very specific and complete case study of the

Sandia National Laboratory's long-term project called JCEAP, for Joint Computational/ Experimental Aerodynamics Program, on hypersonic wind tunnels. Aeschliman and Oberkampf (1997) also highly recommend the AGARD (1994) compendium as "by far the most detailed prescription for dealing with systematic and random errors in wind tunnel data when the systematic errors have been previously identified and estimated," and state that this reference "identifies virtually every conceivable source of experimental error in wind tunnel testing and greatly improves the art of wind tunnel test data uncertainty analysis."

9.5 EXPERIMENTAL ERRORS IN PHYSICAL PROPERTIES

Physical property values (e.g., viscosities, specific heats) can be a source of error in both computations and *experiments*. If a CFD code is based on constant properties, and this causes a deviation of the predictions from the experimental values, the blame is clearly on the (conceptual) modeling. However, it may also be true that a better average value would move the predictions into acceptable agreement with experiment. Also, if the problem is not variability *per se*, but just uncertainty in the correct value, it is hard to blame the computation, especially since the uncertainty is an *experimental* uncertainty (although not the *same* experiment), and since the same uncertainty often leads to experimental error in the data reduction. Coleman (1996) gives an example of a (significant) 2% uncertainty in computed specific impulse due to the range of available reaction rate data. The distinction between "uncertainty" in properties and "variability" over a parameter range can become murky. Coleman (1996) uses the word "uncertainty" in the following sense. "The uncertainty U associated with a measured quantity or a predicted quantity defines the $\pm U$ interval within which we expect the true (but unknown) value of that quantity to lie 95 times out of 100." That is, this "uncertainty' is *defined* to be the 2σ, or approximately 20:1 odds, expectation.

9.6 BOUNDARY CONDITIONS, CONTINUUM AND NUMERICAL

When modelers speak of errors in boundary conditions, they usually are thinking of errors in the numerical modeling, i.e. errors that are *ordered*, either in a discretization measure Δ or in distance to some far-field boundary L_b. These are the subject of Verification, i.e. a purely mathematical problem. As Oberkampf et al (1995) point out, there are also plenty of physical modeling problems associated with even the apparently simplest of boundary conditions in CFD, that of a wall. These include velocity slip and temperature jump in rarefied flow, boundary conditions for chemical species, vibrational nonequilibrium effects, porous walls (continuous porosity or discrete jets?), fidelity of surface shape, surface roughness for some turbulence studies, etc. These are problems in the description of the *continuum* boundary conditions, and are therefore the proper subject of Validation rather than Verification.

9.7 TRENDS, COMPUTATIONAL AND EXPERIMENTAL

In Chapter 8, we have already noted the false optimism often associated with predicting trends, rather than absolute values, from computations. Specifically, we may not be justified in assuming that correct trends (of flow solutions with parameter variations) will always be predicted with even *qualitative* accuracy in under-resolved grid simulations. (That is, counter-examples exist wherein under-resolved grid calculations have predicted the wrong *directions* for trends, let alone the wrong quantitative slopes.) If one attempts a Validation of a code for the prediction of trends rather than absolute values, one encounters the parallel situation for *experimental* errors, as discussed perceptively by Coleman (1996).

Consider an experiment (and corresponding computation) to predict the increment Δ in some flow functional C_D (which might be a drag coefficient, a discharge coefficient, or some other functional) with a change in geometry (or some other parameter). The ΔC_D from two experimental tests would be compared to the ΔC_D determined by two simulations in order to Validate the Code. In treating the uncertainty of the physical experiment, Coleman states that "consideration of correlated bias uncertainties is crucial," and that "many misconceptions have appeared in the engineering literature regarding the effects of bias uncertainties" in such comparative tests, including two ANSI/ASME (1984, 1986) standards! These publications claim that the uncertainty in the (difference) increment measurement "will be composed only of precision (random) errors, because the bias (systematic) errors will cancel out." Coleman shows that these statements are incorrect in general, both for difference tests (Coleman, 1996) and for ratio tests (Coleman et al, 1995). Fortunately, the systematic uncertainty in the increment *can* be significantly less than the systematic uncertainty in the absolute value uncertainty from either test, especially if the same test apparatus and instrumentation are used in the two tests. The minimum increment that can be distinguished in a comparative experimental test must certainly be larger than its (total) uncertainty. This also applies to computational simulations, and as a practical matter demands tighter iterative convergence criteria for comparative computations of increments Δ, more so when Δ is small; otherwise, the true increment Δ will be lost in the noise of incomplete iterative convergence.

In a related article, Steele et al (1996) discuss subtleties of determining confidence intervals for experiments in which some measured variables have asymmetric systematic uncertainties. Again, the ANSI/ASME (1984, 1986) standard methods are found to be lacking. The method of Steele et al would also be applicable to numerical uncertainties.

Van Wie and Rice (1996) give a good example (their Figure 14) of simulations predicting trends well, and absolute values not so well, for mass flows in hypersonic airbreathing engines. They also state that "Even if perfect agreement between experiment and calculation is achieved at one point, the code is nearly useless if the experimental and computational results trend in opposite directions." This is not an adequate condemnation, in my opinion. Rather than "nearly useless," I would say that such agreement is either totally useless or pernicious, since it has such potential to mislead if the trend discrepancy goes unnoticed.

Benek et al (1996) describe the experimental methodology associated with measuring the increment of total aircraft drag caused by engine inlets. They note that experimental accuracy of this [physical] increment is frequently limited by the insufficient number of pressure measurements (i.e., *experimental* resolution) and that the advantage of a CFD simulation is the higher resolution. Then, "the CFD inlet drag [physical] increment can be used as either a correction [numerical increment] to the measured [physical] increment or as a replacement for it." This increment-of-an-increment process, i.e. a numerical increment of a physical increment, would seem to be a powerful capability for simulation, with uniquely tolerant requirements for experimental Validation.

9.8 FALSE NEGATIVES AND FALSE POSITIVES

Aharoni (1995) discusses the pitfalls of using physics experiments to prove or disprove physical theories, but the experience appears to be applicable to CFD simulations as well. He claims that "sometimes theory and experiment are both correct but do not agree with each other; sometimes a wrong theory agrees with experiment. One must therefore be careful not to jump to conclusions."

His examples are from magnetism, and the article is worth reading, although flawed, in my opinion. The *practice* recommended is good, but the semantics are inadequate and misleading. In a not uncommon pattern, the word "theory" is used ambiguously in the above summary. Essentially, the reader should separate his word "theory" into "theory A" and "theory B." The author then says that theory B is responsible for the disagreement with experiment and we (the theoreticians) erroneously attribute the disagreement to theory A. This is somewhat of a shell game with words. Another example involves a (possibly) correct theory that disagrees with experiments, but the details show that it is simply a matter of (perhaps insurmountable) experimental control. Thus it is not fair to say that the (possibly correct) theory does not agree with the experiment, but only that all experiments to date are flawed, and are incapable of validating or invalidating the theory. This is an "un-measurable theory" at present (an increasingly familiar theme in modern physics). The fault is with the experiment, and so should not be described as a "disagreement with theory" but just as an experimental failure.

However, Aharoni's examples are significant to practical science, especially in view of the comments in the previous Section 9.3 e.g., *"Every observation is laden with theory."* As Aharoni states it, "a theoretical value is rarely compared with a *directly* measured value. More often than not, there are hidden assumptions in the analysis of the experimental data." In Aharoni's example, the assumption was one of one-dimensionality near a wall, which turned out to be incorrect.

The principal message is again the difficulty of comparisons with experiments, which should not lightly be taken as correct.

9.9 "NEARBY" PROBLEMS

As noted in Chapter 2, in a meaningful though overly scrupulous sense, a "Code" cannot be Validated, but only a Calculation (or range or calculations with a code) can be Validated. However, it is clear that physical problems and their solutions present more-or-less continuum responses in their parameter spaces. Although parameter "transition" boundaries do occur, at which solution properties can change discontinuously or rapidly, these parameter transition boundaries are at least countable, and are usually few. The determination of the parameter transition boundaries is the task of the entire professional community (experimental, theoretical, computational) working in the subject area. A few examples of such parameter transition boundaries include the following.

Examples Of Parameter Transition Boundaries

- In aerodynamics:
  - the appearance of local supersonic flows
  - boundary layer transition from laminar to turbulent flow
  - appearance of separated flows
  - vortex shedding in the dynamic stall cycle, etc.

- In ocean calculations:
  - the encounter of the continental shelf break
  - thermocline modeling
  - occurrence of salt fingers, etc.

- In groundwater flow and transport calculations:
  - loss of Darcy flow conditions due to significant fracturing or high Reynolds number effects ("high" in this context meaning $Re \sim 1$)
  - appearance of vadose zone or two-phase flow effects, etc.

- In any modeling:
  - the appearance of geometric singularities like sharp corners.

Away from such parameter transition boundaries, and for most of the allowable parameter range variation, solution response is more or less continuous, and Validation for one particular parameter combination can be used to impute confidence to the computation of a "nearby" problem. Thus, most Validation exercises are not viewed as being restricted scrupulously to a single parameter combination but as having some value for nearby problems.

Nevertheless, the "nearby" condition is often harder to achieve than one might think. For example, a NACA 0012 airfoil is very close geometrically to a an NACA 0015 airfoil, yet their dynamic stall behavior can be significantly different. (For other testimony in aerodynamics, see Cosner, 1994.) The methodical determination of the class of "nearby" problems leads to flow taxonomies for Validation particular to each branch of science and engineering, examples of which will be given in Chapter 10. In this sense, we can meaningfully speak of Validation of a Code, with the understanding that the Validation is restricted to the class of nearby problems, the parameter range for which ideally is demarcated as part of the Validation procedure.

9.10 DIFFICULTY OF THE OPTION TREE

The difficulty of the "option tree" growth, discussed previously for Verification of Codes, applies also to Validations, which are often inconclusive. For example, the option trees can include choice of turbulence models, wind tunnel wall effects, wind tunnel flow angularity effects, etc. (Melnik et al, 1996). Tuning of parameters in such models may be good for one set of experiments but bad for another. This regrettable situation is easy to criticize as poor science, but it is a fact of life for many engineering endeavors.

9.11 DATA SPARSITY AND LACK OF SYNCHRONICITY: GROUNDWATER, OCEAN/LAKE, AND METEOROLOGY MODELING

Westerink and Roache (1995) discussed some Validation issues in ocean modeling and other geophysical flows. A geophysical flow modeler typically encounters three major types of error to be accounted in comparing a numerical solution to the physical system.

1. Formulation errors are the result of missing physics and/or more often due to inadequate constitutive relationships.
2. Measured data errors due to inaccuracies in data acquisition and/or analysis techniques as well as data interpretation.
3. Numerical solution errors are incurred due to time and space discretization steps, boundary placement, boundary condition implementation, etc.

Of course, any physical measurement involves some error, and all fields of science and engineering are affected, but areas of geophysical modeling routinely encounter special difficulties with regard to

- data sparsity (groundwater, ocean/lake, and meteorological modeling), and
- lack of synchronicity (ocean/lake and meteorological modeling)

The data sparsity difficulty for groundwater modeling includes data on properties of the medium, which is usually heterogeneous, with properties often varying over orders of magnitude in a region of interest. The more universal difficulty (common to groundwater, ocean/lake, and meteorological modeling) is sparseness in *initial conditions*. If time-dependent simulations were started with only measured data, the PDE's would be grossly under-determined. This sparseness means that initial condition data must be "made up" by interpolation of the experimental data, which is typically sparse and non-uniformly distributed in space. This interpolation problem is usually treated by the family of interpolation methods called *kriging*. Kriging is a geostatistical procedure that uses a weighted moving average interpolation to estimate values at a large number of nodes (the computational grid) from sparse and irregularly spaced data. In a least-squares sense, kriging is the best linear *unbiased* estimator and reproduces the measurements exactly. (E.g., see Bras and Rodriguez-Iturbe, 1985; Englund and Sparcs, 1991.) On the other hand, if the data itself has significant uncertainty (both in regard to the *value* of the variable and to its *location*), it may be preferable to use techniques that do not identically reproduce all data point values, but effectively smooth out a spatial distribution to reduce the effect of outliers.

Besides the data sparsity difficulty, ocean/lake and meteorological modeling involve another thorny problem of *lack of data synchronicity*. Not only are initial condition values sparsely distributed in space, they also are not all available at the same time. To further complicate matters, there often is no hard and fast *time* = 0 for the simulation. As new meteorological data is continually received, it must be incorporated into the model as a kind of continually updated initial condition data set. Note that one cannot simply inject new (later time) data into a computational grid, since it will undoubtedly be incompatible with the solution of the discretized equations at that point in space and time. This continual incorporation of new data into the simulation is a most challenging problem. For a readable overview of geophysical (meteorological) modeling problems, see Somerville (1996).

These three types of errors (formulation errors, measured data errors, and numerical solution errors) all affect the Validation exercise. It is common practice in geophysical problems to lump together all three errors and to only consider how well the numerical solution compares to the measured data. Physically based parameters are subsequently adjusted in the model to obtain "improved" results. However, this approach can adversely affect the representation of the physical system.

Lumped error minimization can in fact give the modeler a false sense of model accuracy since it improves the comparison to the often limited measured data for a few variables, but can actually degrade the accuracy of all variables away from the data points or for variables which are not included in the error minimization. For example, in a tidal computation, tuning the model to fit free surface elevations can lead to a significant misrepresentation of the computed velocities. Furthermore, while tuning exercises can lead to satisfactory results for a select portion of the response spectrum, the lumped error/tuning process does not in general lead to good results over the entire spectrum. Again, this indicates that the basic physics of the process are not correctly represented. For example, in a depth averaged tidal model based on the shallow water equations, the bottom friction factor is often tuned to obtain a "best fit" for the dominant tidal constituent, yet secondary astronomical and nonlinear overtide and compound tidal constituents can fare relatively poorly (Westerink et al, 1989; Grenier et al, 1995).

Thus it is vital for the modeler to obtain separate estimates for numerical and data errors, allowing an estimate of the formulation error, assuming the total error is known. Only then can a reasonable assessment be made as to the appropriateness of the physics incorporated into the model. This will provide a basis for making adjustments in the physically based model parameters and/or the form of the constitutive relationships incorporated into the model. Making modifications in the formulation in this way will certainly lead to a much improved solution for all variables throughout the domain instead of only a few variables at select points within the domain.

Separately defining the numerical error also allows the modeler to define an improved spatial and/or temporal discretization, thus reducing the total model error. This can be especially useful in designing unstructured graded grids (e.g. finite element based grids). Finally, separate error assessments permit the modeler to compare the level of numerical and data errors and to judge what level of numerical accuracy and grid refinement is appropriate.

9.12 THE EFFECT OF PARAMETER RESOLUTION ON GRID CONVERGENCE

As the level of grid resolution is increased, the level of parameter resolution can also change, which complicates the Validation. See previous discussion in Chapter 8, Section 8.1.3.

9.13 SCALE OF UNSTEADINESS

In the previous chapter, we addressed the question of Benek et al (1996). "Let us assume that CFD can accurately compute the unsteady flow. If the steady state approximation is then used, will the same time averaged quantities be predicted?" The answer is "no" in general, but it is a question of scale, and the error bar for the approximation can be determined by CFD calculations.

We can hardly expect more of the simulation than of the physics. For example, it is well known that pitot pressure probes, which give a single physically-averaged pressure reading, do not produce the average pressure in a fluctuating field, but a reading higher than the average; e.g., see Dolling (1996).

Even if one never used the steady-state form of the governing equations, but always used the time-dependent form, the question of scale of unsteadiness remains important, because of filtering effects of Δt and Δx. In principal, these ambiguities would be removed by grid refinement studies, but as a practical matter, small temporal or physical solution scales can be missed. It is probably not an overstatement to claim that virtually all practical flow problems are unsteady at some scale, even in the mean flow. (This even presumes a clear distinction between turbulence unsteadiness and mean flow unsteadiness.)

In regard to the present subject of experimental difficulties, the point is that these small scales are also missed in many *experimental* techniques. As Dolling (1996) states: "A large fraction of the experimental database used for Validation of CFD predictions consists of time-averaged measurements. If meaningful conclusions are to be drawn from comparisons of these time-averaged data with computations, it is important to understand how the data are generated physically." Dolling's examples and his discussion are worth detailed consideration, as they bear significantly on very general questions of Validation.

As already noted, pitot pressure sensors effectively smooth a pressure in time, and the signal is not the same as the simple arithmetic average of the true instantaneous pressures. The pressure is also averaged in space (over the pressure orifice), introducing a similar error that can be significant in regions of flow with small-scale features, e.g. boundary layers. Flow visualization techniques involve a smoothing in space and time scales; for example, a schlieren photograph of a shock wave-boundary layer interaction will appear steady in a photograph with a common light source duration, but will disclose high frequency oscillations of the shock "foot" penetration (and associated high frequency velocity perturbations) when taken with short duration spark photography. Dolling (1996) describes experiments in compression ramp hypersonic flows which ostensibly appear to be steady, but in which mean flow unsteadiness (i.e., below the fluctuation scale of turbulence models) is a dominant phenomenon. The oscillations of the separation shock cause major intermittency in the pressure signal (~ 50% of the mean). Furthermore, the boundary layer separation line indicated by surface patterns (the kerosene - lampblack technique) is *not* near the center of the spread in the intermittent separation positions, but is at or close to the downstream boundary of the region of intermittent separation. The discrepancy is significant, as measured in the appropriate scale of boundary layer thickness. Three-dimensional tests on swept surfaces show that the separation line indicated from surface patterns moves upstream (in the intermittent region) as sweep is increased.

A major qualitative discrepancy between such experiments and the many computations performed over years is that the (steady) simulations always show a much steeper pressure rise (in the area of the shock-boundary layer interaction) than the experiments. Dolling points out that this is a natural and *unavoidable* outcome of the assumption of steady state in the simulations: there is *no* steady shock position that will produce the smoothed-out pressure rise of the experiments, which are due to temporal smoothing of a sharp shock position.

Although these examples of Dolling's have the common element of explaining computational-experimental discrepancy on the basis of the assumed steadiness in the computations, there is a fundamental distinction to be made in regard to the responsibility, between experimenter and computational modeler. First, it would have been preferable if all the experimental documentation had warned explicitly and emphatically that the data were time-averaged and that instantaneous temporal variations were in fact large. Still, the CFD predictions should be expected to be able to model the unsteady flow, and then time-average the CFD results to compare with experiments. It would seem to be the responsibility (though a difficult one) of the computational modeler to reproduce (within some useful tolerance) the averaged positions of the separation shock, and thus reproduce the time-averaged rise to the pressure plateau. The same is true of the time-averaged separation lines. However, there is one significant proviso: *that the unsteadiness is quantitatively inherent to the problem definition.*

The point is this: if the unsteadiness is actually caused by, or significantly modified by, un reported experimental artifacts (such as fluctuations in tunnel operating conditions, free-stream turbulence, or model vibration), then the CFD prediction cannot be expected to anticipate these uncontrolled and unreported experimental variables. The responsibility for the failed Validation in this case would rest with the experimenter. With some knowledge of fluid dynamics, most would agree that it is probable that the unsteadiness of shock position and separation position are in fact inherent to this problem, so that we would expect the steady state assumption used in the CFD simulation to be at fault, but it is also possible that experimental artifacts contribute something to the discrepancy. More pointedly, the problem Dolling describes in regard to pitot pressure measurements is clearly an experimental artifact. Perhaps the CFD simulation could account for the spatial smoothing of pressure data in the boundary layer, but it would seem that the experimenter could more properly account for this in the data reduction technique. For the temporal smoothing (and "pumping" of the pitot pressure above the temporal mean) the CFD practitioner cannot be expected to account for all the relevant experimental artifacts, including (in one cited experiment) the "several feet of plastic tubing" connecting the wall pressure tappings to the scani-valve! This is experimental error, and the failure of the Validation exercise is not an indictment of the code or its assumptions.

For physical airfoils, there is always some scale of unsteady flow (evidenced by small-scale vortex shedding) yet steady-state equation solutions can be acceptably Validated. However, at higher angles of attack, the scale of unsteadiness increases, and perhaps because of this, the accuracy of Validation deteriorates (Jameson and Martinelli, 1996). Whether this discrepancy is viewed as a shortcoming of the conceptual modeling or as a criticism of the code, the Validation of the conceptual model/code combination has failed in this parameter range.

These and other examples from many areas of science demonstrate that the assumption of steadiness is not always justified and is scale dependent. Significantly, the assumption is probably made (implicitly) as often in experiments as in computations, and even more so in theoretical analyses. Recall that in most cases, wind tunnel flow is not the ultimate interest; rather, the wind tunnel flow is itself a "model" for an engineering flow of interest. Likewise, Darcy flow in a laboratory is a "model" for groundwater flows in nature, etc.

In either case, the assumption may be invalid at the scale of interest, making Validation impossible or ambiguous.

Even if the intended application of an experiment (e.g., free flight) produces a steady state flow (at an acceptable scale), the experiment may not, since many experimental ground-based facilities are transient or border-line steady, especially for high enthalpy flows. Van Wie and Rice (1996) note that "In general, more accurate measurements can be made in long-duration facilities as compared to pulse facilities." When purportedly "steady" experimental data really contain artifacts from tunnel or shock tube transients, a literally impossible burden is placed on the computational modeler.

9.14 SPATIAL SCALES, SCALING UP, AND DIMENSIONALITY

Similar considerations apply to assumptions of physical scales and of dimensionality. Turbulence modeling involves a separation of scales, as do virtually all macroscopic concepts. Laminar viscosity works as a concept only when the scale of molecular mean free motion is far removed from the physical size of the aerodynamic body, and fails in high altitude, high speed flows. At the other end of the flow spectrum, the measurement of permeability in a porous medium involves spatial averaging; if the scale of the variability is extreme, it may not be possible to adequately define an "average" permeability. Even qualitative features may not be accounted properly; as computational experiments show, a "scaled up" averaged permeability in strong velocity fields will require a tensor form, even though the underlying fine-scale permeability is a scalar (Durlofsky, 1991; Russell, 1996). Likewise, an elementary thought experiment will quickly disclose that alternating layers of isotropic material with differing heat conductivity will result in a non-isotropic (i.e., tensor property) material when "scaled up," i.e. when computations *or lumped physical measurements* are made at a scale larger than the layer size.

The assumption of dimensionality can be equally misleading. The following examples of simple *laminar* flows demonstrate the problem. Williams and Baker (1997) show that the often-reported under-estimation of reattachment length for a backward-facing step in incompressible flow above $Re \sim 400$ is due to the assumption of two dimensionality; their fully 3-D calculations are Validated by experiments. Likewise, the following examples are provided by Oberkampf et al (1995). Pironneau (1990) notes that nominally 2-D channel flow with a infinite periodic array of cylindrical obstructions exhibits a critical Reynolds number Re (based on channel half width) of $Re \cong 150$, above which the flow remains laminar and 2-D but becomes unsteady; above $Re \cong 600$ the laminar flow becomes 3-D and unsteady. Rudy et al (1991) note that laminar hypersonic flow over a 2-D compression corner with a large deflection becomes 3-D. [It is not necessary to consider hypersonic flows-the classical problem of Goertler vortices (e.g., Saric, 1994) is a strong example of 3-D flow in an 2-D geometry.] Mittal and Balachnadar (1995) have performed "impressive computational work" to shed light on another classical and deceptively simple flow configuration, that of low Re flow perpendicular to a long circular cylinder. For $Re \lesssim 49$, the flow is steady and 2-D. For $49 \lesssim Re \lesssim 180$, only 2-D but unsteady flow exists. For $Re \gtrsim 180$, experiments had suggested that only 3-D unsteady flow exists. Mittal and Balachnadar (1995) shed light on this issue with simulations, which of course have the great advantage over physical experiments of controlled time-dependence and dimensionality. They computed the flow at $Re = 525$ using both a 2-D and a 3-D simulation. "They found that both solutions converged to a periodic solution, but the mean drag coefficient for the 2-D simulation was 1.44 and the 3-D simulation produced a value of 1.24. Experimental measurements yield a value very near their 3-D simulation value." Oberkampf et al (1995) note the lesson involved, that the 2-D assumption might seem to be appropriate, and reasonable computational results are produced, but the results for drag are inaccurate *even for this very simple flow*. "With the change of one parameter (Reynolds number), over a relatively small range, three fundamentally different flow fields, i.e., solutions to the Navier-Stokes equations, emerge."

Even when a flow *can* be 2-D, it is a well-recognized fact of experimental life that it is difficult to achieve accurately 2-D experimental flows in planar geometries. (Axisymmetric 2-D flows are easier.)

Some problems do not scale up well, e.g. engines (Benek et al, 1996). Thus, the aerospace "vehicle performance modeling process becomes a decomposition of effects: the propulsion system characterized by full scale simulations and the aerodynamics characterized by scaled simulations."

In Aeschliman et al (1995), there are many points of agreement with the present book, notably their recommendation for physical experiments specifically designed for CFD Validation. Experiments designed specifically for Validation of codes are quite different from those designed for other practical uses of the data. They are easier in some ways, but more difficult in others. The reason for additional difficulty is the measurements necessary for CFD calculations. The typical wind tunnel approach is to quantify small effects like flow angularity, and perhaps apply correction to data for lift, drag, etc. usually involving a shift in the measured angle of attack. However, this is not sufficient for point values of velocity components or pressures. What is needed is full measurement of inflow properties.

On the other hand, experiments for code Validation are easier in some respects. A very important point is that, for purposes of Code Validation, high quality wind tunnel flow characteristics are often *not* required, as long as they are measured and reported in detail. For example, consider non-uniform flow properties at the beginning of the test section of a wind tunnel. Variations in flow angularity of 5–10° would be unacceptable for almost any engineering measurement, if the purpose were to predict free-flight quantities like lift, drag, etc. But if the purpose were to Validate a panel code (using inviscid equations), such flow angularity might well be acceptable, but *if and only if* the experimental map of flow angularity was given. Once so Validated, the code could then be used with some confidence for the prediction of the free-flight case, the assumption being that the wind tunnel flow, although too distorted to provide quantitative predictions of the free flight case, was still "nearby" and therefore covered the class of Validation problems for the code. Concerns with model blockage effects in wind tunnel tests are thus greatly reduced, if not virtually eliminated, for experiments designed only for CFD Validation. For similar statements on this approach, see Oberkampf et al (1995) and Marvin (1995, p. 1784).

Aeschliman et al (1995) present a "Case Study for CFD Code Validation Methodology" which is recommended reading for anyone serious about design of experiments for CFD code Validation. Also, good case studies are provided by AGARD (1988).

9.15 ASSUMPTIONS OF PERIODICITY

As noted earlier, the fundamental considerations regarding the solution qualities of spatial scales, scaling-up, dimensionality, and steadiness are not unique to simulations, but are equally significant to laboratory and field measurements and to theoretical analyses. Precise quantitative experimental techniques that give only local values often carry with them assumptions about these qualities. Global experimental techniques such as flow visualization are often less precise or simply qualitative, but are less susceptible to gross errors in assumptions. An example from experimental boundary layer transition studies (Knapp and Roache, 1968) illustrates a common assumption that is one level more subtle than an assumption of steadiness: the assumption of (regular) periodicity.

The most studied path to boundary layer transition involves the development Tollmien-Schlichting waves, which are a clearly unsteady phenomenon. However, the assumption is commonly made that they develop in a regular fixed pattern, i.e., the flow exhibits (regular) periodicity. In fact, some erratic behavior had been noted from the earliest studies, and experiments typically were designed to regularize this periodicity, through the introduction of small-amplitude periodic disturbances via single-frequency sound or vibrating ribbons. Even in quasi-natural (unforced) transition, flow visualization with still photographs appeared to show regular periodic traveling waves. Occasional photographic images showing no waves in the transition region could easily be attributed to experimental error due to drift of the smoke flow. But high-speed cinematography showed clearly that the natural phenomenon was *not* regularly periodic, but in fact displayed a bursting of quasi-periodic waves. Predictions of skin friction and surface heating which would assume regular periodicity could be in significant disagreement.

9.16 OTHER DIFFICULTIES OF VALIDATION IN AEROSPACE

As noted in previous chapters, Verification for unsteady flows is not qualitatively different from steady flows; it necessarily involves more data, but is just more of the same process. However, Validation is much more difficult for unsteady flows just because of the paucity of data, e.g. Benek et al (1996). They also note a clearly non-scientific impediment to Validation. "The major stumbling block to Validation of wind tunnel correction methodology is that the majority of data is either classified or proprietary, and therefore, not generally available." The same issues will occur in other than the aerospace industry.

9.17 UNIVERSAL TURBULENCE MODELS vs. ZONAL MODELING

A fundamental difficulty with Validation of turbulence modeling is the *essential* lack of a universal turbulence model. By "essential," I mean that the problem is not going to be solved. Not everyone agrees, and some continue the quest, but many turbulence researchers believe it is not going to succeed. (For example, see discussions in Kline et al, 1981.) Experience has shown that turbulence models that are moderately successful for a *very* limited class of flows (e.g., 2-D incompressible steady boundary layers in zero pressure gradient with no-slip walls) fail miserably and cannot be salvaged by tuning when the class is slightly expanded (e.g. the addition of an adverse pressure gradient to the k-ε turbulence model). The reason for the failure is fundamental, though probably not provable nor universally agreed upon. The "universal turbulence model" is already known, namely, the unsteady and small-scale Navier-Stokes equations. It is asking much of mathematics to devise a different and radically simpler set of equations that provide approximately the same solutions for all classes of flows, i.e. for all flow parameters and boundary conditions. The nearest thing to a viable universal turbulence model is the large eddy simulation (LES) approach, which requires so many fine-scale details that it still requires zonal turbulence modeling to be nearly practical.

In my opinion (as I expressed in Kline et al, 1981) the root of the problem is semantic; "turbulence" is just too broad a term to be meaningful. By creating the words "laminar" and "turbulent" and using them in juxtaposition, we can deceive ourselves into thinking that they have equal weight and meaning. In fact, "laminar" means something, but "turbulent" does not define the condition. If instead we used the term "non-laminar," the lack of specificity would be more obvious. Consider the folly of describing certain flows as "lacking laminarity" and expecting that condition to define the appropriate governing equations! There is just not enough information there to define the problem and to create alternate approximate equations. Using the word "turbulent" vs. "laminar" is like using the words "color" or "non-gray" vs. "gray." We do not expect the words "color" or "non-gray" to adequately define an artist's palette; we need words like green, mauve, fuchsia, ...

Conferences, symposia, and workshops on turbulent modeling that involve Benchmark comparisons with experimental data, especially if the data is not published beforehand (i.e., true "prediction" vs. "postdiction") invariably have come to the following conclusions, perhaps stated more delicately.

1. Sloppy numerical work confuses the entire meeting.
2. "Every dog has his day," i.e. virtually every turbulence model can predict something correctly for some problem.
3. No single turbulence model predicts all features of a moderately complex flow.

This last evaluation is not the result of overly stringent accuracy standards. For examples, see Kline et al (1981), de Vahl Davis (1994), Pollard (1995), and European efforts (ECOFTAC, GAMM, BRITE/EURAM, EUROVAL) mentioned by Pollard (1994).

The alternative to the hopeless quest for a universal model of non-laminarity is the "zonal modeling" approach espoused by Kline and others; see pro and con discussions in Kline et al (1981). In this approach, turbulence models appropriate to each zone of a complex flow are used, e.g. a different model for an attached and approximately zero-pressure gradient boundary layer, another for a separated shear layer, another for swirling flow, etc. This approach leads to significant numerical and coding difficulties (as noted by Rizzi and Vos, 1996), e.g. specification of boundary conditions between zones, and the transfer of information through zone boundaries when different dependent variables are used in each zone, requiring transition/blending functions, etc. Ideally, a zonal modeling code would have a built-in "expert system" for adapting the zonal models to the developing flow solution, rather than requiring the user to define zone boundaries and select zonal models *a priori*.

However it is implemented, a zonal modeling approach greatly complicates each of the following:

1. the act of Verification of code,
2. the act of Verification of calculations, and
3. the act of Validation, even if the individual turbulence models have previously been Validated on single-zone flows.

9.18 STRONG AND WEAK MODEL DEFINITIONS AND MODEL VALIDATION

In Chapter 2 and elsewhere, I noted that, in a meaningful sense, one cannot Validate a Code but only a particular calculation or at least a range of "nearby" calculations defined over some limited parameter range. However, we can also speak of Validating a Code and its underlying conceptual model in a more general sense and still be meaningful, provided that we make some distinctions. The paper by Leijnse and Hassanizadeh (1994) is important for unraveling the semantics associated with the word "model," in particular, whether "model" includes parameter values. They make the worthwhile distinction of "strong" and "weak" definitions of "model" (based on whether or not it includes parameters or is only a conceptual model), and corresponding strong and weak senses of Validation. (They also discuss the related terms of "prediction models" and "analysis models.") They assert (and I concur) that not only is there no clear and unambiguous meaning of "Validation" (neither in the general public nor in the scientific community), but also that there is no unanimity within the scientific community (in their case, the groundwater modeling community) on what constitutes a "model." In their distinction, a "weak" definition of model involves only the conceptual modeling assumptions (e.g., incompressibility, symmetry, Darcy flow, etc.), but a "strong" definition of model also includes data input to the code (e.g., geometry data, parameter values, and boundary conditions and initial conditions).

Validation of a model in the weak definition is a less daunting task than Validation in the strong definition. For example, one may demonstrate that Darcy flow assumptions are valid for a geologic site, and even obtain quantitative prediction to an acceptable level of accuracy. However, if the strong definition of model is used, and the input includes uncontrollable factors (e.g., climate effects such as rainfall into the groundwater model) then one could argue [as did Konikow and Bredehoeft (1992), without benefit of this distinction of weak and strong definitions] that one could never hope to Validate the (predictive) model in this strong sense. Even then, I would argue that Validation in a meaningful sense is conceivable, depending on error tolerances, and especially if a single predictive calculation is replaced by more meaningful Monte Carlo simulations that sample rainfall over historical ranges, etc.

In any case, whether comparisons of simulations with physical data are judged to be acceptable or not, and thus either Validate or invalidate a model, the analyst would do well to acknowledge this distinction between strong and weak definitions of "model", and to define or describe terms with sufficient precision. As Leijnse and Hassanizadeh (1994) assert, these considerations "do not support abandoning the use of the term 'model Validation'. It only makes it apparent that the modelers have to carefully specify what their models constitute and what is being Validated." They quote McCombie and McKinley (1993); if confusion arises about Validation, "use of the term Validation is much less to blame than poor science and sloppy documentation."

9.19 THE MYTH OF THE TOTALLY VALIDATED CODE

In discussing Verification in Chapter 8, I noted the "myth of the converged solution," in the sense that different variables can converge at different rates, and that it would always be possible to devise some error measure that is exquisitely sensitive to discretization error, so that this measure is far from converged even when other, more benign measures are well converged. Thus, in the pursuit of Verification of a Calculation, the concept of a "converged solution," ascertained to be so *independent of the intended error measure*, is a myth.

There is an obvious analog in Validation. Although we want to maintain that it is meaningful to speak of a "Validated Code" (either a commercial code or a specialized scientific code) it has to be understood in the context

1. of a class of nearby problems,
2. for specified variables, and
3. for a specified level of accuracy.

Given a set of variables known to be of interest for an engineering project, one can follow the directive of Tjonneland (1988). "If you want a Validated Code you should look for the most critical part of the flow and for the toughest variable to get right and look at that." In the terminology defined in Chapter 2, this would certainly be appropriate for a Certification exercise, i.e. a project-oriented evaluation of a code, with understood variables and accuracy. But more generally speaking, the above 3 restrictions on the context must be recognized and honored.

For a real-world example, consider Validation of Code based on a k-ε turbulence model. Many Validation exercises have demonstrated that k-ε is good for attached boundary layer flows in zero and favorable pressure gradients, and is inadequate for adverse pressure gradients, swirl, etc. However, "good" is defined in the context of engineering needs, primarily friction drag (i.e., shear stress at the wall τ). While the k-ε model *predicts* turbulent kinetic energy k and dissipation rate ε, Validation is usually based on τ. But suppose someone now wants to predict not just τ at the wall but the ε distribution throughout the boundary layer, say for applications to optical propagation through boundary layers (e.g., Truman and Lee, 1990). Validation for τ is encouraging, but it cannot be assumed that there is such a strong correlation that this can be taken for Validation of $\varepsilon(y)$ throughout the boundary layer. The calculation would have to be Verified for $\varepsilon(y)$ (first for iterative convergence, if that variable had not previously been tested in the iterative convergence criteria, and second for grid convergence) and then Validated for $\varepsilon(y)$. In actual fact, the k-ε model prediction of $\varepsilon(y)$ is inadequate.

Thus, the concept of a "Validated Code," ascertained to be so *independent of the intended variable measured*, is a myth.

REFERENCES FOR CHAPTER 9

Aeschliman, D. P., Oberkampf, W. L., and Blottner, F. G. (1995), "A Proposed Methodology for Computational Fluid Dynamics Code Verification, Calibration, and Validation," *Proceedings 16th International Congress on Instrumentation in Aerospace Simulation Facilities*, WPAFB, Ohio, 95–CH3482–7, July 1995, pp. 27.1–27.13.18–21. See also Aeschliman and Oberkampf (1997)

Aeschliman, D. P. and Oberkampf, W. L. (1997), "Experimental Methodology for Computational Fluid Dynamics Code Validation," SAND95–1189, Sandia National Laboratories, Albuquerque, New Mexico, September 1997. Also, *AIAA Journal*, to appear.

AGARD (1988), *Validation of Computational Fluid Dynamics*, Lisbon, Portugal, 2–5 May 1988, NATO Advisory Group for Aeronautical Research and Development, AGARD CP 437, Dec. 1988. Volume 1: Symposium Papers and Round Table Discussion. Volume 2: Poster Papers.

AGARD (1994), *Assessment of Wind Tunnel Data Uncertainty*, NATO Advisory Group for Aeronautical Research and Development, Advisory Report AGARD–AR–304, 1994.

Aharoni, A. (1995), "Agreement Between Theory and Experiment," *Physics Today*, June 1995, pp. 33–37.

ANSI/ASME (1984), *Measurement Uncertainty for Fluid Flow in Closed Conduits*, MFC–2M–1983, ASME 1984, American National Standards Institute/American Society of Mechanical Engineers, New York.

ANSI/ASME (1986), *Measurement Uncertainty*, PTC19.1–1985, ASME 1986, American National Standards Institute/American Society of Mechanical Engineers, New York.

Barber, T. J. (1996), "The Role of Code Validation and Certification in the Design Environment," AIAA Paper 96–2033, 27th AIAA Fluid Dynamics Conference, New Orleans, Louisiana, 17–20 June 1996.

Benek, J. A., Kraft, E. M., and Lauer, R. F. (1996), "Validation Issues for Engine/Airframe Integration," AIAA Paper 96–2031, 27th AIAA Fluid Dynamics Conference, New Orleans, Louisiana, June 17–20, 1996.

Bras, R. L. and Rodriguez-Iturbe, I. (1985), *Random Functions and Hydrology*, Addison Wesley, New York.

Brinster, K. (1995), personal communication.

Coleman, H. W. and Steele, W. G. (1995), "Engineering Application of Experimental Uncertainty Analysis," *AIAA Journal*, Vol. 33, No. 10, October 1995, p. 1888.

Coleman, H. W., Steele, W. G., and Taylor, R. P. (1995), "Implications of Correlated Bias Uncertainties in Single and Comparative Tests," *ASME Journal of Fluids* Engineering, Vol. 117, December 1995, pp. 552–556.

Coleman, H. W. (1996), "Uncertainty Considerations in Validating CFD Codes with Experimental Data," AIAA Paper 96–2027, 27th AIAA Fluid Dynamics , New Orleans, Louisiana Conference, June 17–20, 1996.

Cosner, R. R. (1994), "Issues in Aerospace Application of CFD Analysis," AIAA Paper 94–0464, AIAA 32nd Aerospace Sciences Meeting, Reno, Nevada, January 10–13, 1994.

de Vahl Davis, G. (1994), "Going for Quality in CFD Codes," *Proceedings 2nd World Conference in Applied Computational Fluid Dynamics*, World User Association–CFD, Basel, Switzerland, 1–5 May 1994, pp. 18.1–18.6.

Dolling, D. S. (1996), "Considerations in the Comparison of Experimental Data with Simulations–Consistency of Math Models and Flow Physics," AIAA Paper 96–2030, 27th AIAA Fluid Dynamics Conference, New Orleans, Louisiana, June 17–20, 1996.

Durlofsky, L. J. (1991), "Numerical Calculation of Equivalent Grid Block Permeability Tensors for Heterogeneous Porous Media," *Water Resources Research*, Vol. 27, 1991, pp. 699–708.

Englund, E. and Sparcs, A. (1991), *Geostatistical Environmental Software User's Guide*, U. S. EPA 600/8–91/008.

Grenier, R. R., Luettich, R. A. and Westerink, J. J. (1995), "A Comparison of the Nonlinear Frictional Characteristics of Two-Dimensional and Three-Dimensional Models of a Shallow Tidal Embayment," *Journal of Geophysical Research*, Vol. 100, No. C7, pp. 13719–13735.

Jameson, A. and Martinelli, L. (1996), "Mesh Refinement and Modeling Errors in Flow Simulation," AIAA Paper 96–2050, 27th AIAA Fluid Dynamics Conference, New Orleans, Louisiana, June 17–20, 1996.

Jankowski, D. (1985), personal communication.

Johnson, L. T. (1996), *The Real Jesus: The Misguided Quest for the Historical Jesus and the Truth of the Traditional Gospels*, Harper, San Francisco, p. 172.

Knapp, C. F. and Roache, P. J. (1968), "A Combined Visual and Hot-Wire Anemometer Investigation of Boundary Layer Transition," *AIAA Journal*, Vol. 6, No. 1, January 1968, pp. 29–36.

Kline, S. J., Cantwell, B. J., and Lilley, G. M. (1981), Eds., *Objectives, Evaluation of Data, Specifications of Test Cases, Discussion, and Position Papers. The 1980–81 AFOSR/HTTM-Stanford Conference on Complex Turbulent Flows*, Vol. 1, ThermoSciences Division, Mechanical Engineering Department, Stanford University, Stanford, California.

Konikow, L. F. and Bredehoeft, J. D. (1992), "Groundwater Models Cannot be Validated," *Advances in Water Resources*, Vol. 15, 1992, pp. 75–83.

Leijnse, A. and Hassanizadeh, S. M. (1994), "Model Definition and Model Validation," *Advances in Water Resources*, Vol. 17, 1994, pp. 197–200.

Marvin, J. G. (1995), "Perspective on Computational Fluid Dynamics Validation," *AIAA Journal*, Vol. 33, No. 10, October 1995, pp. 1778–1787.

McCombie, C. and McKinley, I. (1993), "Validation-Another Perspective," *Ground Water*, Vol. 31, 1993, pp. 530–531.

Mehta, U. (1996), "Guide to Credible Computational Fluid Dynamics Simulations," *AIAA Journal of Propulsion and Power*, Vol. 12, No. 5, October 1996, pp. 940–948.

Melnik, R. E., Siclari, M. J., Marconi, F., Barber, T. and Verhoff, A. (1995), "An Overview of a Recent Industry Effort at CFD Code Validation," AIAA Paper 95–2229, 26th AIAA Fluid Dynamics Conference, San Diego, California, June 19–23, 1995.

Mittal, R. and Balachnadar, S. (1995), "Vortical Structures in Bluff Body Wakes," AIAA Paper 95–0867, 33rd AIAA Aerospace Sciences Meeting, Reno, Nevada, January 1995.

Moffat, R. J. (1981), "Contribution to the Theory of Uncertainty Analysis for Single-Sample Experiments," *The 1980–81 AFOSR/HTTM–Stanford Conference on Complex Turbulent Flows*, Vol. 1, ThermoSciences Division, Mechanical Engineering Department, Stanford University.

Naylor, T. H. and Finger, J. M. (1967), "Verification of Computer Simulation Models," *Management Science*, Vol. 14, No. 2, 1967, pp. B92–B101.

Oberkampf, W. L., Blottner, F. G., and Aeschliman, D. P. (1995), "Methodology for Computational Fluid Dynamics Code Verification/Validation," AIAA Paper 95–2226, 26th AIAA Fluid Dynamics Conference, San Diego, California, 19–22 June 1995. See also Oberkampf and Blottner (1997).

Oberkampf, W. L. and Blottner, F. G. (1997), "Issues in Computational Fluid Dynamics Code Verification and Validation," SAND95–1352, Sandia National Laboratories, Albuquerque, New Mexico, September 1997. Also, *AIAA Journal*, to appear.

Pironneau, O. (1990), Ed., "Summary and Conclusions for Test Case T2," in *Numerical Simulation of Unsteady Flows and Transition to Turbulence*, Cambridge University Press.

Pollard, A. (1994), "Validation of CFD Codes," *CFD Bulletin*, CFD Society of Canada, Winter 1994, pp. 2–3.

Pollard, A. (1995), "Validation Exercise of Turbulent CFD Codes," Proceedings CFD Society of Canada Meeting, Banff, July 1995.

Porter, J. L. (1996), "A Summary/Overview of Selected Computational Fluid Dynamics (CFD) Code Validation/Calibration Activities," AIAA Paper 96–2053, 27th AIAA Fluid Dynamics Conference, New Orleans, Louisiana, June 17–20, 1996.

Rizzi, A. and Vos, J. (1996), "Towards Establishing Credibility in CFD Simulations," AIAA Paper 96–2029, 27th AIAA Fluid Dynamics Conference, New Orleans, Louisiana, June 17–20, 1996.

Rudy, D. H. et al (1991), "Computation of Laminar Hypersonic Compression-Corner Flows," *AIAA Journal*, Vol. 29, July 1991, pp. 1108–1113.

Russell. T. (1996), personal communication.

Saric, W. S. (1994), "Goertler Vortices," in *Annual Review of Fluid Mechanics*, Vol. 26, 1994, pp. 379–409.

Somerville, R. C. J. (1996), *The Forgiving Air: Understanding Environmental Change*, University of California Press, Berkeley.

Steele, W. G., Maciejewski, P. K., James, C. A., Taylor, R. P., and Coleman, H. W. (1996), "Asymmetric Systematic Uncertainties in the Determination of Experimental Uncertainty," *AIAA Journal*, Vol. 34, No. 7, July 1996, pp. 1458–1463.

Tjonneland, T. (1988), in Round Table Discussion in AGARD (1988).

Truman, C. R. and Lee, M. J. (1990), "Effects of Organized Turbulence Structures on the Phase Distortion in a Coherent Optical Beam Propagating through a Turbulent Shear Flow," *Physics of Fluids A*, Vol. 2, No. 5, 1990, pp. 851–857.

Van Wie, D. M. and Rice, T. (1996), "Quantification of Data Uncertainties and Validation of CFD Results in the Development of Hypersonic Airbreathing Engines," AIAA Paper 96–2028, 27th AIAA Fluid Dynamics Conference, New Orleans, Louisiana, June 17–20, 1996.

Westerink, J. J., Stolzenbach, K. D., and Connor, J. J. (1989), "General Spectral Computations of the Non-Linear Shallow Water Tidal Interactions Within the Bight of Abaco," *Journal of Physical Oceanography*. Vol. 19, 1989, pp. 1350–1373.

Westerink, J. J. and Roache, P. J. (1995), "Issues in Convergence Studies in Geophysical Flow Computations," Joint JSME–ASME Fluid Mechanics Meeting, Hilton Head, South Carolina, Session F137, August 14–18, 1995. See also Proceedings of the EPA Workshop on *Next Generation Environmental Models Computational Methods*, G. Delic and Mary F. Wheeler, Eds., Society for Industrial and Applied Mathematics, Philadelphia, PA, 1997, pp. 95–107.

Wilcox, D. (1994), personal communication.

Williams, P. T. and Baker, A. J. (1997), "Numerical Simulations of Laminar Flow over a 3-D Backward-Facing Step," *International Journal for Numerical Methods in Fluids*, Vol. 24, 1997, pp. 1159–1183.

10

METHODOLOGIES AND EXAMPLES OF VALIDATIONS, CALIBRATIONS, AND CERTIFICATIONS

The examples of Validations, Calibrations, and Certifications in this Chapter are each presented with a philosophy or "moral of the story" in mind. Sometimes the moral is just the high quality of the work and/or the early date (i.e., the high quality obtained with earlier and therefore comparatively primitive computer power). More often, the example is chosen to elucidate some aspect of Validation exercises.

10.1 SOURCES OF PHYSICAL MODELING ERRORS IN AERODYNAMICS CFD

Oberkampf et al (1995) give the following taxonomy of sources and list of non-exclusive sub-categories of physical modeling errors in CFD. Their emphasis is in aerodynamics, and the original paper includes worthwhile discussion based on their extensive experience.

- Physical Modeling Errors for Aerodynamics CFD
- Partial Differential Equations for Fluid Dynamics
  - Inviscid Flow
  - Viscous Flow
  - Incompressible Flow
  - Gas with Vibrationally Excited Molecules
  - Inert Gas Mixtures
  - Chemically Reacting Gas
  - Turbulent/Transitional Flow
  - Additional Physical Phenomena (e.g. thermal nonequilibrium, ionized flows, radiative transfer in gases, multi-phase mixtures)
  - Temporal Nature Assumptions
  - Spatial Dimensionality Assumptions
- Auxiliary (or Closure) Physical Models
  - Equations of State
  - Thermodynamic Properties
  - Transport Properties
  - Chemical Model, Reactions, and Rates
  - Turbulence Model
- Boundary Conditions for the Partial Differential Equations
  - Wall Boundary Conditions
  - Open Boundary Conditions
  - Free Surface Boundary Conditions

In addition, simple specification of geometry (even smooth surfaces) can introduce significant errors. Cosner (1995) states "Serious issues are unresolved in transferring geometry efficiently from CAD systems to the CFD grid generators."

10.2 ACCURACY LEVEL FOR VALIDATION

As previously discussed in Chapter 2, any concept of Validation of a code must involve an error tolerance or accuracy level, and this must be determined "from the perspective of its intended uses" (Mehta, 1995). It follows that a code could be convincingly Validated for one use but shown to be unacceptably inaccurate for another use, even if the physical problems are identical. (For example, see Van Wie and Rice, 1996.) Also, note that Validation will tend to be more precise, or have higher standards of accuracy level, for a purely scientific project that for an engineering project. As Rizzi and Vos (1996) note, "in industry this [level of acceptable accuracy] usually is in the direction of 'just enough' certainty for the problem at hand." In considering the accuracy level appropriate for a Validation, it is also worthwhile to keep in mind that a CFD simulation can have various expectations, depending on the application. In increasing level of accuracy requirements, the simulation may be expected (Benek et al, 1996) to

- only provide diagnostic information,
- supply only incremental data, or
- generate "base line data" for the performance model data base.

10.3 GENERIC MODELS VS. REALISTIC MODELS FOR VALIDATION AND CALIBRATION: PHASES OF VALIDATION

Marvin (1995) notes somewhat different Validation experimental needs for different stages of aerospace CFD code development: research codes, pilot codes, and production codes, referring to "building block and benchmark experiments." He notes the historical development of Validating computations (and theoretical approaches) "first, by performing small-scale experiments in ground-based facilities to Validate the approaches and any related extrapolation techniques and, later, by data from flight experiments with design hardware or prototypes." He notes that research organizations (such as NASA) tend to view Validation from a more fundamental level, stemming from a long history in CFD research and pilot-code development, whereas industry tends to view Validation in a broader sense, having to additionally consider the total design process in which the code is used.

Although design engineers would prefer to see codes Validated and Calibrated for complex geometries, these increase experimental uncertainties. It is often preferable to perform Validation/Calibration exercises for simpler generic models, especially for new areas of investigation. Marvin (1995) cites an example of approaching the formidable challenges of Validation in hypersonic flow. A generic hypersonic vehicle model was tested to provide Benchmark data for Validating forebody codes developed for use in integrating air-breathing propulsion systems with vehicle airframes (Lockman et al, 1992). Marvin (1995) also notes the attraction of flight experiments designed with Code Validation as an integral part of their motivation, though he notes these are often difficult and expensive.

Marvin (1995) also provides an excellent example of a success story for experiments specifically designed for Validation/Calibration, rather than "practical" designs, in the development of the Johnson-King (1985) turbulence model. The experiment (Bachalo and Johnson, 1979) that provided much of the data for guidance of the development and Validation for transonic wing applications did not remotely resemble a wing. The test configuration was an axisymmetric cylinder with a circular arc section, aligned parallel to the flow; the motivation was to avoid extraneous 3-D effects. "The turbulence model derived made use of experimental evidence on the development of the turbulent shear stresses through shock waves of varying strength that developed on the circular arc section as the freestream Mach number was varied." The resulting Johnson-King non-equilibrium turbulence model has an intermediate level of complexity, more so than simple algebraic models (equilibrium models which have no "history" dependence) but less so than two-equation models like k- ε and k-ω because it uses a single additional ordinary differential equation (i.e., 1-D) rather than multiple partial differential equations. For a small increase in computing burden and complexity over algebraic models, the Johnson-King model provided better predictions of pressure recovery, separation location and extent, and velocity profile development. Improvements to the original model (Johnson and

Coakley, 1990) have been made, and the Validation has been extended to additional experiments (Marvin, 1995).

Similarly, Marvin (1995) cites the example of Menter's (1992) in developing a variation of the Wilkox *k*-ω model (e.g., see Wilcox, 1993) based on the reliable "building block experiments" on non-design geometries of Driver (1991).

Sindir et al (1996), like Marvin (1995), Melnik et al (1995), and Rizzi and Vos (1996), see Validation as a multi-phase process. Their code "Validation" process is perhaps closer to what others call Certification, but the meaning is clear, and Validation as used herein is certainly involved at all phases.

Their four phases (see Figure 10.3.1) progress from

- Unit Problems (single flow features), to
- Benchmark Cases (more than one flow feature, simple to moderate flow physics), to
- Simplified/Partial Flow Path (multiple flow features, moderate to complex flow physics), to
- Actual Hardware (complete flow physics).

As the phases progress in geometric and flow complexity, the authors note a tendency for decreasing availability and accuracy of experimental data.

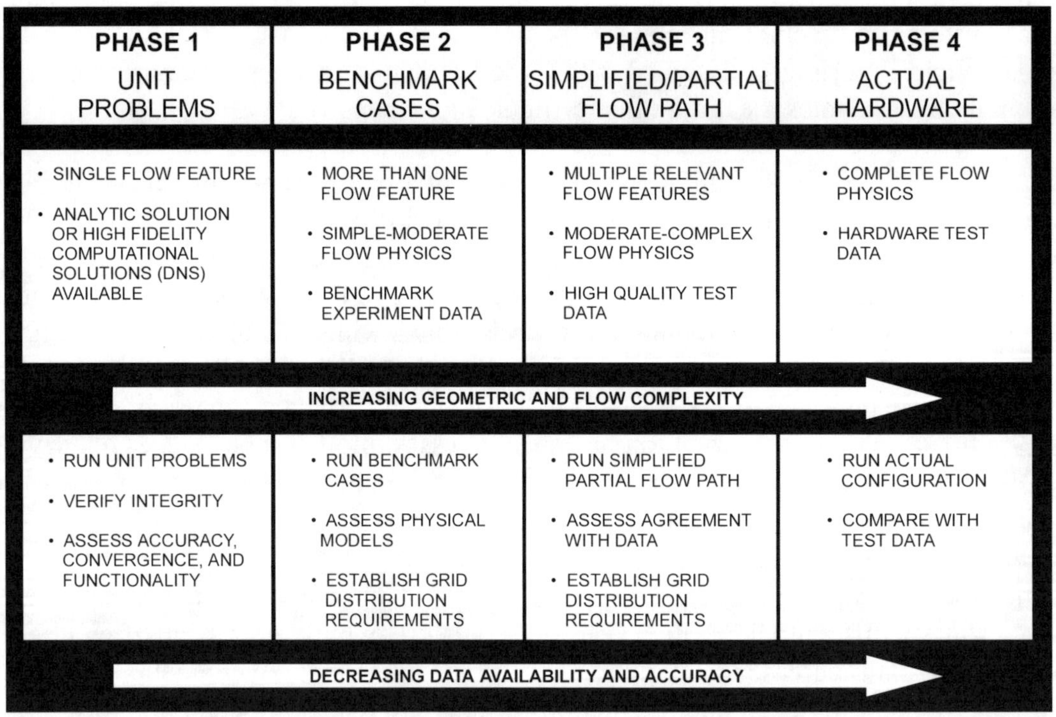

Figure 10.3.1. A Four-Phase View of Code Validation. (From Figure 1 of Sindir et al, 1996.)

Sinder et al (1996) also note three levels of "Validation" depending on the use:

- Conceptual Design – Validated Code,
- Preliminary Design – Validated Code, and
- Detail Design – Validated Code,

with increasingly high standards of accuracy and specificity. Here, it is clear that their use of "Validation" corresponds more to other's use of "Certification." They give a case study of this Validation process in the application of their UniFlo code to a turbomachinery dynamic loading problem.

Similarly, Marvin (1995) cites the example of the experiments by Olsen and Seegmiller (1993) specifically designed for code Validation. The problem was transonic flow over a low aspect ratio wing. Rather than try to simulate free-flight conditions as well as possible, they tested the model in a solid wall wind tunnel, so that issues of corrections for transonic tunnel wall ventilation and interference were eliminated. The key feature is that conditions at the tunnel walls and free-stream inflow values were measured and included in the code Validation computation. (The modeling included viscous calculations at the tunnel walls by way of boundary layer calculations and a displacement thickness included in the CFD code being Validated.) In such exercises, once the basic code is Validated, practical calculations can be obtained with better far-field boundary conditions (which themselves must have been Verified and/or Validated).

10.4 CFD AND EXPERIMENTAL FACILITY CORRECTIONS

Benek et al (1996) give an overview of CFD successes in providing corrections (Calibrations) for experimental facilities. (See discussion of "Trends, Computational and Experimental" in Chapter 9, Section 9.7.) "Sickles and Erickson (1990) demonstrated excellent agreement between measured and computed corrections to an aircraft configuration." The calculations of Martin et al (1993) of increments of normal force and pitching moment on the space shuttle launch configuration "agreed well with measurements at several model scales." Willhite et al (1995) computed sting interference corrections that "compared well over a wide range of flow conditions." Hinkleman (1995) made "unpublished comparisons of drag increments accounting for a mismatch of wind tunnel and flight Reynolds number" that were "in excellent agreement with the measured values" from a "conventional" vs. a "high Reynolds number wind tunnel." Benek et al (1996) note the potential for CFD to "make practicable the evaluation of increments that can not be readily obtained experimentally," such as scale model geometric compromises (like oversized boundary layer gutters in wind tunnels), and Reynolds number corrections for inlet swirl and total pressure maps. However, they also note "The major stumbling block to Validation of wind tunnel correction methodology is that the majority of data is either classified or proprietary, and therefore, not generally available."

The use of CFD in the operation of *adaptive* wind tunnel walls is discussed in Section 10.6.4 below.

10.5 VERIFICATION MUST BE INDEPENDENT OF VALIDATION: AIRFOIL CALCULATIONS

In Chapter 2 and elsewhere, the importance of performing Verification of Calculations *before* Validation was stressed. Logically speaking, the actual time sequence would be immaterial, of course; the point is that Verification is required regardless of agreement with experiments. But humanly speaking, people can get led astray with fortuitous agreement with experiments, and are tempted to not bother with Verification of Calculations.

Zingg (1992) has pointed out the importance of this precept in his airfoil calculations involving thorough and systematic grid refinement studies (previously cited in Chapter 6 as examples of good Verification practice). In 4 out of the 7 flow cases calculated, agreement with experiment was better with coarse grid calculations than with fine grid calculations. This would have given a "false positive" Validation, caused by inadequate grid resolution, had not the systematic Verification been performed independent of agreement with experiments.

Rizzi and Vos (1996) note the same phenomenon, and identify the reason. The physical model used includes the Baldwin-Lomax turbulence model. Being algebraic (a "zero-equation" model), there is no lag in adjustment to new boundary layer edge conditions. This may be an adequate approximation for slowly varying edge conditions, but is not for transonic flows with shocks. Physically, these flows require a small distance aft of the shock for turbulence equilibrium to be attained. Quite by accident, spatial discretization provides such a lag in adjustment. The longitudinal discretization increment Δx thus provides a *non-physical* parameter that behaves something like a physical relaxation length. The agreement with experiment occurs only from this numerical artifact, the computational error happening to partially cancel the physical modeling error in these cases. As Rizzi and Vos say, "Issues related to the interaction of these two sources of errors have been of constant concern to the CFD and turbulence modeling communities because fortuitous cancellation of errors can lead to a very erroneous conclusion about Validation."

Clearly, to jump to a conclusion about Validation without thorough Verification is to build a house on sand.

10.6 SYNERGISM BETWEEN COMPUTATION AND VALIDATION EXPERIMENTS

Aeschliman and Oberkampf (1997, p. 14) note that beneficial synergism arises when experiments are designed specifically for CFD Validation. "By a 'synergism', we mean an activity whose primary intent is to meet a requirement for one approach, whether CFD or experiment, but which generates improvements in capability and/or accuracy of the other, such that both computational and experimental methods benefit."

Presented below are instances of such synergism.

10.6.1 Artificial Heart Valves

In pioneering work on simulation of artificial heart valves, Mueller and colleagues (Mueller, 1974; Underwood and Mueller, 1977,1979) showed the advantages of an integrated numerical and experimental program. Over the multi-year program, four different experimental flow visualization techniques (electrochemical, dye, hydrogen bubble, and nylon micro-spheres) were used to Validate the numerical solutions. Even using 1972 vintage computers such as the UNIVAC 1107 and the IBM 370/155, this group managed to perform reasonable grid convergence testing, i.e. Verification as well as Validation. "The overall flow patterns as well as the occurrence, location and extent of separated flow regions obtained numerically agreed very well with the experimental results for the laminar flow cases." (Underwood and Mueller, 1977.) The advantage of the integrated experimental/numerical program worked both ways; "the numerical results were found to be very helpful in positioning the hot-wire and hot-film anemometer probes in later physical experiments." (Underwood and Mueller, 1979.)

10.6.2 Transonic Flow

Marvin (1995) cites another interesting example of synergism between computations and Validation experiments when cooperative efforts are made.

The related papers by McDevitt et al (1976), Levy (1978) and Marvin et al (1980) involved experiments and simulations of transonic flow over a circular arc airfoil. Airfoil pressures, tunnel wall pressures, and flowfield velocity profiles were measured experimentally with a (non-intrusive) laser anemometer system. Depending on the experimental Mach number M, steady flow or unforced buffet flow developed. The buffet onset condition displayed hysteresis; the buffet M domain was different for experiments during which M was increased or decreased. Computations using a time-accurate Navier-Stokes code and a simple turbulence model were performed at three M. At the intermediate $M = 0.755$ the computations were not converging. The experimentalists pointed out that the flow was unsteady at this M for experiments wherein M was decreased once buffet occurred, and that perhaps the computation was in fact reproducing the correct physics. With this motivation, the computational modelers continued the simulation for a time equivalent to several airfoil chords of flow travel, and indeed periodic buffet developed in the simulation. The magnitude and frequency of the computed buffet "compared well with the experiment, although the frequency was found to be about 20% less than the experimental value." According to Marvin (1995), "These studies confirmed for the first time that time-accurate solution methods used to solve the Reynolds-averaged Navier-Stokes equations had the potential for predicting buffet."

10.6.3 Simple-to-Complex Geometry Flows

Aeschliman and Oberkampf (1997, p. 14) also present an example of beneficial synergism between CFD and experimentation in the Validation activity in their long-duration hypersonic facility.

"If in a wind tunnel experiment the wind tunnel model is designed for easy modification from geometrically simple to complex, it becomes possible to produce a wide range of flow conditions. The geometrically simple flows could possibly be calculated with high confidence, while the complex geometry flows may exceed the current computational state of the art. As an example, for attached, perfect gas, laminar flow over a slender sphere/cone at low angle of attack, confidence in the computed solutions for flow over the simple model with simple flow physics can be at such a high level that the results are usable for an *in-situ* calibration of the freestream wind tunnel flow. This type of calibration can provide new, and sometimes surprising, information about the facility. For flow over more complex geometries, the measurements can be used to Validate the code."

10.6.4 Operation of Adaptive Wind Tunnel Walls

The use of CFD in the operation of *adaptive* wind tunnel walls and in correcting for wall and model support interference is described by Aeschliman and Oberkampf (1997, p. 14) as "a synergism that has a large potential payoff. It is desirable to test aircraft configurations at the largest possible scales to maximize Reynolds number, a goal which is in immediate conflict with minimizing interferences." That is, the larger the wind tunnel model, the better is the Reynolds number, but the worse is the tunnel interference. The interference effects can be calculated (and removed from the experimental results) by CFD. These authors and others note that the "CFD capability required to compute interference corrections must advance in concert with the testing requirements."

Likewise, Aeschliman and Oberkampf (p. 14) state "In a similar vein, advances in the use of CFD to compute flows in perforated-wall wind tunnels are retarded by a lack of well characterized wall boundary conditions. Detailed measurement of the actual wall boundary conditions as a function of test section location and given tunnel operational parameters would directly improve wind tunnel data accuracy, in addition to providing the needed [boundary conditions] for a CFD calculation."

10.6.5 Boundary Layer Transition

The paper by Haynes, Reed and Saric (1996) presents an excellent example of Validating a reduced set of equations, the nonlinear Parabolized Stability Equations, for 3-D crossflow-dominated transition by direct comparison with extremely carefully performed wind tunnel experiments. This paper contains many insights and suggestions on the general topic of Validation, and especially provides an excellent instance of synergism between computation, theory, and Validation experiments.

As noted in Chapter 2, a major opportunity for computational simulation technologies like CFD is to contribute to experimental work being used to predict the importance of difficult-to-measure quantities like wind tunnel flow angularity, non-uniform stagnation enthalpy, surface waviness, etc. (See Aeschliman et al, 1995.). A premier example of this given by Haynes et al is the sensitivity of boundary layer transition to free stream vorticity, which is "an unusually difficult experiment" but which can be modeled by CFD. Surprisingly, when streamwise vorticity is not included in the simulations, "the subharmonic mode is present as predicted by theory but not seen experimentally. When streamwise vorticity (as is present in the flow from the turbulence screens upstream of the nozzle) is also included, the subharmonic mode is overshadowed by the fundamental mode (as in the experiments!)"..."Here is a case in which the computations have explained [long standing] discrepancies between theory and experiments."

Haynes et al (1996) also demonstrate conclusively that, although linear theory predicts many features, a nonlinear calculation is required to obtain complete Validation for the growth rates for stationary cross-flow waves.

Other examples cited in Haynes et al (1996) are the computations by Fasel (1990) in which the CFD simulations isolated the significant effect of a small streamwise pressure gradient, and the CFD simulations by Joslin and Street (1992) in which subharmonic modal discrepancies with experiments were eliminated by including a small adverse pressure gradient and a small effective frequency variation in the input disturbance.

Haynes et al (1996) also suggest an innovative and synergistic use of CFD simulations in conjunctions with experiments. Their simulations have shown that free-stream vorticity has a major effect on transition, yet free-stream vorticity cannot be measured! "At first glance, the situation seems hopeless. How do you quantify an effect you can't even measure? How do you ever hope to establish [an experimental] database for CFD Validation when you can't provide precise initial and boundary conditions?" Their solution is to systematically apply CFD (spatial DNS) simulations to study the effect of different conditions along edges of the computational domain, "identifying those quantities in the boundary layer that are indicative of the forcing." The key is that "amplified quantities in the boundary layer are often larger than their predecessors in the freestream and therefore measurable, thus providing a database." They call for "CFD leadership in the identification, cataloging, and modeling of the effects of freestream disturbances."

The subject area of direct spatial simulation of boundary layer transition is one of the few that proves difficult from the aspect of computer round-off error. From Haynes et al (1996): "Because of the long fetch from the onset of instability to breakdown and the large amplitude ratios associated with this process [$O(e^{10})$ and larger], resolution and bit accuracy limit how far into breakdown a spatial simulation can go. Because of the large differences in amplitudes throughout the domain and the large growth rates known to exist near breakdown where smaller scales appear, truncation and round-off errors can easily contaminate the solution. Consequently, spatial simulations are currently unable to proceed completely through transition and into turbulence."

This area of fluid dynamics (boundary layer transition) can also be sensitive to outflow boundary conditions, which (as noted in Chapter 2) might appear to be a "gray area" in the distinction between Verification and Validation, but, in the final analysis, is best considered part of (the purely mathematical area of) Verification. Haynes et al (1996) suggest five exercises for testing [Verifying] a CFD boundary layer transition code:

a. grid refinement studies,
b. solving test problems for which the solution is known,
c. changing the 'far-field' boundary conditions systematically and re-solving,
d. comparing linear growth rates, neutral points, and eigenfunctions with linear stability theory,
e. running the unsteady code with time-independent boundary conditions to ensure that the calculations remain steady.

They also require that the "manufactured solutions should be chosen with topological qualities similar to those anticipated for the solution to the 'real' problem (e.g. gradients close to the wall)." As noted in Chapter 3, I do not believe this is a necessary feature mathematically, although it is desirable from the viewpoint of intuition and confidence building.

These studies from Haynes et al (1996) are powerful examples, not only of Validations of CFD codes, but also of the power of simulations when performed very carefully, with rigorously Verified codes, in conjunction with high-level theory and highly controlled experiments. These results could not have been attained by haphazard application of CFD commercial codes by unqualified engineering personnel; such exercises could only serve to pollute the literature, and degrade the reputation of CFD. Likewise, sloppy experiments do much more harm than good. As Haynes et al (1996) note for boundary layer transition (but the observation holds in many other areas of fluid dynamics, heat transfer, and other areas of applied science), "Validation requires comparison with careful archival experiments, but few such experiments have been performed." Computations can contribute to the "determination of relevant Validation experiments."

10.7 THE DIFFICULTY OF DEFINING A "NEARBY" PROBLEM

As noted in Chapter 2, we would not want to be so scrupulous as to insist that only a single Calculation can be Validated. However, the concept of Validation of a Code, rather than just a single calculation, is meaningful *provided* that the evaluation of "Validated" is applied only to a range of "nearby" problems, close to the discrete set of experimental Validation parameter space. It was also noted that the concept of a "nearby" problem will fail near parameter "transition" boundaries, which should be identifiable from knowledge of the general field (see list in Chapter 9, Section 9.9.). However, the following example shows how surprisingly difficult this can be.

Consider the effect of chemistry modeling on aerodynamic coefficients. Without trying to decide *a priori* by theoretical considerations whether or not chemistry is important, one might reasonably assume that, if chemistry could be shown to be insignificant in a given parameter range for one aerodynamic coefficient, it would also be insignificant for another coefficient at the same parametric conditions, i.e., surely this would constitute a "nearby" problem. Unfortunately, this is not always the case. Marvin (1995) gives an example of experimental data on sharp and blunt cone models fired in a hypervelocity gun range, compared to two sets of calculations, one assuming a perfect gas, and one using non-equilibrium chemistry. High enthalpy, reacting but laminar flow was established over the model, so chemistry was an issue, but turbulence modeling was not. For a 5° blunt cone, perfect gas calculations for drag were within the experimental uncertainty and similarly close to the nonequilibrium chemistry calculations. However, the data for drag are insensitive to chemistry modeling only because drag is composed mainly of forebody pressure drag, whereas the flowfield generally, and the pitching moment C_m especially, prove to be sensitive to chemistry. The error in C_m for the non-equilibrium air calculations (full Navier-Stokes and seven species air) was within the estimated experimental error of ±3% over the entire range of angle of attack α from 0° to 8°, but the error for the perfect gas calculation at $\alpha = 5°$ is an order of magnitude larger, ~ 30–35%. This is indeed a sobering example of the difficulty of defining a "nearby problem" for Validation of a Code.

10.8 MISSING EXPERIMENTAL DETAILS

Barber (1996) provides a review of several code Validation/Certification studies in the aerodynamics literature, and cites examples wherein missing experimental details have lead to poor Benchmark calculation comparisons. The details he discusses are

a. the geometrical definition,
b. the data reduction procedures, and
c. the dominant physics.

Advances in experimental techniques can now supply details of flow fields that were impossible in the past. For example, Eklund et al (1995) performed Validation of the SPARK Navier-Stokes code predictions of Mach 2 flow over a rearward-facing step. This Validation was unique in that the experimental techniques used (Laser-Induced Iodine Fluorescence and Laser Doppler Anemometry) produced not only velocity data but also in-stream static temperature measurements.

10.9 ONSET OF 3 DIMENSIONALITY IN BACKSTEP FLOW

The experiments of Armaly et al (1983) show the abrupt onset of 3 dimensionality in the nominally 2-D laminar incompressible flow over a backstep. These trustworthy experiments have been used as the basis of comparison for many CFD Validation exercises, mostly for the 2-D flow regime ($Re < 400$). The 2-D CFD predictions for reattachment length are in good agreement with the experiments up to $Re \sim 400$, but predict a longer than observed reattachment length for $Re > 400$, evidently due to the onset of 3-D effects in the experiments. Conjectures on the causes of the 3-D flow (besides possible experimental flow or other errors) included Taylor-Goertler instability associated with the concave streamlines near reattachment, and tunnel side-wall boundary layer growth, the latter being rejected due to discrepancy between the proposed mechanism and experiment as Re is increased (Ghia et al, 1989).

The actual mechanism was determined by 3-D calculations by Williams and Baker (1997). This carefully performed study (involving 7 levels of grid refinement, and evaluation of sensitivity to entrance length) first of all showed good agreement with experiments for both 2-D and 3-D flows, thus Validating the numerical solutions (using the finite element Continuity Constraint Method). More interestingly, the detailed flow calculations disclosed a previously unrecognized flow feature, that of a wall jet at the backstep plane that grows in strength with increasing *Re*. This wall jet is the source of 3-D vortices, and the Lagrangian particle tracks of the simulation "reveal a fascinating picture of very complex three-dimensional flow structures."

This study by Williams and Baker (1997) shows the power of carefully performed CFD simulations applied to trustworthy experimental data, and guided by sound theoretical reasoning, to convincingly explain complex phenomena.

10.10 GRAY AREA: "VALIDATION" FROM A CALCULATED BENCHMARK

A gray area exists in using a DNS (Direct Navier-Stokes) computational data base for Benchmarking "Validation" of simpler turbulence models. Should this be called Verification or Validation? Apparently, confidence in the governing equations and the solution (obtained with highly accurate spectral or mixed spectral and high-order FDM) is so high that agreement is taken as a Validation. I am reluctant to corrupt the mathematics vs. science distinction of Verification vs. Validation, but I have no trouble with the word "Benchmark."

10.11 GRAY AREA: "VALIDATION" OF AN EXPERIMENTAL TECHNIQUE BY A COMPUTATION

In Newling et al (1997), a comparison is made between experiment and computation with the object not of Validating the CFD computation, but "Validating" the experimental technique!

The experimental technique is velocity determination by Magnetic Resonance Imaging (MRI). In fact, it is difficult to determine what is the cause for discrepancies—the CFD code, the grid, the experimental flow (especially symmetry problems), the "seeding" of the flow with particles that respond to MRI, the MRI data acquisition, or the MRI data reduction, which involves significant calculations. The authors take the CFD computation as the standard, but in such a case there is no clear-cut "Benchmark" and no certainty of what is being Validated or invalidated. Nevertheless, the overall agreement (at least qualitative) builds some confidence in the experiment, the MRI technique, and the simulation, at a discrepancy level appropriate for a new technology.

Another approach to Validating a new experimental technique by way of simulations, with more easily defensible results, is that of Jumper and Hugo (1995); see also Hugo et al (1995) and Jumper et al (1994). The paper describes the small-aperture beam technique, which is a sparse-instrument method for quantifying the instantaneous optical wave front distortions caused by propagation of a laser beam through an optically active (aberrating) turbulent flowfield. The method takes advantage of the fact that the structures on the wave front *convect* through the viewing aperture. A numerical simulation of a 2-D heated jet, using a discrete vortex model, is used to "explore the validity" of using beam-jitter signals from multiple probe beams to obtain a measure of the distortion (optical path difference, or OPD) in a flow region where eddy production constitutes the major character of the turbulent flowfield. The concept involved is that the numerical simulation provides a Benchmark of high-resolution space and time data for the numerical quadrature of the optical propagation. In this situation, high fidelity between the simulation and a physical experiment is not strictly required. Since "the essential character of the experimental flow appears to be captured by the numerical flow," the numerical data base can be used to "Validate" the experimental technique.

Although these types of studies cloud the distinction between Validation and Verification, their value is indisputable, semantics aside.

10.12 THE MADE-2 EXPERIENCE: CAN GROUNDWATER FLOW MODELS BE VALIDATED?

In Chapter 2, Section 2.2, the paper by Konikow and Bredehoeft (1992), "Groundwater Models Cannot be Validated," was criticized for its excessive pessimism. However, it is true that error tolerance in groundwater flow and transport modeling is huge compared to other fields. Rigorous Validation sometimes seems to be an insurmountable problem for field data on groundwater flow, and some sites are worse than others. A well known difficult case is the MADE-2 experiment (Boggs et al, 1992; Adams and Gelhar, 1992; Rehfeldt et al, 1992; Boggs and Adams, 1992; MacIntyre et al, 1993; Stauffer et al, 1994).

MADE-2 (for MAcroDispersion Experiment) was a natural-gradient experiment in which water and dissolved contaminants (tritium and four organics) were injected into an unconfined aquifer in a two-day pulse and monitored for 15 months. The tritium provided a conservative tracer (its decay with a 12.26 year half life was easily corrected) while the non-conservative organics (realistic components of fuel spills) exhibited sorption and proved to experience significant natural (probably biological) attenuation, which is highly beneficial to prospects for natural remediation techniques. MADE-2 was unique in two respects: the density of the experimental resolution in both space and time, and the heterogeneity of the site. A 3-D sampling network tracked contaminant plumes, and other extensive field measurements make it one of the most completely characterized sites in the open literature (Boggs et al, 1990). Compared to other experimental field sites, this one is highly heterogeneous. Significantly, it is not known at the present time if this is non-representative of nature in general, or just of the instrumented field sites. Which is more representative of a typical contamination site is hard to say. What is clear is that, given the limits of present field test measuring capability, no combination of Tuning of model parameters, reasonably constrained by in-situ field measurements of properties, and within the conceptual model limits of isotropic Darcy law flow, will reasonably match the field measurements, without introduction of *ad hoc* assumptions.

As dense as the field measurements were (e.g., there were 2500 measurements of hydraulic conductivity in 77 different boreholes), there is room for improvement with conceivable developments in instrumentation. Although permeability measurements were dense, porosity measurements were scarce. (Only one measurement was made of specific yield.) These affect transport calculations for time-dependent flow. Also, the field measurement technique used to determine permeabilities is considered to be state of the art, but nevertheless has a significant limitation.

The borehole flowmeter technique involves injection of water through a probe, and measurement of resistance to infer permeability. Vertical resolution is achieved, so the data is 3-D. This is important for geological formations, since geological layering causes significant heterogeneity (as well as isotropy) in the vertical direction. Of course, there is some inevitable smoothing associated with the data, being inherently averaged over some volume of space of the scale of the probe itself. The more fundamental limitation is the inherent averaging in the circumferential direction. This means that only an effective scalar (non-directional) permeability can be measured. If the site actually had significant tensor properties (i.e., directional preference in hydraulic conductivity) this could not be detected because of instrumentation limitations. Recall the statements of Chapter 9, Section 9.3: "A conceptual model is needed prior to experimental data gathering" and "Every observation is laden with theory." The step of converting the raw data into permeabilities *assumes* a Darcy Law formulation with scalar (non-directional) permeability. Nor could the data be re-interpreted with a different assumed form of the law, since the experimental technique has no directional sensitivity.

Likewise, the inference of dispersivities is obviously dependent upon the assumed conceptual and mathematical model, and upon the scale of definition, which clouds the border between continuum definition and discretization (Roache et al, 1997).

The well known vertical directional properties (due to geological layering) are probably not a major problem, because the horizontal resolution was adequate. In the absence of vertical flow data, groundwater modelers typically use a factor of 10 vertical heterogeneity, i.e. a tensor property with vertical permeability assumed to be 1/10 that of the measured or otherwise modeled horizontal permeability. For the MADE-2 site, an average value of vertical anisotropy was calculated as 0.18. The results are usually not very sensitive to the ratio, since vertical flow is a secondary effect. In fact, most groundwater flow simulations are performed with a 2-D code in which the vertical layer is represented with a single control volume, and no vertical flow is calculated. ("Vertical" is also a simplification; the small angle approximation is used, so that no flow is calculated through the top and bottom of the layer, but the layer may be inclined slightly to the horizontal.) For the MADE-2 experiments, we found (Roache and Rucker, 1996; Roache et al, 1997) that in order to obtain good fit with the experimental plume data, it is necessary to use highly anisotropic horizontal conductivities (by factor of 15). These were not substantiated by field measurements, nor can they be because of the circumferential averaging implicit in the field instrumentation. However, the possibility of anisotropy was suggested by aerial photographs of the site that show some indication of previous streambeds in the area, and by some well draw-down tests. Clearly something is going on, since at the scale of measurement, the observed flow paths (inferred from contaminant tracers) did not follow the head (pressure) gradients; no amount of Tuning with scalar properties can match this behavior.

Whereas improvement is conceivable with new instrumentation that would allow measurement of tensor properties, it is also clear that there are limits of what can be predicted with mathematics. If standards for precision are excessive, it is true that "groundwater models cannot be Validated," but neither can anything else. As stated previously, the concept of Validation must include an acceptable error tolerance or level of precision. It is also true that groundwater flows will always have an error tolerance large compared to most areas of science and engineering.

This does not mean that groundwater modeling is useless, only that it must be used with commensurate tolerance. The "predictions" may be of very coarse precision, but still useful compared to no predictions at all. Even very long term simulations can be useful, e.g. the 10,000 year simulations *required by law* for low-level radioactive waste disposal (WIPP PA Dept. (1992; Helton et al, 1995, 1996). Such exercises properly are not conceived as "predictions" at all, but as rational bounding exercises. No one knows if or when some luckless future wildcat operator will drill into the repository; nevertheless, a rational probabilistic study (including geostatistics) can be made to bound the consequences and rationally and quantitatively (vs. emotionally) evaluate the risks and costs, and allow comparisons to alternative disposal concepts.

As quoted in Chapter 2, Konikow and Bredehoeft (1992) assert that "In the end, action concerning waste disposal will be a judgment; a professional judgment by the scientific community and a judgment by society." As discussed in Chapter 2, Section 2.2, such replacement of modeling by "judgment by the scientific community" or scientific expertise is nonsense. Decisions must be made quantitatively; we need numbers, whether or not the numbers come from codes. If the decision is based on a classic "back of the envelope calculation" based on a simple theory, then that is the *model*! The only difference between such a calculation and a computer code simulation is that the "back of the envelope" model is incredibly limited, and makes much *less* defensible approximations and assumptions than the computer model. Once the knowledge of the geophysics scientist is put into usable quantifiable terms, it *is* a *model*. And decisions cannot be made on non-flow-model geological data. The type of information dear to the hearts of geologists—the age of the rocks, or the color of the rocks, or the chemical composition, or the description of earth processes that formed the rocks—even if these data were quantitatively accurate, would not answer the question of the suitability of the site or of a proposed remediation technique.

Groundwater flow modeling is not a choice, it is the *only* approach. The only remaining question is the scale of the "modeling," from a suite of thousands of 3-D time-dependent computer simulations with $O(10^6)$ cells over 10,000 years, to a "back of the envelope" single-cell steady-state "model." In any case, the models must be Validated in some sense, at some level of accuracy, no matter how crude, in order to be quantitatively useful.

10.13 DYNAMIC STALL WIND TUNNEL DATA: WHO DOES THE TWEAKING?

In Salari et al (1994), we presented simulations of 2-D dynamic stall using the Wilcox k-ω turbulence model and comparisons with experiments. The experimental data were unpublished. Agreement with the experimental data initially was poor. It became clear from inspection of the data itself (without regard to simulation results) that a problem existed, namely, there was significant flow angularity in the tunnel. (This was evident from the behavior of the static data on the symmetric airfoil.) We (i.e., the computer modelers) estimated a flow angularity correction from the static tunnel data, not using our simulations, and applied the corrections to the input parameters (angle of attack) of our simulation. The result was greatly improved agreement with experiment.

This kind of data reduction activity creates a gray area—it "muddies the water" of Validation. If such corrections are done by the CFD user (or worse, by the code developer in a Validation exercise) they could be subject to snickers and knowing accusations of "tweaking the data to fit the computations." But if the experimenter had done the same calculation and called it "data reduction" instead of tweaking, no one would criticize it. The folklore observation is close to literally true: *no one* believes a CFD calculation except the person who performed it, and *everyone* believes the experimental data and reduction except the person who performed it. Cooperation and team work between computer modelers and experimentalists is often and properly called for, yet it does open the door to criticism.

10.14 CONSORTIUM EFFORT AT CFD CODE CERTIFICATION

Melnik et al (1995) describe a collaborative effort in CFD code assessment that covers Verification, Validation, and Certification. The work assessed the usefulness of CFD codes in a production design environment. The project was conducted by a U. S. industry/university team working within a consortium called MADIC (the Multidisciplinary Analysis and Design Industrial Consortium) with funding provided by NASA.

The Code Certification plan was developed by a MADIC/NASA review panel and exercised on five NASA codes in six Certification projects defined by the configuration tested. (Only one configuration was calculated using all five codes, and another using two codes; the remaining three configurations were each calculated with a single code.) The panel developed the overall organizational structure and plan for the Certification process, and facilitated formation of the Certification teams. Although the term QA (Quality Assurance) was not used in the paper, this Certification involved some standardization of metrics, evaluation forms, and log forms, separately published in the open literature (Melnik et al, 1994) and was essentially a QA exercise.

The focus was not on fundamental flow quantities (like velocities, pressures, etc.) but on design performance data; this distinguishes Certification from Validation as the terminology was used by the authors. (See my comments in Chapter 2.) Herein, I use the more general term Validation in some cases where Melnik et al used "Certification." Also, when they use the term "Verification" or even "Verification of Code," they are referring to what I have described (see Chapters 2 and 4) as "Verification of Calculations." (They do not address "Verification of Codes" in the sense used herein in Chapters 2 and 3, i.e. demonstration of the order of discretization convergence to a known exact solution.)

Aircraft design performance data include lift, drag, moment, maximum lift and loads, flap effectiveness, stall prediction, buffet onset, internal flow losses, etc. The authors stated that, despite many *ad hoc* studies in the literature, it was not possible to draw broad conclusions, e.g. the fundamental question of how well maximum lift of a simple transport model can be predicted "remains largely unanswered."

The six classes of flow configurations simulated were as follows. (Only four of these will be synopsized herein.)

- Isolated Wing
- Nozzle/Boattail Flows
- Axisymmetric Nozzle
- Wing-Body Combination
- Turbine/Compressor
- Flexible Wing (steady and unsteady aeroelastic problems)

The experimental data used was restricted to wind tunnel data rather than flight data, because "there is little flight data available that was suitable for Code Certification." This is an impressive statement on the paucity of reliable and complete flight data, coming as it does from a consortium including NASA and major aerospace firms. In addition to performing simulations, the teams also assessed the experimental data with respect to four "factors/issues": "measurements, test conditions, data accuracy, and wind tunnel installation effects." The panel noted that experimental error estimates should be provided for:

- model and support geometry dimensions
- wind tunnel wall and support system interference
- instrumentation
- free stream and operating conditions
- model distortions due to aerodynamic loads.

Corrections for wind tunnel interference should include

- Mach number
- incidence
- forces and moments.

They noted that "it is also highly desirable that pressure measurements on all wind tunnel walls and model support systems be provided. These can be used either to develop wind tunnel interference corrections or to provide boundary conditions in CFD computations that include the wind tunnel configuration in the computational model." They also pointed to "the continuing need for high quality experiments that are designed specifically for CFD Code Certification" [Validation].

The panel developed a set of criteria and metrics for the assessments "to add uniformity to the process and because Certification of a Code as a design tool requires specific criteria and metrics." The criteria are in four broad categories: "numerical, physical, operability, and range of applicability." The criteria are listed in Figure 10.14.1; see Melnik et al (1995) for further discussion.

A sampling of interesting results from Melnik et al (1995) is now given for various configurations.

10.14.1 Isolated Wing C

This was the only configuration for which all five NASA codes were applied. The wing is highly three dimensional, with aspect ratio = 2.6, leading edge sweep angle = 45°, tip chord/root chord = 0.3, and twist = 8.17°.

The Wing C configuration was simulated for three flow conditions of Mach Number M and angle of attack α, all at Reynolds Number = 10 million: $M = 0.82$, $\alpha = 5°$; $M = 0.84$, $\alpha = 4.62°$; $M = 0.85$, $\alpha = 5°$. Only an algebraic turbulence model was used because only this was common to all five codes.

Grid sensitivity [Verification of the Calculation] was investigated using two single-block C-O topology grids, one with $49 \times 241 \times 49$ (578,641) nodes and the other with $65 \times 321 \times 65$ (1,356,225) nodes. "Solutions on both grids agreed well with each other" [presumably using only one code] and so the coarse grid was selected for code-to-code comparisons. Note this is an example of the methodology described in Chapter 5, common especially in such cost-conscious design environments, of Verifying a *coarse* grid calculation by comparison with a fine grid calculation for a single "nearby" problem. In this case, "nearby" signifies not only small changes in flow conditions M and α but also a "nearby" *code*. The coarse grid is then used for the sequence of simulations including parametric variations and, in this study, five different codes.

Although the stated "Certification" goal was design parameters, comparisons were made for surface pressure coefficients (Figure 11 of Melnik et al, 1995). The assessment was that "all [five] solutions agree quite well at the inboard station. The cause of discrepancy between ADPAC [code] solutions and others at the outboard stations needs to be further investigated."

Numerical	Physical	Operability	Range of Applicability
• Accuracy – Global, local parameters – Spatial accuracy through grid refinement estab. • Computer requirements – RAM, CPU, disk – Portability • Grid dependent sensitivity • Convergence characteristics • Grid types – Single block – Multiple block – Overset – Unstructured – Adaptive • Portability – Mainframe, mpp, workstations – Distributed Workstations, Clusters	• Accuracy – Performance data – Requirements for design data • Model adequacy – Transition model – Turbulence mode	• Reliability – Sensitivity to numerical – Tuning parameters & grid robustness • Credibility – Established limitations – Known error estimates • Responsiveness – Wall times – Design environment • Geometry/grid/solver interfaces • Ease-of-use – Grid automation – Set-up time – User friendly features – Pre & post processor integration – User skill requirements • Maintainability – Modularity – Support – Documentation	• Flow regime – Incompressible – Compressible Steady/unsteady – Transonic, supersonic, hypersonic – Perfect gas/ real gas • Geometric flexibility • Grid flexibility • B.C. flexibility

Figure 10.14.1. Criteria /Metrics for CFD Code Assessment for Production Design Environments.
(From Figure 7, Melnik et al, 1995.)

	C_L	C_D	C_N	C_M
Experiment	–	–	0.5400	–0.0393
Codes:				
CFL3D	0.5256	0.04097	0.5272	–0.0726
TLNS3D	0.5254	0.04290	0.5271	–0.0736
ENSAERO	0.5212	0.0423	0.5230	–0.0739
ADPAC	0.5217	0.03453	0.5227	–0.0686
OVERFLOW	0.5267	0.04422	0.5285	–0.0768

Table 10.14.1.1. Comparisons of five code predictions and the experimental values for aerodynamics coefficients of a highly three dimensional isolated wing for the case $M = 0.85$, $\alpha = 5°$. (From Table 1 of Melnik et al, 1995.)

Comparisons of the five code predictions and the experimental values for aerodynamics coefficients for the case $M = 0.85$, $\alpha = 5°$ are given in Table 10.14.1.1. [The normal force coefficient C_N is of course close to the lift coefficient C_L at this small α, so the experimental agreement for C_N is indicative of the agreement for C_L and is fairly good, 3.2% discrepancy in the worst case. The discrepancy in drag coefficient C_D might be expected to be worse, but no experimental values were given. However, see the M100 wing-body combination results below. The discrepancy in moment coefficient C_M is large, 75% in the best case, 88% in the worst.] See Melnik et al (1995) for comparison of code economics; total computer times ranged from 3440 to 22000 seconds, and memory required ranged from 17.3 to 39 mega-words.

10.14.2 Nozzle/Boattail Flows

This geometry consisted of a 3D aft end with a 2D converging/diverging nozzle. A seven zone grid with a total of 989,595 nodes was used. Verification involved calculations with a grid with 75% grid density and another with 125%. Two codes were used, with three turbulence models: an algebraic or "zero-equation" model (Baldwin-Lomax), and the Baldwin-Barth and Spalart-Allmaras one-equation models. Both codes confirm the now well-documented inadequacy of algebraic turbulent models in separation, wakes, shear layers, etc., flows for which they were not designed and where they require *ad hoc* adaptations. A representative comparison of surface pressures is given in Figure 10.14.2.1.

10.14.3 Wing-Body Combination

Melnik et al (1995) provide an excellent lesson on the pitfalls of inadequate initial Verification of Calculations, misleading Tuning of experimental corrections, and mixing of Verification and Validation, in their case study of the Wing-Body Combination.

The configuration is a transport type wing (designated M100) mounted on a simple circular cross section fuselage. The mean-chord Reynolds Number was 3.2 million, and boundary layer transition was forced at 7% chord in the experiments. The thin-layer Navier-Stokes code TLNS3D was used with two turbulence models: the one equation Spalart-Allmaras (S-A) model

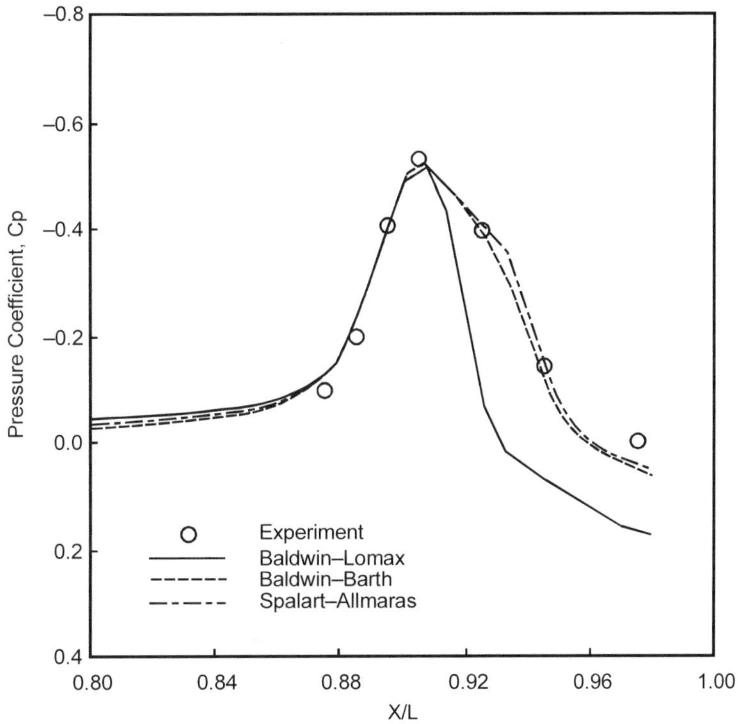

Figure 10.14.2.1. Comparison of Surface Pressures at Mid Sidewall for an Aft-End/2D Converging-Diverging Nozzle Flow at *M* = 0.938. (From Figure 13b, Melnik et al, 1995.)

and the blended k-ω/k-ε model of Menter (M–T). Original simulations were done assuming full turbulent flow everywhere; later calculations enforced transition at the experimental location. Also, the code had options for the strictly numerical explicit artificial dissipation using either a scalar form of dissipation (SD) or a full matrix form (MD). Although "extensive grid studies were conducted in this project" details were given on two grids, $321 \times 57 \times 49$ (896,553) nodes and one doubled in the k-direction to $321 \times 57 \times 97$ (1,774,809).

Important to the tale is the fact that an angle of attack correction (for tunnel flow angularity and interference) of $\Delta\alpha° = -0.2636\ C_L$ was supplied with the experimental data. (Note the point made in Chapters 2 and 9, that experiments as well as simulations require computations and therefore include numerical errors.) "The one problem noted in the initial phase of the work was attributed to an aeroelastic effect; the first computations made predicted a 15% higher lift than the experiment. In particular, the section lift at the wing tip section was significantly higher than the experiment," suggesting that the culprit was aeroelasticity, i.e. the wing model was twisting significantly under the aerodynamic load. The experimental report stated "measurements on a similar model indicated maximum aeroelastic wing tip deflection and twist of about +25mm and 0.25° nose down." So the modelers included the correction for twist in their computations, and this "significantly improved the lift comparisons."

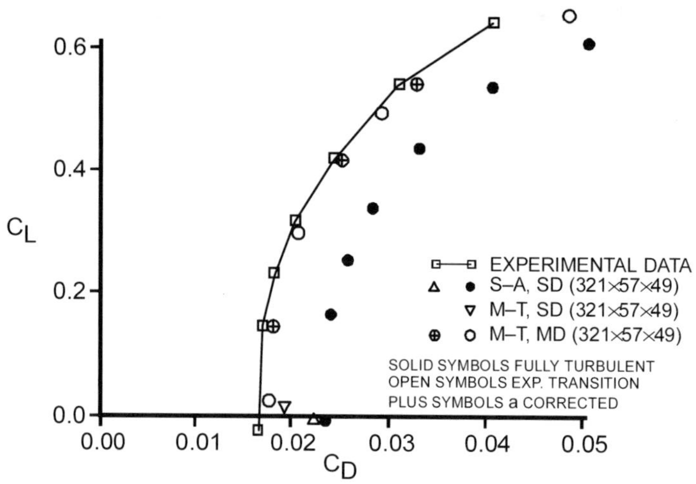

Figure 10.14.3.1. Drag polar for M100 Wing-body Combination; Experimental Data and Simulations Using Code TLNS3D. S-A = Spalart-Allmaras turbulence model, M-T = Menter blended *k*-ω/*k*-ε model. SD = scalar artificial numerical dissipation, MD = matrix dissipation. (From Figure 19 of Melnik et al, 1995).

After this twist correction, the drag polar (the plot of C_L vs. C_D, parameterized by α, Figure 10.14.3.1) remained shifted about 69 counts, i.e., the drag coefficient C_D was 0.0069 higher than the experimental data. Originally, this was attributed to inaccuracy in the computations, but the issue was revisited in Melnik et al (1995).

The next effort to reconcile the simulations with the experiments involved Tuning of both numerical parameters and physical models [i.e. an ill-advised mix of Verification and Validation, in the present terminology]. The base line simulation on the $321 \times 57 \times 49$ grid had used scalar dissipation (SD), fully turbulent flow, and the Spalart-Allmaras (S–A) turbulence model, and had produced drag higher than experimental data by 69 counts. A 10 count reduction in this discrepancy was achieved by including another physical feature in the simulation, eliminating fully turbulent flow by including boundary-layer transition at the experimentally fixed location. (see the upward-pointing triangle in Figure 10.14.3.1). Another 30 count reduction was achieved by changing the constitutive equations, switching the turbulence model (from S–A to M–T). After these *conceptual model* changes, purely *numerical* changes were employed, doubling the spanwise grid to $321 \times 57 \times 97$, and switching from scalar artificial dissipation (SD) to a matrix form of dissipation (MD). These purely numerical changes achieved another 20 count reduction in the drag at zero lift C_{D0} (along the abscissa of Figure 10.14.3.1), giving the best match to experiment, high by only 12 counts instead of the original 69.

This set of simulation parameters were taken as the new base-line, and attention shifted from the drag polar, in which angle of attack α parameterizes C_L vs. C_D (Figure 10.14.3.1), to generating a new lift curve of C_L vs. α (Figure 10.14.3.2). Here, the pitfalls of Tuning become evident. The old baseline set of simulation parameters had predicted C_L vs. α fairly well, but had produced the 69 count discrepancy in the drag polar. The new baseline set improved the zero-lift

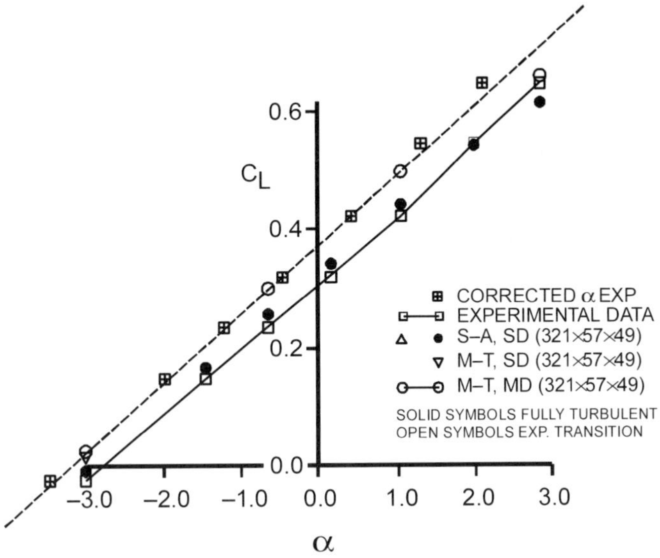

Figure 10.14.3.2. Lift Curve for M100 Wing-body Combination; Experimental Data and Simulations Using Code TLNS3D. S-A = Spalart-Allmaras turbulence model, M-T = Menter blended *k*-ω/*k*-ε model. SD = scalar artificial numerical dissipation, MD = matrix dissipation. "Corrected α Exp." refers to the new base-line set of simulation parameters, which had produced improved agreement with the experimental drag polar. (From Figure 20 of Melnik et al, 1995).

drag discrepancy to 12 counts, but now (Figure 10.14.3.2) produced C_L vs. α curves nearly 20% high!

The resolution of the paradox begins with the recognition that the agreement in C_L vs. α of the original baseline was somewhat illusory. The aeroelastic effect (0.25° twist at the wing tip) had been "substantiated by matching the experimental lift, so that it is not surprising that the original base-line computation (solid circles) matched the experimental lift curve." The sleuthing then continues (see Melnik et al, 1995 for details) to conclude that the source of the problem is the under-estimation of aeroelastic correction to angle of attack provided in the experimental data. This correction had been based on under-resolved simulations, i.e. on *inadequate Verification of the calculation inherent in the experiment.*

The team forcefully concluded that Verification of the Calculation should precede Validation and Certification. Especially, this study provides an example of the error and confusion propagated by faulty experimental data reduction calculations, including the pernicious practice of judging grid refinement adequacy simply by agreement with experimental data. "This experience clearly indicates the importance of the three steps (i.e. Verification [of the Calculations], Validation and Certification) in the assessment of a code as outlined in this paper. Before Certifying a code for design, its numerical accuracy must be established (i.e. Verify that the grid used is adequate) and the turbulence/transition model must be Validated. Only then can the Certification [Validation] process begin."

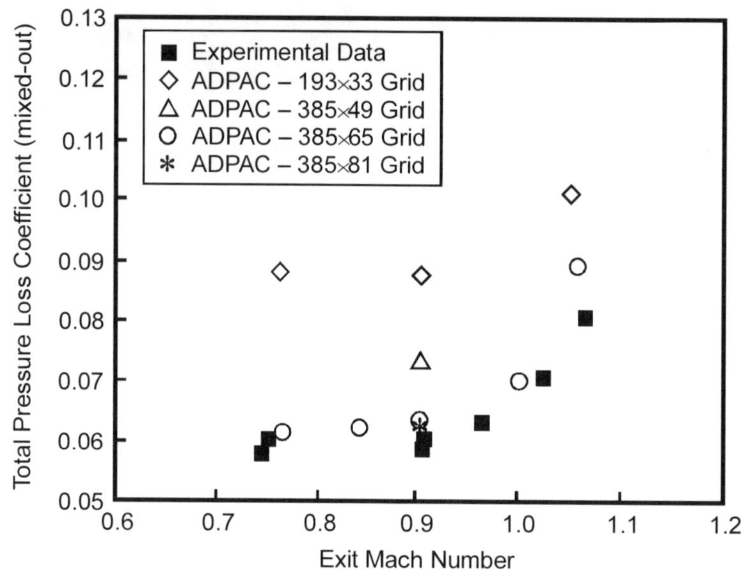

Figure 10.14.4.1. Loss Coefficient for a Turbine/Compressor Flow as a Function of Exit Mach Number, Showing the Effect of Grid Density. (From Figure 23 of Melnik et al, 1995).

10.14.4 Turbine/Compressor

The study in Melnik et al (1995) of Turbine/Compressor flows involved eight different configurations. The grid convergence study involved 2D slices with grid density ranging from 193 × 33 to 385 × 81. Even so, the results (Figure 10.14.4.1) show "the strong dependence of the total pressure loss coefficient (mixed-out) on the grid spacing at an exit Mach number of 0.90."

The "fine mesh prediction of 94.1% peak efficiency agrees fairly well with the measured value of 93.4% and occurs at essentially the same mass flow relative to the choking flow (9.990 predicted vs. 0.989 measured)."

The team's assessment of the single code used noted that "ADPAC is relatively easy to apply by a first-time user," and gave this curious, virtually universal statement that "Moderate CFD experience is suggested to achieve optimal results." [I should hope so.] Less obviously, they noted the need for "serious debugging of less used options" [a Verification of Code issue, in present terminology], for better than an algebraic turbulence model [a Validation issue, in present terminology], and for "more consistent version control" [a Certification or Code QA issue, in present terminology].

10.14.5 Selected Observations of Melnik et al (1995)

A brief selection of observations from Melnik et al (1995) follows. Some are repetitious, but deserve to be.

- "Certification is an expensive and time consuming process that is currently carried-out in piece-meal or ad-hoc fashion in the [aerospace] industry. Our experience in the project has demonstrated, at least to us, that a consortium or other type of teaming approach is a very effective way to accomplish code Certification. It is cost effective because it reduces duplication of effort, it accelerates technology transfer to the designers who are the ultimate users and it promotes code developers to do the right thing through the generation of consistent and broad based industry feedback."
- "Code assessment (at the Verification [of Calculations] step) requires computations on grids that are fine enough to reduce numerical errors below those from other sources. Our studies in the project clearly established that this is a necessary first step in the assessment process. Failure in the Verification step will, at best lead to inconclusive results, and at worst to misleading or erroneous assessments."..."In this project, because we were able to use grids that were sufficiently fine, we were able to observe first-hand the pitfalls of working with inadequately Verified codes [Calculations]." From our experience on this project it is clear that proper Verification [of Calculations] is an absolutely necessary first step in the code Certification process."
- The complete Certification process cannot be done in general or once-and-for-all but must be carried through for each general class of configurations...and/or physical phenomena...Each of these situations involves differing grid and physical modeling requirements that must be addressed in the [Verification of Calculations, Validation, and] Certification process."... "Certification has to be thought of as an ongoing activity..."
- "Data archiving is an important aspect of the code Certification process..."..."The data should be archived in a standardized form..."..."Lack of a process of archiving Certification data is hindering the Certification process and the transfer of CFD technology to the industrial design community."
- All the studies pointed to the inadequacy of simple algebraic turbulence models.
- "The prevailing view seems to be that CFD cannot be trusted for absolute values of drag. Our results, at least for the simple M100 wing/body combination, indicate that the TLNS3D [code] can predict drag to within 5% on a relatively crude 1.7 million point grid."..."This conclusion was reached only after a careful grid study in the Verification stage on the assessment."
- "All teams pointed to the critical need for additional experimental data for CFD code Certification."
- "Larger computing resources will be needed to treat problems of greater complexity..."
- "Support is needed for the acquisition of suitable experimental data as well as for the Certification process itself. Unfortunately, this NASA supported project is the exception and not the rule."

10.15 SIMULATION TEAM RESPONSIBILITIES IN VALIDATION

Rizzi and Vos (1996) note that the engineering activity of CFD simulations involves the interaction of three distinct teams.

1. Code Developers–Software engineers, mathematicians, and physical model developers.
2. Physical model experts who carry out Calibration and Validation.
3. Users, who are [generally] not experts in either the code nor modeling, but are application specialists.

Rizzi and Vos state that, of the four steps (Verification, Validation, Calibration, Certification) required to build credibility, Verification [of the Code] and Certification are the responsibility of Team 1, the Code Developers. (Recall that, as described in Chapter 2, Rizzi and Vos do not include Validation in their concept of Certification.)

"The work of the second and third teams are interrelated because Calibration and Validation are specific to a particular design application and user community. It is the task of the second team to Calibrate a code, and through Validation determine its range of applicability, which then has to be transferred to the end users. Validation and Calibration are related. In Calibration you improve the results of your Validation, at the expense of a general Validation because you modified or specialized the physical model to make it work for a certain type of generic flow. In Validation you identify all the generic flow types that make up the complex flow and then choose the appropriate model to make the predictions. The reasoning behind this approach is inductive, and therefore must be Validated. The key activity, which is also the most demanding and expensive, is Validation."

10.16 SHIFTING RESPONSIBILITIES AND GRAY AREAS

Discussion of shifting responsibilities is a practical consideration, even though it might appear confrontational (vs. team-oriented). Consider the common situation (e.g., cited in Barber, 1996) wherein a Validation exercise shows discrepancy between CFD simulation and experiment, and the cause is agreed to be probe interference. That is, the experimental flow is perturbed significantly by the introduction of the experimental measuring apparatus.

Should the CFD modeler volunteer to model the perturbed flow about the probe?

If he does, the responsibility for the discrepancy shifts from the shoulders of the experimentalist (whose job it is to minimize probe interference effects, calibrate them, and if possible account for them in the data reduction process) to the shoulders of the modeler. Unfortunately, the probe flow introduces another scale of resolution, and possibly another dimension into the modeling problem. (For example, the undisturbed flow may be axisymmetric 2-D, but a boundary layer probe makes it fully 3-D.)

This is a QA and engineering management issue. The modeler's decision will be influenced by the professional environment, levels of trust (will "responsibility" become "blame"?) and personal values and goals.

More difficult assignment of responsibility will occur in gray areas. For example, in a comparison with wind tunnel data, if the experimenter claims uniform free-stream flow conditions, then any deviation from this condition is the responsibility of the experimenter. But it would not be unreasonable to expect a CFD code to accommodate some non-uniformity in free-stream conditions. (Unlike the probe example, this would not introduce another scale of flow variation, although it might destroy axisymmetry.) If the CFD code cannot include variable free-stream conditions (due to limitations on input data simplicity, memory requirements, or the fundamental formulation of the equations) then the responsibility (blame?) shifts towards the modeler.

10.17 WUA BENCHMARKS IN 1994 AND 1996

The World User Association in CFD sponsored conferences in 1994 and 1996 (Muller, 1994; Muller and Loffler, 1996) that included benchmark calculations by commercial code vendors. The problems were defined by a WUA committee, with suggestions solicited from true industrial users. In spite of much work and expertise, difficulties were experienced with inadequate problem definition. These included undefined geometry and undefined inflow boundary conditions needed for turbulence models. By commitment and attention to feedback from the first conference (de Vahl Davis, 1994), the committee succeeded in correcting many difficulties for the next meeting.

The test cases provide good range of industrial fluid dynamics, including a generic automobile model, flow in a chemical smoke flue, impinging turbulent jet flow with heat transfer, and a stirred reactor tank. The biggest difficulty with the WUA comparisons, common to all such exercises, is that any vagueness in problem specification (e.g., inflow turbulence quantities), no matter how realistically representative of the industrial modeling environment, means that the exercise becomes a comparison of the skills of analysts, when what is desired is a comparison of codes.

Note again the false criticism of turbulence models and CFD, previously discussed. No reasonable turbulence model requires more information (boundary conditions and initial conditions) than the physics. (In fact, all turbulence models are simplifications of the true physics, i.e. the full Navier-Stokes equations, and therefore may be expected to require less information than required to rigorously define the problem.) Thus, if a CFD turbulence model shows that the results of interest (e.g., distance to reattachment for a backstep separated flow problem) are sensitive to some parameters (such as the upstream turbulence kinetic energy distribution) and if this model has been Validated on this point, then the fact that some experiments have not measured this quantity indicates that the experimenters have been running a test with uncontrolled variables. (No wonder agreement between facilities is poor!) Nevertheless, these undefined variables from incomplete experiments impede Validation, making the agreement with experiment or lack thereof dependent more on engineering guesswork and intuition than on reliable data.

10.18 CFD TRIATHLONS

The CFD Triathlons organized by Freitas (1993) and sponsored by the ASME were designed to apprise potential customers and users of the capabilities of CFD commercial codes. Rather than simulate complex industrial fluid dynamics problems with the difficulties noted above (Section 10.17), the vendors were invited to submit their simulations for fairly simple, unambiguously defined, somewhat academic problems. Freitas did not evaluate the results but just let the papers stand for themselves. Even though the problems were simple, some of the discrepancy can be attributed to the skill of the modelers as well as codes.

These observations come to mind.

1. Verification of Calculations were generally inadequate, with most of the papers reporting results from only one grid on all or some of the problems, or minimal non-quantified reference to other studies.
2. Often, CFD commercial codes show poor accuracy, usually due to the sacrifice of accuracy to robustness.
3. A general point, aside from the evaluation of commercial codes or research codes, etc. is that user training is required. (See also Rizzi and Vos, 1996, and Section 10.23 below.)
4. There was a wide range of computer times required to solve the same problem with roughly similar grid densities.
5. Some results were clearly erroneous, and others differed from one paper to another.
6. Although the flows were laminar, Validation (i.e., determination of whether or not the "right equations" were used) was still an issue, because of the assumption of 2-D flow in two of the problems.

10.19 CANADIAN CFD SOCIETY TEST CASE

A. Pollard organized a comparison exercise of commercial and research CFD codes for the 1995 meeting of the CFD Society of Canada (Thibault and Bergeron, 1995). The problem was one of turbulent heat transfer in the vicinity of stagnation point flow. It was intended to be a true prediction exercise (rather than "postdiction"), i.e. the experimental data was not made available to the participants beforehand. However, some of the participants recognized the problem specification and identified the journal paper with the experimental data. Not surprisingly, their results were best. As is often the case, some of the turbulence information required was missing from the experiment, so the problem specification had to be supplemented by the experience and intuition of the modelers.

As is always the case with non-trivial turbulent flow cases, the results were generally unsatisfactory. Agreement with experiment was generally poor, and agreement between different modelers using the same turbulence model was also poor. The exercise was worthwhile, again showing the unsatisfactory state of experimental data, especially for turbulence measurements, and the lack of Verification of Calculations.

10.20 WORKSHOPS

Workshops have been a part of CFD for many years. A workshop for Verification and/or Validation may be described as "a collective learning process where a group of researchers actively working on a sharply defined topic meet to discuss in detail their problems and experiences and make direct comparisons and critical evaluation of their computational and experimental data." (Rizzi and Vos. 1996.) The 1980/81 Stanford Conference widely known as the "Stanford Turbulence Olympics" (Kline et al, 1981a,b) is a premier example. Rizzi and Vos (1996) review several such activities, and note that a strong case exists for collaborative Validation, involving a number of different codes and at least two rounds of analysis. (See also Section 10.23 below.) Another good example is the transonic airfoil workshop of 1987 (Holst, 1987).

Marvin (1995) notes that an important aspect of Validation strategy is the "cataloging, documentation, and assessment of experimental databases. Such information can be used in workshops to assess code development and Validation, by CFD design groups, who must Validate codes used in the design process, and by planners of future experimental programs, who should be aware of existing data." He cites AGARD (1979, 1985), Kline et al (1981a,b), and Settles and Dodson (1991) as precedents of Validation Workshops. Porter (1996) also contains references to Primary Workshops, Databases, etc. The JANNAF Airbreathing Propulsion Subcommittee, Airframe Integration Panel has sponsored a series of workshops on CFD Code Validation/Verification and "has found the subject to be one of major interest and impact on the aerospace community as a whole." Proceedings are published yearly by the Chemical Propulsion Information Agency (CPIA) under JANAFF auspices, e.g. CPIA Publication 551, etc. See Porter (1996) for discussion.

10.21 AGARD 1988 VALIDATION OF COMPUTATIONAL FLUID DYNAMICS

In May 1988, AGARD (the NATO sponsored Advisory Group for Aeronautical Research and Development) hosted a conference/workshop entitled *Validation of Computational Fluid Dynamics* in Lisbon, Portugal. The proceedings (AGARD, 1988) cover 3 invited papers, 34 contributed papers, 16 poster papers, and a Round Table Discussion. The lengthy proceedings provide an overview of the status of aeronautical Validation, including terminology, as of 1988. Session topics included: CFD Validation Concepts; Airfoils; Vortex Flows; Wings/Wing Body; External High Speed Flows; Turbomachinery; and Intakes and Ducts.

The proceedings are worth perusal by anyone active in CFD aerodynamics, yet some deficiencies are noteworthy. The semantic problems go beyond the usual ones, approaching "epistemological chaos." Not all the papers even treat Validation—some just present CFD methods, some present only experimental data that might be used for Validation, but were not designed specifically for CFD Validation. Some are not well done, and few address general concepts. Some common themes emerge from the specifics of grunt work, including these.

- There is real value in using some definitions (though not overly precise) to distinguish concepts.
- Good experiments are hard to find.
- It is best to have experiments specifically designed for CFD Validation.
- As a corollary, cooperation is required (and was often evident) between computationalists and experimentalists.
- There is a wide range of opinion as to what one may honestly describe as "good agreement with experiments"
- Navier-Stokes codes are better than Euler codes for predicting separation.
- Boundary layer transition was never modeled, neither in the CFD computations nor in the wind tunnel "modeling."

The candidate definitions presented have already been discussed in Chapter 2; in AGARD (1988), semantic clarity was lacking. Some of the participants certainly grasped the essential distinction between mathematics and science (e.g. see the remarks by K. Gersten and by L. Roberts in the Round Table Discussion) although no one used the precise terms Verification and Validation. (The conference was held in 1988, and agreed-upon distinctions were still emerging.) On the other extreme, even the standard terms consistency, stability, and convergence were misused. "Consistency" was confused with behavioral errors (Roache, 1972) so that the desirable, but not at all necessary, property of an algorithm identically preserving uniform free-stream flow at all discretizations became confused with recovering of derivatives from differences. Likewise, "stability" was confused with parameter sensitivity. Definitions of "stages of CFD code development" may be expected to be somewhat vague managerial terms, but the criteria of "value exceeds expectations" is irrelevant, having nothing to do with the code *per se* but only with previous managerial prognostications. The claim that [iterative] convergence acceleration schemes constitute a limiting factor to accuracy is not true; to the contrary, faster iterative convergence will make it easier to achieve near steady-state solutions and therefore may be expected to slightly improve accuracy. Empty statements are common, such as "today's computers do not permit total elimination of this error source" related to grid density—as if tomorrow's computers will? There was significant lack of consensus on the difficulty of gird generation, with one participant claiming that "It is now clear that the construction of grids around complex configurations is not a major problem anymore." This is at a conference in 1988! Generally, there were lots of lists with incomplete coverage. Many of the papers note the now familiar need for convincing grid convergence in order to evaluate the experimental comparisons.

In the AGARD Proceedings, Bradley (1988) describes four categories of experiments related to CFD development.

Experiments designed to
1. *understand flow physics*
2. develop *physical models* for CFD codes
3. *calibrate* CFD codes
4. *validate* CFD codes.

Each of these experiments should be designed differently with regards to variables measured, measurement resolution, accuracy required, etc.

On the quality of Validation experiments, the respected experimentalist S. Bogdonoff gave (in the Round Table Discussion) an attention-getting evaluation. "I am appalled at the use of experiments that were made over the past 40 years to do computational 'Validation'. Those experiments were carried out to do specific things for many reasons, but they were surely not designed for computational Validation. I know of very few experiments which were carried out 30 years ago or 20 years ago, or even 10 years ago that have the quality, the basic resolution, and the parameter variations that are required for computational Validation. I would suggest that developing a code and then looking for some experiments to Validate it is unreasonable. The experiments that you need to Validate a code, I believe, have to be designed specifically for that task. You should have computations to tell you what kind of details you need from the experiment, what range of parameters, what kind of instrumentation you need to really test if the code is worth anything or not." Bogdonoff also stated that, as an experimentalist, "the major contribution that I have gotten from computation...is that what I thought was important in the experiment was not the physics that turned out to be critical." This is strong testimony on the effectiveness of simulation, recalling the statement of Hamming (1962): "The purpose of computing is insight, not numbers."

In regard to what constitutes "good agreement," I am not talking about the differing needs for differing end uses, but differing standards of quality for the same end use. The laxness of some computors is simply shocking.

It should come as no surprise that Navier-Stokes codes are better than Euler codes for predicting separation. On the contrary, one would be forgiven for expressing surprise that use of Euler codes was even attempted, since inviscid flows do not separate. What happens is that discretization introduces viscous-like behavior. Often in Euler codes it is an explicit artificial viscosity that is very close to physical viscosity; sometimes it is implicit effects that are only analogous, not strictly equivalent, to viscosity; or sometimes it is not really viscous-like but just some kind of numerical error that allows separation to occur. If the body has a geometrical feature that practically fixes the location of separation (i.e., a sharp leading edge delta wing) then virtually *any* viscosity, physical or otherwise, gives a reasonable answer. Unfortunately, people then push their luck, and try some rounding of the leading edge. If that works, why not a normal airfoil? The reason why not is that the artificial viscosity, and therefore the separation location and the entire solution, depends on the grid spacing. As the grid is refined, the solution does not converge to a physical answer. (This discourages grid refinement studies.) Nevertheless, attempts persist, and ill-conceived Euler solutions of separation are as persistent as 1st-order upstream differencing, or crabgrass. Of course, if Euler codes are built with zonal boundary layer submodels, grid convergence can be meaningful.

Boundary layer transition was never modeled because it is very difficult, both computationally and experimentally. Instead, transition was always induced artificially in the wind tunnel "modeling" with a trip wire or other device. This fixed transition location is then input to the CFD model, often with some blending required (over ~5 cells) to maintain stability. This is a wise Validation approach, because to use "natural" transition would hopelessly confuse the Validation procedure. Transition is known to be devilishly sensitive to wind tunnel idiosyncrasies, even in high-quality tunnels specifically designed for transition studies. In tunnels designed for gross measurement of aerodynamic coefficients, it is a very loosely controlled phenomenon, unrepeatable from one tunnel to another, or even from one run to another in the same tunnel. Also, the CFD codes (certainly those in 1988) did not have transition models in them. This does not mean that the modelers were ignoring the importance of transition, just that they did not know how to handle it. The Validation of a CFD code is then performed with the understanding that it cannot be applied to a new free-flight case without addressing the transition issue. Fortunately, tremendous progress has been made in the last few years in predicting transition (e.g., see Haynes et al, 1996; Reed and Saric, 1989; Reed et al, 1996). It may soon be possible to embed a transition module in full Navier-Stokes codes.

For details on what is involved in the careful design of experiments for CFD Validation, the AGARD (1988) papers by Elsenaar et al (1988), Firmin and McDonald (1988), and Kordulla et al (1988) are recommended.

10.22 A CASE STUDY FOR CFD CODE VALIDATION METHODOLOGY

Aeschliman and Oberkampf (1997) present details of their case study for CFD code Validation methodology. The project is a long term, coupled CFD and experimental effort initiated at Sandia National Laboratories–Albuquerque in 1990, and referred to as JCEAP for Joint Computational/Experimental Aerodynamics Program. The study involved Validation of CFD codes using experiments in long duration hypersonic ($M = 7.84$) wind tunnels, which experiments were specifically designed for the Validation. The methodology will be of interest far beyond hypersonic wind tunnels, however. Such a project represents a major commitment to quality work on the part of all personnel and management. As noted above in Section 10.6, synergism arises when experiments are designed specifically for CFD Validation, and much was learned about both computational and experimental techniques.

The geometry used was a spherically blunted cone with a sliced aft region and flaps at the rear of the slice. With varying flap angles, this relatively simple yet modular geometry generated a wide range of flow complexity, from simple attached flow to very complex flow with massive separation and strong shock wave/boundary layer interactions. The flow was laminar everywhere, as determined by flow visualization techniques. Especially pertinent to the evaluation of tunnel flow angularity and nonuniformity errors was that the model was rolled, to take data at 0, 90, 180 and 270 degrees, and its axial location within the test section was also varied. This unusually thorough approach used in conjunction with 96 pressure ports, various flap angles, and repeat runs produced a massive amount of comparatively high-resolution data. The authors performed a thorough experimental Uncertainty Analysis, obtaining an experimentally based statistical estimate of variance components of surface pressure measurements, including the effects of model geometry uncertainty.

Full details of the statistical methods and conclusions will be found in Aeschliman and Oberkampf (1997), and this report is highly recommended. They determined unambiguously that "the dominant contributor to uncertainty in these surface pressure measurements is due to the nonuniformity of the tunnel test section flow field. Although we had previously suspected this was the case, the present statistical analysis quantitatively demonstrates it." Their suggestion is that this result probably applies to other hypersonic facilities, and perhaps to transonic facilities as well.

10.23 JOINT CONSIDERATION OF EXPERIMENTAL AND SIMULATION UNCERTAINTIES

In Section 10.5, we noted the difficulty of possible cancellation of conceptual modeling errors and numerical errors (primarily inadequate grid convergence). "Issues related to the interaction of these two sources of errors have been of constant concern to the CFD and turbulence modeling communities because fortuitous cancellation of errors can lead to a very erroneous conclusion about Validation." (Rizzi and Vos, 1996.) The importance of performing Verification of Calculations *before* Validation was stressed. (See also Chapter 2.) However, this admonition does not address the question of what to do with the purely numerical error bar during the process of Validation. The safest approach is simply to reduce the purely numerical error bars to a level negligible compared to the experimental uncertainty. This may be uneconomical.

Coleman and Stern (1997) have proposed an approach to a Methodology of Validation that includes joint consideration of both experimental and simulation uncertainties. This detailed approach is thoughtful and thought-provoking, but is unsatisfactory in an important aspect, in my opinion. Nevertheless, the full article should be required reading for anyone who intends to take Validation seriously.

As a preliminary consideration, Coleman and Stern (1) discuss hypothetical data-simulation comparisons as given in their Figure 1 (see also Coleman, 1996), similar to Figure 10.23.1. Parts a and b respectively show the data without and with experimental uncertainty bars. The authors state that one might interpret part a to indicate that Model 1 is superior to Model 2 since it seems to "capture the trend of the data." In contrast, when the experimental uncertainty is considered in part b, "it is obvious that arguing for one method over another based on comparison with the experimental data is wasted effort since the predictions from both methods fall well within the data uncertainty."

In my view, this negative evaluation would be true if and only if the uncertainty distribution within the plotted range of experimental uncertainty (commonly called "error bars") was flat, i.e. a step function. (For example, if the "error bar" covered the data value D to ±10%, this flat distribution would give equal probability that the true value was at D +1%, D-3%, D+9%, etc. or at D itself.) In such a case, our probabilistic knowledge is indifferent to any values within the range of the "error bar" and therefore to the relative merits of Model 1 and 2, and the authors'

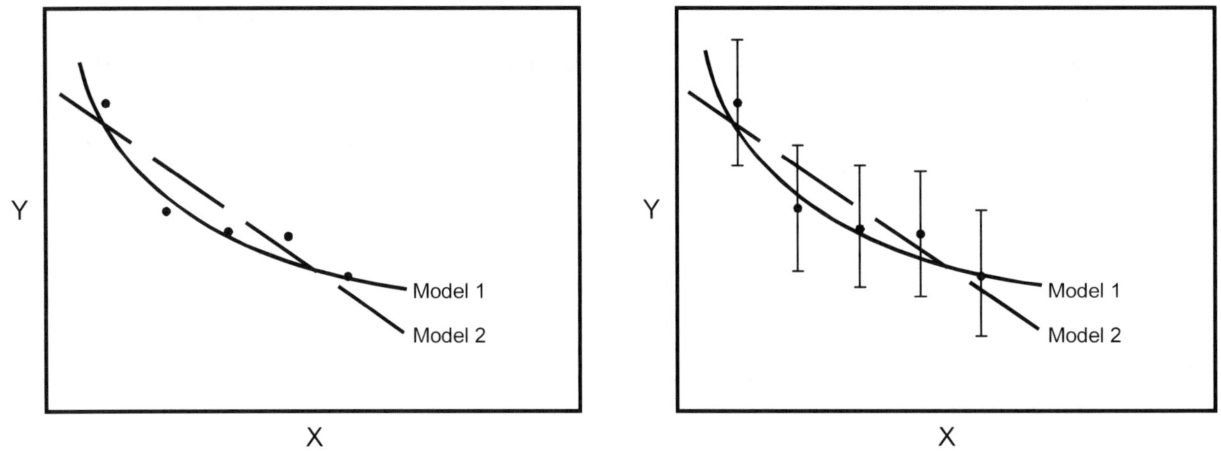

Figure 10.23.1 Comparison of experimental data and model predictions.

argument holds. Of course, we do not know what the appropriate probability distribution is, but most physical experiments surely have some preference for the data values reported, and the default model of probability is a Gaussian distribution. If, in the judgment of the reader, this is the more likely distribution, then there probably is a trend in the true values, and Model 1 does indeed "capture the trend of the data." One could readily express this quantitatively as a weighted integral over the assumed probability distributions within the "error bars", but qualitatively, I believe that the naive preference for Model 1 is justifiable.

The authors then define the "comparison error" E as the difference between the experimental Data set value D and the value produced by the Simulation S.

$$E = D - S \tag{10.23.1}$$

These may be point values (e.g., a drag coefficient C_D at given flight conditions) or 2-D plots (e.g., C_D vs. α) or multi-dimensional plots (e.g., C_D vs. α, M, Re, ...) E is then just what everyone would ordinarily look at in a Validation exercise; is $|E|$ small enough to consider the Validation acceptable for the project? There are experimental uncertainties associated not only with answer C_D but also with the parameters (α, M, Re, ...) so the experimental error "bar" about C_D should really be an n-dimensional error "box" about α, M, Re, etc. With an assumption of independence of error sources (the subtleties of which are not the main concern here) one can denote the uncertainty values U in the usual way, i.e. that the true value lies within $D \pm U$ 95 times out of 100 (95% $\cong 2\sigma \cong$ 20:1 odds). Again assuming independence of the error statistics, the different sources of uncertainty can be combined in an *rms* sense to give

$$U_E^2 = U_D^2 + U_S^2 \tag{10.23.2}$$

where U_E is the uncertainty is the comparison error E, U_D is the uncertainty in the Data set D, and U_S is the uncertainty in the Simulation results S.

Coleman and Stern decompose the simulation uncertainty U_S into three sources: the Simulation Numerical solution uncertainty U_{SN}, the Simulation modeling uncertainty arising from using Previous experimental Data U_{SPD}, and the Simulation uncertainty arising from Modeling Assumptions U_{SMA}.

$$U_S^2 = U_{SN}^2 + U_{SPD}^2 + U_{SMA}^2 \tag{10.23.3}$$

Then Equation (10.23.2) becomes

$$U_E^2 = U_D^2 + U_{SN}^2 + U_{SPD}^2 + U_{SMA}^2 \tag{10.23.4}$$

(In the examples in Coleman and Stern, "the magnitude of U_{SPD} was assumed negligible relative to the other uncertainties.")

The key step in their approach is to define the metric for Validation. Ideally, they would like to use U_E, but they "know of no approach that gives an estimate of U_{SMA}, so U_E cannot be estimated. That leaves a more stringent Validation test as a practical alternative." They then define the "Validation uncertainty U_V as the combination of all uncertainties that we know how to estimate (i.e., all but U_{SMA})," giving

$$U_V^2 = U_D^2 + U_{SN}^2 + U_{SPD}^2 \tag{10.23.5}$$

Their proposal is that this *Validation uncertainty U_V is the key metric in the Validation process*, and is to be used as follows.

If the "comparison error" $E = D - S$ is such that $|E|$ is less than the Validation uncertainty U_V,

$$|E| \le U_V \tag{10.23.6}$$

then the Validation has been achieved at the U_V level.

That is, the combination of all the errors in the experimental data set D and in the results of the simulation S is smaller than the estimated Validation uncertainty U_V, and Validation has been achieved at the U_V level.

The reasoning is that U_V is the key metric because it is the Validation "noise level" imposed by the uncertainties – *all* the uncertainties: uncertainties inherent in the experimental data set *and* uncertainties in the numerical solution (as well as uncertainties due to the use of previous experimental data in the simulation model). A key point is that "one cannot discriminate once $|E|$ is less than this, i.e., as long as $|E|$ is less than this one cannot evaluate the effectiveness of proposed model 'improvements' ". (If my view [given earlier] of the interpretation of Figure 10.23.1 in regards to Model 1 "capturing the trend in the data" is accepted, then their statement would have to be tempered somewhat for an assumed Gaussian rather than flat probability distribution, but basically, their point is well taken.)

The authors note that "Choice of the required level of U_V is associated with the degree of risk acceptable in a program." It is important to stress the sense in which the term "Validation uncertainty" U_V is being used here. It is not a measure of program requirements, but of what can be achieved with a given set of experimental data and simulation results. For example, if programmatic design requires Validation of some prediction (like C_D) to 2%, this sets the *required* level of U_V at 2%. However, if (say) the data set alone has an uncertainty of 5%, then this sets a minimum on U_V of 5%, and the data set simply is not reliable enough to meet the program objectives.

This much is true and undebatable. However, there are possible difficulties of interpretation with the *use* of the definition of U_V, as the authors are aware. Validation, in this definition, requires that the comparison error E lie within the combined error bands of the data set and the simulation. Consider this quote. "The Validation uncertainty U_V sets the level at which Validation can be achieved." This is true for a given Validation exercise including a calculation, but taken out of context, this would seem to suggest that we are stuck with a level of uncertainty. Recall that U_V contains numerical uncertainty, the major component of which is often grid convergence error. Thus, U_V can be improved (lowered) by simply improving the quality of the numerical simulation. Unfortunately, as the authors point out, the more uncertain the data set and the more uncertain the calculation, the *easier* it is to achieve Validation – at the U_V level, because the data/simulation uncertainty make U_V larger. But, it will be more *difficult* to achieve Validation to a program-specified level of *required* U_V. The potential for confusion and misuse is evident.

If the U_V of a Validation attempt (i.e., given data set and a simulation) is greater than that required for a program, then "the required level of Validation could not be achieved without improvement" in the reliability of the data set, the calculation, or both. Likewise, if the simulation uncertainties (U_{SN} and U_{SPD}) are not estimated but $|E|$ is less than U_D, then "Validation has been achieved" [in a sense, and within this framework of definitions] "*but at an unknown level*. Obviously, if *no* uncertainties are estimated, no statements about Validation can be made within the concept of the Validation process considered..."

Although logically true, this approach contains dangerous paradoxes. Besides those already noted by the authors, consider these. Less attention to iterative convergence will increase the numerical component of U_V and therefore make it easier to claim Validation (although at a higher level of uncertainty). Likewise, if your numerical uncertainty estimate for grid convergence error has been calculated using the GCI (Chapter 5) with a factor of safety F_S = 1.25, justified by the execution of a 3- or 4-grid convergence study, consider using only 2 grids and increasing F_S = 3. This may achieve "Validation" with the same experimental data set! (although at a higher level of uncertainty). These paradoxes seem to me to be fraught with danger, and will work against acceptance of this approach. Nevertheless, the article contains important insight and careful work (although some details are missing in the presentation, possibly due to length restrictions on the journal paper).

Most importantly, the proposal of Coleman and Stern for a Validation methodology fails to account for an acceptable error tolerance in the Validation. This is easily demonstrated by the following "thought experiment." Imagine a programmatic need for Validation with a 5% error tolerance (call it TOL_V), i.e. the need is for a model/code that can predict physical reality to within 5%. Further imagine the limit of good experiments and calculations, i.e. vanishingly small experimental uncertainty U_D and simulation uncertainty $U_{SN} + U_{SPD}$ so that the Validation uncertainty U_V is also vanishingly small (by Equation 10.23.5). Finally, imagine excellent agreement between experiment and simulation, say an observation error $|E| = 0.1\%$. Then, the formulation of Equation (10.23.26) does not allow this to be considered a Validation. No matter how good the agreement between experiment and simulation, if it is not as good as the combined uncertainties from experiment and simulation, it cannot be considered a Validation, in this approach.

A modification of Equation (10.23.6) to include a programmatic Validation tolerance TOL_V would avoid this difficulty.

$$|E| \leq U_V + TOL_V \qquad (10.23.7)$$

However, the previously discussed paradoxes remain, e.g. increasing experimental and/or numerical uncertainty still make it easier to achieve Validation, at the level now defined by $U_V + TOL_V$. Further consideration by the simulation community will be required to determine whether or not this modification makes their proposal practical.

I believe that the complexities and paradoxes raised by Coleman and Stern (other than the lack of consideration of a Validation tolerance TOL_V) are not just due to their approach but are inherent in the problem. At the very least, by elucidating the complexity and interactions of the combined experimental and numerical uncertainty issues, Coleman and Stern (1997) have provided motivation for performing thorough grid convergence work and otherwise reducing numerical uncertainty to a level negligible relative to the experimental uncertainty, whenever feasible.

10.24 DYNAMIC DATABASES FOR VALIDATION

Rizzi and Vos (1996) have presented their vision for a process of systematic Code Validation that involves dynamic, interactive databases stored and accessed on the World Wide Web. The particulars come from their experience in the European aerospace community, but the general concepts are applicable to all areas of engineering and science involving the numerical solution of partial differential equations.

As noted earlier, Validation of a Code, if it has meaning at all (as opposed to Validation of a particular Calculation) depends on the concept of a "nearby" problem, and this concept will fail near parameter "transition" boundaries, which should be identifiable from knowledge of the general field. (See list in Chapter 9; see also the cautionary examples of difficulty cited earlier in this Chapter, Section 10.7.) The first requirement in the vision of Rizzi and Vos (1996) is for the establishment of a detailed taxonomy of flows for the subject application field, establishing for the database users sets of flow cases acceptable for Validation of "generic flow cases" (what I have herein described as "nearby" flows). Rizzi and Vos state: "The taxonomy is also the basis upon which the code user transfers credibility from previously Validated generic flow cases to the simulation at hand."

Other elements of their vision for the Dynamic Database for Validation are these.

- Creation of a detailed flow taxonomy for the applications area.
- Establishment of an electronic database on the World Wide Web.
- The database performs these functions:
- Archives the experimental data from trustworthy experiments;
- Provides easy access and full public scrutiny to the experimental data for Validation exercises;
- Provides easy access and full public scrutiny to the previous Validation computational data;
- Provides data manipulation (plotting) software;
- Allows for ongoing full public discussion of the information.

The essential first requirement for a taxonomy of flows follows what Rizzi and Vos denote as the "Stanford Paradigm" (Kline et al, 1981a), which they consider (and most would agree, myself included) to be a "landmark event." They note the following other features of the Stanford Paradigm, which "still stands today as complete in the sense of providing all elements necessary for understanding the state of the science and for carrying our CFD validation."

- Taxonomies that organize
  - flows,
  - methods of modeling,
  - numerics.
- Classification of experimental data and computational data through a review process.
- Comparison of all computational results with experimental data and with other computed results, case by case.
- Examination, analysis, and discussion of the comparisons.
- Overall synthesis of the findings by distinguished experts.
- Storage of all data in specified machine format.

Rizzi and Vos also state that the Stanford Conference was "a very professional attempt to give an overview of the state of the art in computing complex turbulent flows in 1981 using the experimental data base established for this purpose in 1980." While I agree whole-heartedly with this evaluation, I note some limitations of this very professional attempt. (I served on the Evaluation Committee of the Conference.) These limitations have a lesson for future projects as well. The project did not really represent the state of the art of computing turbulent flows because not everyone active in this area participated in the conference; a notable absentee was any representative from the Los Alamos group, who had in fact pioneered the development of two-equation turbulence models (Harlow and Nakayama, 1967), what had by the time of the Stanford Conference been re-named as the k-ε model. Of those who did participate, not all computed all the flows for which they were equipped. The most obvious reason is that participation in such an activity requires much effort, time, and money. Other understandable personal reasons are reluctance to subject one's models and work to so much public scrutiny and inevitable comparison, avoidance of confrontational discussion, disagreement and distrust about the evaluators, reluctance to loose one's competitive advantage by disclosure prior to journal publication, etc. The taxonomies of flows and modeling were excellent, but the attempted taxonomy of numerics proved to be inadequate because many methods are hybrids of the various taxonomic categories (e.g., implicit vs. explicit). Many participants felt the scope of the conference was overly ambitious, and that more narrowly focused problems would have been preferable; in fact, this has been the trend in later workshops (see Rizzi and Vos, 1996 for examples). Finally, as I noted in Chapter 1 and elsewhere, the lack of Verification by the participants made firm conclusions on the relative merits of the turbulence models virtually impossible (Kline et al, 1981b).

Nevertheless, the Stanford Paradigm, as well as its experimental data base, still serves well, a remarkably long-lived success considering the advancements in computers, communications, and modeling in the intervening years. The following major additions in the Rizzi and Vos vision are enabled by advances in technology since the 1981 Stanford Conference.

- A *dynamic* (rather than static) database that
  - can be updated,
  - can be interrogated by the user with graphical tools,
  - can be searched electronically (via keyword files),
  - can provide some generation of synthesis documents [automatic report preparation],
  - can include reports, pictures, videos,
  - can include expert evaluation and advice on
    - appropriate level of modeling complexity for a user's needs,
    - specific computational codes, and
    - specific computational procedures,
    - thus providing on-line user training and
    - serving as an electronic "Handbook on How to Use CFD."

The updating feature is extremely important. As Rizzi and Vos noted, traditional methods of disseminating experimental data are actually quite inadequate. The artificial length restriction on journal articles means that the data published is often incomplete and that precision is greatly reduced because of small Figures. A dynamic, electronic archive removes these restrictions, and further allows access to "raw data" and details of the data reduction (numerical) techniques used to process the raw data into final data. This would allow later refinement of the data reduction;

for example, someone (the original experimenter, a reviewer, or someone attempting a Validation of a Code) could improve upon the ideal gas assumption made in the original data reduction by incorporating real-gas effects. Generally, the dynamic aspect of the database allows for the iterative nature of the scientific process to proceed more naturally, without the artificial constraints of hard-copy publications that sometimes cast errors in concrete. Describing previous static databases, Rizzi and Vos (1996) note that "In some cases the input description is incomplete and the output format for the results [has] been specified imprecisely."

The ASME *Journal of Fluids Engineering*, under the editorship of D. P. Telionis, has established a dynamic database at the Virginia Polytechnic Institute and State University. Selected experimental or numerical papers are accompanied by data deposited in the Data Bank, accessible at either

telnet scholar.lib.vt.edu

or

http://scholar.lib.vt.edu/

This system also contains a bulletin board where readers can read discussions and contribute their own on specific papers.

A significant difficulty, at least in the aerospace sector and likely in other sectors of industrial fluid dynamics, is that some databases are not accessible from the public domain, e.g., the HAEDB system described in Rizzi and Vos, 1996. They note: "In addition, the most sensitive data is saved off-line, and can only be accessed after intervention of the HAEDB manager."

The most important first step, the development of a taxonomy of experimental results, will be specific to sub-fields within the fields of electrodynamics, chemistry, biological systems, laser physics, structural dynamics, fluid dynamics, etc., etc. The fluid dynamics category would itself be much too broad for a single database. There is little in common between the fluid dynamics of ocean current modeling, free-surface wave interactions, groundwater flow and transport in porous media, dusty aerodynamic flows, hypersonics, free convection, etc. Within the category of Industrial Sectors, Rizzi and Vos (1996) note that ERCOFTAC (the European Research Community on Flows, Turbulence and Combustion Association) created an Industrial Advisory Committee. One of their tasks is to define the taxonomies for flows in different industrial sectors, such as Aerospace, Power and Energy, Process Engineering, Automotive, Environment,...

Even restricting the dynamic database to the sub-field of aerodynamics might be to broad. Restricting themselves to the aerospace sector and compressible flows that are either fully turbulent or for which the transition location is known, Rizzi and Vos (1996) offer "a glimpse of what would be a preliminary attempt to categorize features," presented here in Table 10.24.1.

- I. Wall-Bounded Flows
  - I.1 attached
    - I.1.a with pressure gradient
    - I.1.b without pressure gradient
    - I.1.c with shocks
    - I.1.d with or without suction
    - I.1.e with or without blowing
  - I.2 separated
    - I.2.a bubble type
    - I.2.b with reattachment
    - I.2.c without reattachment
- II. Free Shear Layer Flows
  - II.1 vortex flow
    - II.1.a delta wing leading edge
    - II.1.b forebody
  - II.2 ...
- III. Mixing Layers
- IV. Flows with Confluence of Two Shear Layers
  - IV.1 high-lift devices
    - IV.1.a off-body recirculation zones
  - IV.2 ...
- V. Flows In Cavities
- VI. Wake Flows

Table 10.24.1. Preliminary Taxonomy of Compressible Turbulent Flows of Interest in the Aerospace Sector. (Rizzi and Vos, 1996)

Besides this preliminary taxonomy, Rizzi and Vos give another list of phenomena pertinent to classification of fluid dynamics classes: "turbulence, relaminarization, retransition, catalyticity, chemical kinetics, combustion, shock interactions, shear-layer mixing, separation, secondary flows, etc." Obviously, this preliminary attempt is presented to give the flavor of the task; the result is neither unique nor complete, nor would it ever be-the *dynamic* aspect applies to the taxonomy itself as well as the data. As the authors state, "The definition of classes is never completely precise and of necessity imperfect because single classes of features never occur isolated in a complex flow, but the categorization is an aid in establishing credibility and should be seen as such. The process is more difficult and less reliable without it."

Of course, one can easily construct a taxonomic tree and sort through it to the area of interest electronically, even as we send e-mail today. But the point is that creation and maintenance of a dynamic database involves more than just a decision tree. As Rizzi and Vos (1996) point out, such a database requires a substantial investment to develop and maintain, including the cost elements shown in Table 10.24.2.

- Personnel costs of an *evaluation committee* (to assess the quality of each experiment).
- Initial investment in hardware and software.
- Costs to run Validation cases.
- Support for a database *manager*.
- Hardware replacement and/or upgrade.
- Software maintenance.

Table 10.24.2. Cost Elements of a Dynamic Validation Database. (Rizzi and Vos (1996).

Such efforts will require major funding and collaboration from interested groups. Rizzi and Vos (1996) state that "funding on a governmental (or European) level is needed to build and maintain such a system." They give examples of some electronic databases for aerodynamics already in existence.

Also, in the 1996 AIAA "Year in Review" features, the AIAA Fluid Dynamics Technical Committee gave the following electronic databases (Devenport et al, 1996).

DATABASE	WWW URL (http://)
ERCOFTAC Fluid Dynamics Database	fluindigo.mech.surrey.ac.uk/
NAS Data Set Archive	www.nas.nasa.gov/NAS/DataSets/
MADIC/NASA Code Certification	hpccp-www.larc.nasa.gov/ccpintro.html
ASME Journal of Fluids Engineering	scholar.lib.vt.edu/ejournals/JFE/jfe.html
NPARC Alliance Validation Archive	info.arnold.af.mil/nparc/Validation_information.html

Rizzi and Vos discuss some successes and some partial failures, both in the database technology and in the science. In the partial failure category, they note examples of lack of compliance on the part of contributors, including failure to provide keyword files, geometry rescaling and coordinate system irregularities, and incomplete answers. They state "although some of the problems appear trivial, they prohibited the automatic functioning of the data manipulation tools." Also, they note the need for "efforts [to] avoid or at least explain major differences in results obtained by different computors with the same turbulence model." (My impression is that, in fluid dynamics, the communications technology is of better quality than the science. Also, my impression is that the uniformity of results from various structures codes is less than stated by Melnik et al, 1995 and by Rizzi and Vos, 1996.) But even with modern communications technology, it will not work to try to create a "database of everything." (At least, not in the immediate future. Later in the 21st century, a "database of databases" is feasible in principle.)

REFERENCES FOR CHAPTER 10

Adams, E. E. and Gelhar, L. W. (1992), "Field Study of Dispersion in a Heterogeneous Aquifer 2. Spatial Moments Analysis," *Water Resources Research*, Vol. 28, 1992, pp. 3293–3307.

Aeschliman, D. P., Oberkampf, W. L., and Blottner, F. G. (1995), "A Proposed Methodology for Computational Fluid Dynamics Code Verification, Calibration, and Validation," *Proceedings 16th International Congress on Instrumentation in Aerospace Simulation Facilities*, July 1995, WPAFB, Ohio, 95–CH3482–7, pp. 27.1–27.13.18–21. See also Aeschliman and Oberkampf (1997)

Aeschliman, D. P. and Oberkampf, W. L. (1997), "Experimental Methodology for Computational Fluid Dynamics Code Validation," SAND95–1189, Sandia National Laboratories, Albuquerque, New Mexico, September 1997. Also, *AIAA Journal*, to appear.

AGARD (1979), "Experimental Data Base for Computer Program Assessment," NATO Advisory Group for Aeronautical Research and Development, Fluid Dynamics Working Group 04, AGARD AR–138, May 1979.

AGARD (1985), "Test Cases for Inviscid Flow Field Methods," NATO Advisory Group for Aeronautical Research and Development, Fluid Dynamics Working Group 07, AGARD AR–211, May 1985.

AGARD (1988), *Validation of Computational Fluid Dynamics*, Lisbon, Portugal, 2–5 May 1988, NATO Advisory Group for Aeronautical Research and Development, AGARD CP 437, Dec. 1988. Volume 1: Symposium Papers and Round Table Discussion. Volume 2: Poster Papers.

Armaly, B. F., Durst, F., Pereira, J. C. F., and Schonung, B. (1983), "Experimental and Theoretical Investigation of Backward-Facing Step Flow," *Journal of Fluid Mechanics*, Vol. 127, p. 473.

Bachalo, W. D. and Johnson, D. A. (1979), "An Investigation of Transonic Turbulent Boundary Layer Separation Generated on an Axisymmetric Flow Model," AIAA Paper 79–1479, July 1979.

Barber, T. J. (1996), "The Role of Code Validation and Certification in the Design Environment," AIAA Paper 96–2033, 27th AIAA Fluid Dynamics Conference, New Orleans, Louisiana, 17–20 June 1996.

Benek, J. A., Kraft, E. M., and Lauer, R. F. (1996), "Validation Issues for Engine/Airframe Integration," AIAA Paper 96–2031, 27th AIAA Fluid Dynamics Conference, New Orleans, Louisiana, June 17–20, 1996.

Boggs, J. M., Young, et al (1992), "Field Study of Dispersion in a Heterogeneous Aquifer. 1. Overview and Site Description," *Water Resources Research*, Vol. 28, 1992, pp. 3281–3291.

Boggs, J. M. and Adams, E. E. (1992), "Field Study of Dispersion in a Heterogeneous Aquifer. 4. Investigation of Adsorption and Sampling Bias," *Water Resources Research*, Vol. 28, 1992, pp. 3325–3336.

Bradley, R. G. (1988), "CFD Validation Philosophy," in AGARD (1988).

Coleman, H. W., and Stern, F. (1997), "Uncertainties and CFD Code Validation", *ASME Journal of Fluids* Engineering, Vol. 119, December 1997, pp. 795–803.

Cosner, R. R. (1995), "CFD Validation Requirements for Technology Transition," AIAA Paper 95–2227 and the oral presentation, AIAA 26th Fluid Dynamics Conference, San Diego, California, 19–22 June 1995.

de Vahl Davis, G. (1994), "Evaluation of Benchmark Problems ," in Muller, A. (1994), ed., *Proceedings of the Basel World CFD User Days 1994, Second World Conference in Applied Computational Fluid Dynamics*, Basel, Switzerland, 1–5 May 1994.

Devenport, W., Powell, K and Power, G (1996), "Fluid Dynamics," *Aerospace America*, December 1996, pp. 16–17.

Driver, D. M. (1991), "Reynolds Shear Stress Measurements in a Separated Boundary Layer," AIAA Paper 91–1787, June 1991.

Eklund, D. R., Fletcher, D. G., Hartfield, R. J., Northam, G. B., and Dancey, C. L. (1995), "A Comparative Computational/Experimental Investigation of Mach 2 Flow over a Rearward-Facing Step," *Computers and Fluids*, Vol. 24, No. 5, pp. 593–608.

Elsenaar, A., Hjelmberg, L., Butefisch, K., and Bannink, W. J. (1988), "The International Vortex Flow Experiment" in AGARD (1988), Paper No. 9.

Fasel, H (1990), "Numerical Simulation of Instability and Transition in Boundary Layer Flows," in D. Arnal and R. Michel (Eds.), *Laminar-Turbulent Transition*, Springer-Verlag, Berlin.

Firmin, M. C. P., and McDonald, M. A. (1988), "The Design of the GARTEUR Low Aspect-Ratio Wing for Use in the Validation of Shear Layer and Overall Flow Prediction Methods," Paper P5 in AGARD (1988), Vol. 2.

Freitas, C. J. (1993), Ed., *The CFD Triathlon: Three Laminar Flow Simulations by Commercial CFD Codes*, FED Vol. 160, ASME Fluids Engineering Conference, Washington, DC, June 20–24, 1993.

Ghia, K. N., Osswald, G. A., and Ghia, U. (1989), "Analysis of Incompressible Massively Separated Viscous Flows using Unsteady Navier-Stokes Equations," *International Journal for Numerical Methods in Fluids*, Vol. 9, p. 1025.

Hamming, R. W. (1962), *Numerical Methods for Scientists and Engineers*, McGraw-Hill Book Company, New York. "Chapter N + 1. The Art of Computing for Scientists and Engineers."

Harlow, F. H. and Nakayama, P. I. (1967), *Physics of Fluids*, Vol. 10, 1967, p. 2323.

Haynes, T. S., Reed, H. L., and Saric, W. S. (1996), "CFD Validation Issues in Transition Modeling," AIAA Paper 96–2051, 27 AIAA Fluid Dynamics Conference, New Orleans, Louisiana, 17–20 June 1996.

Helton, J. C., Anderson, D. R., Roache, P. J., et al (1995), "Effect of Alternative Conceptual Models in a Preliminary Performance Assessment for the Waste Isolation Pilot Plant," *Nuclear Engineering and Design*, Vol. 154, 1995, pp. 251–344.

Helton, J. C., Anderson, D. R., Roache, P. J., et al (1996), "Uncertainty and Sensitivity Analysis Results Obtained in the 1992 Performance Assessment for the Waste Isolation Pilot Plant," *Reliability Engineering and System Safety*, Vol. 51, 1996, pp. 251–53–100.

Holst, T. L. (1987), "Viscous Transonic Airfoil Workshop Compendium of Results," AIAA Paper 87–1460, 1987. See also *Journal of Aircraft*, Volume 25, No. 12, December 1988, pp. 1073–1087.

Hinkleman, M. (1995), Private Communication cited as Ref. 18 in Benek et al (1996).

Hugo, R. J., Jumper, E. J., Havener, G, and Stepanek, S. A. (1995), "Time-Resolved, Aero-Optical Measurements of a Wavefront Aberrated by a Compressible Free Shear Layer," AIAA Paper 95–1979, AIAA 26th Plasmadynamics and Lasers Conference, San Diego, California, June 19–22, 1995.

Johnson, D. A. and King, L. S. (1985), "A Mathematically Simple Turbulence Closure Model for Attached and Separated Turbulent Boundary Layers," *AIAA Journal*, Vol. 23, No. 11, November 1985, pp. 1684–1692.

Johnson, D. A. and Coakley, T. J. (1990), "Improvements to a Nonequilibrium Algebraic Turbulence Model," *AIAA Journal*, Vol. 28, No. 11, November 1990, pp. 2000–2003.

Joslin, R. D. and Street, C. L. (1992), "Crossflow Disturbance Evolution on Swept Wings Computed by Spatial DNS," *Bulletin of the American Physical Society*, Volume 37, p. 1778.

Jumper, E. J., Hugo, R. J., and Cicchiello, J. M. (1994), "Turbulent-Fluid-Induced Optical Wavefront Dynamics: Near- and Far-Field Implications," AIAA Paper 94–2547, AIAA 18th Aerospace Ground Test Conference, Colorado Springs, Colorado, June 20–23, 1994.

Jumper, E. J. and Hugo, R. J. (1995), "Quantification of Aero-Optical Phase Distortion Using the Small-Aperture Beam Technique," *AIAA Journal*, Vol. 33, No. 11, November 1995, pp. 2151–2157.

Kline, S. J., Cantwell, B. J., and Lilley, G. M. (1981a), Eds., *The 1980–81 AFOSR/HTTM–Stanford Conference on Complex Turbulent Flows*, 3 Volumes, ThermoSciences Division, Mechanical Engineering Department, Stanford University, Stanford, California.

Kline, S. J., Cantwell, B. J., and Lilley, G. M. (1981b), eds., *Objectives, Evaluation of Data, Specifications of Test Cases, Discussion, and Position Papers. The 1980–81 AFOSR/HTTM-Stanford Conference on Complex Turbulent Flows*, Vol. 1, ThermoSciences Division, Mechanical Engineering Department, Stanford University, Stanford, California.

Konikow, L. F. and Bredehoeft, J. D. (1992), "Groundwater Models Cannot be Validated," *Advances in Water Resources*, Vol. 15, 1992, pp. 75–83.

Kordulla, W., Schwamborn, D. and Sobieczky, H. (1988), "The DFVLR-F5 Wing Experiment," in AGARD (1988), Paper No. 13.

Levy, L. L. Jr. (1978), "Experimental and Computational Steady and Unsteady Transonic Flows About a Thick Airfoil," *AIAA Journal*, Vol. 16, No. 6, 1978, pp. 564–572.

Lockman, W. K., Lawrence, S. L. and Cleary, J. W. (1992), "Flow Over an All-Body Hypersonic Aircraft: Experiment and Computation," *Journal of Spacecraft and Rockets*, Vol. 29, No. 1, 1992, pp. 7–15.

MacIntyre, W. G., Boggs, J. M., Antworth, C. P. and Stauffer, T. B. (1992), "Degradation Kinetics of Aromatic Organic Solutes Introduced Into a Heterogeneous Aquifer," *Water Resources Research*, Vol. 29, 1993, pp. 4045–4051.

Martin, F. W. Jr., Sickles, W. L. and Stanley, S. A. (1993), "Transonic Wind Tunnel Wall Interference Analysis for the Space Shuttle Vehicle," AIAA Paper 93–0420, January 1993.

Marvin, J. G., Levy, L. L. Jr., and Seegmiller, H. L. (1980), "On Turbulence Modeling for Unsteady Transonic Flows," *AIAA Journal*, Vol. 18, No. 5, May 1980, pp. 489–496.

Marvin, J. G. (1995), "Perspective on Computational Fluid Dynamics Validation," *AIAA Journal*, Vol. 33, No. 10, October 1995, pp. 1778–1787.

McDevitt, J. B., Levy, L. L., and Deiwert, G. S. (1976), "Transonic Flow About a Thick Circular-Arc Airfoil," *AIAA Journal*, Vol. 14, No. 5, May 1976, pp. 606–613.

Mehta, U. (1995), "Guide to Credible Computational Fluid Dynamics Simulations," AIAA Paper 95–2225, June 1995.

Melnik, R. E., Siclari, M. J., Barber, T., and Verhoff, A. (1994), "A Process for Industry Certification of Physical Simulation Codes," AIAA Paper 94–2235, AIAA 25th Fluid Dynamics Conference, Colorado Springs, Colorado, June 1994.

Melnik, R. E., Siclari, M. J., Marconi, F., Barber, T., and Verhoff, A. (1995), "An Overview of a Recent Industry Effort at CFD Code Certification," AIAA Paper 95–2229, AIAA 26th Fluid Dynamics Conference, San Diego, California, 19–22 June 1995.

Menter, F. R. (1992), Improved Two-Equation *k*-ω Turbulence Models of Aerodynamic Flows," NASA TM 103975, October 1992.

Mueller, T. J. (1974), "On the Fluid Dynamics of Prosthetic Heart Valve Flow—A Preliminary Numerical and Experimental Study," Technical Note 101, Von Karman Institute, Rhode Saint Genese, Belgium, June 1974.

Muller, A. (1994), Ed., *Proceedings of the Basel World CFD User Days 1994, Second World Conference in Applied Computational Fluid Dynamics*, Basel, Switzerland, 1–5 May 1994.

Muller, A. and Loffler, B. (1996), Eds., *Proceedings of the Basel World CFD User Days 1996, Third World Conference in Applied Computational Fluid Dynamics*, Basel, Switzerland, May 1996.

Newling, B. et al (1997), "Comparisons of Magnetic Resonance Imaging Velocimetry with Computational Fluid Dynamics," *ASME Journal of Fluids Engineering*, Vol. 119, March 1997, pp. 103–109.

Oberkampf, W. L., Blottner, F. G., and Aeschliman, D. P. (1995), "Methodology for Computational Fluid Dynamics Code Verification/Validation," AIAA Paper 95–2226, 26th AIAA Fluid Dynamics Conference, San Diego, California, 19–22 June 1995.

Olsen, M. E. and Seegmiller, H. L. (1993), "Low Aspect Ratio Wing Code Validation Experiment," *AIAA Journal*, Vol. 31, No. 10, 1993, pp. 1744–1752.

Porter, J. L. (1996), "A Summary/Overview of Selected Computational Fluid Dynamics (CFD) Code Validation/Calibration Activities," AIAA Paper 96–2053, 27th AIAA Fluid Dynamics Conference, New Orleans, Louisiana, June 17–20, 1996.

Reed, H. L. and Saric, W. S. (1989), "Stability of Three-Dimensional Boundary Layers," *Annual Review of Fluid Mechanics*, Vol. 21, 1989, pp. 235–284.

Reed, H. L., Saric, W. S., and Arnal, D. (1996), "Linear Stability Theory Applied to Boundary Layers," *Annual Review of Fluid Mechanics*, Vol. 28, 1996, pp. 389–428.

Rehfeldt, K. R., Boggs, J. M., and Gelhar, L. W. (1992), "Field Study of Dispersion in a Heterogeneous Aquifer. 3. Geostatistical Analysis of Hydraulic Conductivity," *Water Resources Research*, Vol. 28, 1992, pp. 3309–3324.

Rizzi, A. and Vos, J. (1996), "Towards Establishing Credibility in CFD Simulations," AIAA Paper 96–2029, 27th AIAA Fluid Dynamics Conference, New Orleans, Louisiana, June 17–20, 1996.

Roache, Patrick J. (1972), *Computational Fluid Dynamics*, First Edition, Hermosa Publishers, P. O. Box 9110, Albuquerque, New Mexico 87119-9110, USA. http://www.hermosa-pub.com/hermosa. Revised Printing 1976. See also Roache (1998).

Roache, P. J. and Rucker, D. (1996), "SECO Code Development for MADE-2 Simulations," Proceedings of the 3rd Army Groundwater Model Users Workshop, Albuquerque, New Mexico, 31 July 1996.

Roache, P. J., Knupp, P., Salari, K., et al (1997), "High Accuracy Robust Groundwater Flow and Transport Codes for Environmental Engineering: Phase II," Final Report, F08637–94–C6049, Armstrong Laboratory, Tyndall AFB, Florida, AL/EQ–TR–1996–0044, May 1997.

Roache, P. J. (1998), *Fundamentals of Computational Fluid Dynamics*, Hermosa Publishers, P. O. Box 9110, Albuquerque, New Mexico 87119-9110, USA. http://www.hermosa-pub.com/hermosa. To appear.

Salari, K., Roache, P. J., and Wilcox, D. C. (1994), "Numerical Simulation of Dynamic Stall Using the Two-Equation K-ω Turbulence Model," ASME Fluids Engineering Division Summer Meeting, Lake Tahoe, Nevada, 20–23 June 1994, ASME FED Vol. 196, K. Ghia and M. Hafez, Eds., pp. 1–7.

Settles, G. S. and Dodson, L. J. (1991), "Hypersonic Shock/Boundary-Layer Interaction Database," NASA CR–177577, April 1991.

Sickles, W. L. and Erickson, J. C. Jr. (1990), "Wall Interference Corrections for Three-Dimensional Flows," AIAA Paper 90–1408, June 1990.

Sindir, M. M., Barson, S. L., Chan, D. C., and Lin, W. H. (1996), "On the Development and Demonstration of a Code Validation Process for Industrial Applications," AIAA Paper 96–2032, 27 AIAA Fluid Dynamics Conference, New Orleans, Louisiana, June 17–20, 1996.

Stauffer, T. B., Antworth, C. P., Young, R. G., MacIntyre, W. G., Boggs, J. M., and Beard, L. M. (1994), "Degradation of Aromatic Hydrocarbons in an Aquifer During a Field Experiment Demonstrating the Feasibility of Remediation by Natural Attenuation," U. S. Air Force, Armstrong Laboratory, AL/EQ–TR–1993–0007, 1994.

Thibault, P. A. and Bergeron, D. M. (1995), Eds., *Proceedings of the Third Annual Conference of the CFD Society of Canada*, Volumes I and II, June 25–27, 1995, Banff, Alberta.

Underwood, F. N. and Mueller, T. J. (1977), "Numerical Study of the Steady Axisymmetric Flow Through a Disk-Type Prosthetic Heart Valve in a Constant Diameter Chamber," *ASME Journal of Biomechanical Engineering*, Vol. 99, May 1977, pp. 91–97.

Underwood, F. N. and Mueller, T. J. (1979), "Numerical Study of the Steady Axisymmetric Flow Through a Disk-Type Prosthetic Heart Valve in an Aortic-Shaped Chamber," *ASME Journal of Biomechanical Engineering*, Vol. 101, August 1979, pp. 198–204.

Van Wie, D. M. and Rice, T. (1996), "Quantification of Data Uncertainties and Validation of CFD Results in the Development of Hypersonic Airbreathing Engines," AIAA Paper 96–2028, 27th AIAA Fluid Dynamics Conference, New Orleans, Louisiana, June 17–20, 1996.

Wilcox, D. C. (1993), *Turbulence Modeling for CFD*, DCW Industries, La Cañada, California, 1993.

Willhite, P., Jasper, D., and Romer, W. (1995), "A Critical Evaluation of CFD Predictions of Full Aircraft Drag Increments," AIAA Paper 95–2289, June 1995.

Williams, P. T. and Baker, A. J. (1997), "Numerical Simulations of Laminar Flow over a 3D Backward-Facing Step," *International Journal for Numerical Methods in Fluids*, Vol. 24, 15 June 1997, pp. 1159–1183.

WIPP PA Dept. (1992), *Annual Performance Assessment for the Waste Isolation Pilot Plant, Dec. 1992, Vol. 1, Preliminary Comparison with 40 CFR 191, Subpart B, Vol. 2: Technical Basis*, Performance Assessment Department, Sandia National Laboratories, Albuquerque, New Mexico.

Zingg, D. W. (1992), "Grid Studies for Thin-Layer Navier-Stokes Computations of Airfoil Flowfields," *AIAA Journal*, Vol. 30, No. 10, October 1993, pp. 2561–2564. See also AIAA Paper 92–0184.

PART IV
BROADER ISSUES

Parts II and III of this book covered Verification and Validation, which (to oversimplify) involve mathematics and science, respectively. The broader issues in this Part IV assume Verification and Validation, and involve (to oversimplify again) engineering and management practice.

11

CODE QUALITY ASSURANCE AND CERTIFICATION

What the hell is Quality? What *is* it?

Zen and the Art of Motorcycle Maintenance: An Inquiry into Values
Robert M. Pirsig, William Morrow Publishing, New York, 1974

11.1 INTRODUCTION

This Chapter covers Code Quality Assurance (QA), Code Certification, characteristics of good coding, and some other aspects of QA relevant to engineering projects. See Chapter 2 for definitions and connotations. The context in this chapter, as in all this book, is that of non-real-time numerical solution of partial differential equations. Following the discussion of quality and QA system characteristics, a template for a QA system will be given.

11.2 QUALITY ASSURANCE (QA) VS QUALITY WORK

As noted in Chapter 2, Section 2.11 (repeated here), it is worth making the distinction between the technical term "Quality Assurance" or QA vs. "quality work." QA boosters like to talk as though the two were equivalent, but formal QA is largely a system of paperwork (e.g., see ISO, 1991) run by managers. A project can meet all the formal QA requirements and still be low quality (or flatly erroneous). On the other hand, high quality work can and has been done without a modern and formal QA program. QA can, of course, be helpful for quality work. (If nothing else, it can be used to encourage management to support quality work.) Even the formality and paperwork are helpful in issues like version control. But if allowed to run amuck, formal QA can mire real work in forms and procedures and definitions, impeding genuine quality rather than fostering it. For example, formal QA procedures may require expensive re-running of an entire suite of confirmation problems whenever a code is modified; this can discourage minor code fixes and improvements. I have seen QA reviewers complain and require written justification (with committee approvals, documentation of resolution, archiving of correspondence on approved forms, etc.) for a change in input data that produced numerical value changes from 3.0 to 2.999998. And this, in a geophysical problem wherein the parameter ranges were sampled over three orders of magnitude!

11.3 QA VS CREATIVITY

Early and excessive insistence QA issues (e.g., early documentation) will smother creativity. It is analogous to the situation of true mathematicians insisting on rigorous proofs and theorems for every stage of development in an applied mathematics arena. A quote from an old paper by Biot (1956) on the relation between pure and applied mathematics is equally applicable to the relation between QA and creative algorithm and code development.

One could understand the feelings of the artist who undoubtedly benefits from the scientific study of colors but who would be constantly reminded of proceeding with rigorous conformity to the dictates of physics and psychology.

11.4 QA AND TEMPERAMENT TYPES

It is this smothering by paperwork that creates tension and gives QA a bad name among creative scientists and engineers. There are sound management reasons for QA, yet there are also deeply rooted personality and temperament issues involved as well. In the Myers-Briggs temperament classification system (Myers, 1962; Keirsey and Bates, 1984), the typical QA temperament type is SJ, whereas the creative scientist type is typically NF, diagonally across the matrix from SJ. These two types do not work well together; they get on each others' nerves. Yet they need each others' traits. Big trouble occurs when one or the other type does not recognize his or her own temperament traits as simply idiosyncrasies, and tries to enforce them on others. QA types will even prefer crude and inaccurate algorithms, just because they are easier to document clearly and to understand. (This statement is *not* an exaggeration on my part. I have experienced strong requests from regulators, unqualified in either the science involved or in the simulation algorithms, to drop sophisticated algorithms, no matter that they are efficient and accurate, in favor of childish and inaccurate spread-sheet type of algorithms which any accountant could understand.)

In spite of these statements, and my own temperament type (which, as the reader may have guessed by now, is NF), I do believe in the usefulness of a formal QA system for large projects, provided that the QA system is designed and applied with some common sense, and that personal power trips are avoided.

11.5 THE PREVALENCE OF ERRORS IN SCIENTIFIC SOFTWARE: USE OF STATIC ANALYZERS

Hatton (1997) performed a huge suite of tests over 5 years on commercial, government and university scientific software to evaluate the prevalence of errors. The results were horrifying.

11.5.1 The "T" Experiments of Hatton

Hatton's projects were in two categories. The "T1" experiment measured defects without running the codes, by performing static "deep flow analysis" that "looks for inconsistent or undefined use of language." The software analysis programs QAC and QA Fortran were used to gauge "formal consistency," e.g., inconsistency in variable types and/or numbers between code components (functions, subroutines). The T1 experiment covered over 3 million executable lines of scientific software written in C and Fortran, from "many industries and applications areas." The "T2" experiments covered a single applications area and involved running the codes. The results of the T2 experiments were terrible but, I believe, not so representative—the applications area was seismic data processing, a notoriously difficult area involving vary large ill-conditioned matrices and fine-grained input data. On the other hand, I believe the poor results of the T1 experiments to be more generally representative of scientific software.

"More than 100 codes in some 40 applications areas" were tested in T1. No details could be given on the types of codes, although the applications areas included "graphics, nuclear engineering, mechanical engineering, chemical engineering, civil engineering, communications, databases, medical systems, and aerospace," earth science, and control. We may presume that many of these codes involved our interest in this book, numerical solution of partial differential equations. "The age of the codes was evenly spread between 1 and 20 years." All were "mature," i.e. "in regular use by their intended end Users." Significantly, "internationally standardized systems of quality control" [presumably ISO 9000 or something like it] "were comprehensively represented."

Once again, it is necessary to introduce semantic distinctions. In Hatton's terminology, the "defects" measured in T1 are called *faults*, while those defects found in T2 by actually running the codes are called *failures*. Failures can arise through a simple defect (i.e., a programming mistake) but *could* also be caused by non-programming defects such as an inappropriate [or just plain incorrect] algorithm, or variations in floating-point implementations. However, Hatton's results indicate that for his T2 tests, in which a common algorithm was used across codes, the only significant source of such inaccuracy *failures* was in fact “simple defects” or *faults*. Hatton notes that “the terms ‘defect’, ‘error’, and ‘bug’ are often used as a generic term for either fault or failure, dependent on context.”

Briefly, Experiment T1 showed that “C and Fortran codes, independent of the application area, are riddled with statically detectable faults. For example. calling-sequence inconsistencies (also known as interface inconsistencies) occurred at the average rate of one in every 7 interfaces in the Fortran codes we tested, and one in every 37 interfaces in C.”

What can we do with such numbers?

First or all, Hatton does *not* suggest using C instead of Fortran. As he says, “In C, the delights of pointers add many new ways of getting it all wrong. In C++, even more pitfalls await the unwary...” About half of the higher incidence of interface errors in the Fortran codes is easily normalized away, since the Fortran components on average were about 2.5 times as large and had about 2.5 times as many arguments. I venture the opinion that the higher prevalence of errors in Fortran than in C (and the longer components) results from simple skewing in time; more old code was written in Fortran. (And there is nothing as ugly as really old Fortran code.) Hatton states that “The net result of changing languages appears to be that the overall defect density appears to be about the same. In other words, when a language corrects one deficiency, it appears to add one of its own.”

Second, the presence of these *faults* does not always indicate failures. Most codes with many options were built with the intention of performing a rather limited subset of the options. It is this limited subset that has been given the most attention and exercised. Often, code authors will build in an option more with a view towards future expansion, rather than immediate application, and test it only cursorily. Even with a commercial code, a potential client will usually be advised as to whether the features of interest have been well-exercised or not. Also, some *faults* are simply not errors at execution time, e.g. failure to initiate variables *can* be correct if restricted to computer systems or compiler options that automatically zero all variables. (It is still not good coding practice.) Even inconsistencies in interfaces may not be a true error; I have written correct code which only requires the first few arguments for some function options, and no error results from not including a list of unused arguments in some calls. It could be argued that such a call with a full list of unused variables is less readable. (I do not argue this, anymore, and have added dummy zero’s, properly floating point or integer, to the argument list, just to avoid questions and flags from code analyzers.) However, Hatton addressed this in a statistical fashion, asking experts in the code applications areas to weight the severity of the faults, from 5% to 100%. A 5% severity rating represented a low probability that the *fault* would result into a *failure*, “for example, casting a pointer to an integral type in C.” A 100% severity weighting meant “effective certainty that failure will take place in a reasonable software life-cycle,” but no guarantee. Briefly, their evaluation was that the situation was serious. Some fluids engineering codes happily were on the low end of the weighted fault distributions, while one *nuclear engineering* code was off-scale with an “awe-inspiring 140 weighted static faults per 1,000 lines of code..." [If the mean weighting for a fault was about 50%, this would indicate one coding fault every 4 or so lines of code!] As Hatton says, this package "amounted to no more than a very expensive random number generator."

Third, Hatton's results gives evidence that (e.g., see Chapter 3) there will always be a continuing requirement for Confirmation exercises.

11.5.2 Use of Static Analyzers

Fourth, and most important, Hatton's survey indicates the importance of performing static tests on codes, and raises an obvious question. If Hatton and his collaborators could perform these static tests, why did not the code developers do it first?

Good question. One might try to argue that the static analysis tools were not available when the old code was developed. However, this applies only to the oldest Fortran code, since such static analyzers have been commonly available at least since the early 1980's. Then why is not the analyzer applied after the fact, *now*, to old code? Because (I believe) while it is simple to turn on the analyzer, it is difficult to go through the results, especially for old code. In my limited experience, static analyzers typically present many "false positives," flags that formally *could* be a problem, but are not. The most common (in my experience) has been the flag that states "tests for equality between floating point variables may not be meaningful." Of course, they may *be* meaningful, and are probably not even dangerous when the test is not for simple equality but for "less (greater) than or equal," which construction still elicited flags. (Similar flags occur with some compiler options.) There are other false positive flags, or flags that are true positive but *intended*. For example, one might build a code data structure to accommodate future expansions, including more generality in a key sub-program than is presently required. This can result in error flags of un-initialized variables, "dead" code segments that can never be accessed, etc. In a *good* QA program (which *must* include such static tests) all these flags have to be addressed, either by changing the code (with subsequent headaches of re-running of test cases, new documentation, etc.) or by explaining on QA forms why they are *false* positives, which is perhaps as big a headache, and often leaves a residue of doubt. In the template QA system outlined later, scientific codes at the C level (Candidate for "A," or QA-approved, level) are required to be processed by a static analyzer. However, because these runs typically produce many false positive warnings, and consequently require significant effort and judgment from the Code developer to process the results, its use is not to be *required* at all stages of code development, although early use is recommended (see below).

These static analyzers can also output various indices of the code complexity. For old and ugly Fortran code, the numbers are high, and indicate the terrible state of the art at the time the codes were written. However, I do not believe in arbitrary limits on these complexity indices, nor on the validity of code comparisons based on them. If you want a simple (non-complex) code, you use a simple algorithm. It is a sad but true fact that very efficient algorithms tend to be inherently complex, and that general purpose codes will have many option combinations. I have had QA aficionados suggest dropping a complex but efficient 3-D semi-coarsening multigrid algorithm in favor of simple point relaxation, at a cost of 2–3 orders of magnitude in efficiency for large problems. This is too high a price to pay for the aesthetic of coding simplicity.

The static analyzers can also determine adherence to structured programming and other programming standards, if anyone is interested. Hatton notes that "experiment T1 proved conclusively that attempts to maintain programming standards were risible," i.e. laughable, hilarious. One can only imagine the output of error flags!

In spite of these sad stories of additional work required to use static code analyzers, I believe that

- such static tests should be required in any formal QA program,
- static tests should be used by any code builder or User interested in true quality, and
- static tests should be used early and throughout the lifespan of a code.

It may take a day of drudgery to wade through the analyzer output for a moderate size scientific package, but the discovery of one genuine bug will be relatively painless compared to releasing and using faulty scientific code. As in the case of Verification of mathematical correctness of codes and of calculations, there is a time to put aside creativity, and just think like an accountant for awhile.

11.6 FACTORS OF CODE QUALITY ASSURANCE

Definitions of "Quality Assurance" are not very informative. The following *description* is adapted from Rechard et al (1991,1992).

QA as related to computer software means ensuring that the software consistently does what it is supposed to do to meet the expectations of the customer (i.e., recipient, purchaser, or beneficiary of the work). For the type of projects considered in this book (non-real-time applications), the primary customer expectation is that the software not deceive the User (either by the software producing a meaningless answer or by the developer overstating software capabilities). Hence, QA means that a software User has a reasonable degree of assurance that the software will correctly perform the stated capabilities (e.g., provide a satisfactory solution to the mathematical model) when given the proper input. The degree of assurance must be commensurate to the complexity, importance, and risk of failure of the analysis.

Beyond this very general description, many factors of code QA could be listed. The list in Table 11.6.1 follows Mehta (1996). Some definitions are obvious and commonplace, others might require elaboration to be legalistically specific, but all are at least suggestive of the elusive term "quality."

Performance Factors
- Reliability
- Integrity
- Usability
- Efficiency
- Robustness

Design Factors
- Correctness
- Testability
- Maintainability
- Modularity
- Traceability
- Documentation Adequacy

Adaptability
- Reusability
- Flexibility
- Portability
- Inter-operability

Table 11.6.1. Factors of Code Quality. (After Mehta, 1996, augmented.)

11.7 SOME COMPONENTS OF PROJECT CODE CERTIFICATION

If we define Project-oriented Code Certification to be broader than Code or Software QA, the following components of Certification are suggested (Table 11.7.1).

As noted in Chapter 2, my inclination is to include in Code Certification the activity of Code Verification independent of the code authors, i. e., Code Confirmation, other aspects of code QA including robustness, documentation (internal, external, users manuals), version control, the QA

- Verification
- Validation (for a project- oriented range of parameters)
- Confirmation exercises
- Robustness exercises
- Documentation
- Version Control Procedures
- QA System
- Perhaps, predictions (vs. "postdictions") by an independent review team

Table 11.7.1. Some Components of Project Code Certification.

system itself (including such clearly non-mathematical but practical considerations as the selection process for review committees, signature authority for change orders, forms for bug reports, conflict resolution procedures, etc.). For project-level Certification, one would also need Code Validation for a project-oriented range of parameters—for several authors, a principal distinction is that "Validation" involves comparisons to experiments primarily on model problems while "Certification" involves comparisons to experiments on more realistic geometries and range of operating parameters (e.g., Melnik et al, 1995; Barber, 1996).

11.8 ENGINEERING TEAMS AND THE DIVISION OF LABOR

For many senior workers in CFD, their orientation towards the field was formed during their graduate school studies, and in most cases one person was responsible for everything. This paradigm is not adequate for present practice. As Blottner (1990) opined (and I agree), it is asking too much of one person to be the expert (conceptual) turbulence modeler *and* the expert in numerical algorithms *and* the expert in code development.

For significant scale engineering projects, the division of labor must be more extensive. Rizzi and Vos (1996) describe a useful model for the division of labor in an engineering simulation (CFD) activity, divided over three teams.

- The code developers, perhaps consisting of software engineers, mathematicians, and physical modelers
- Physical modelers who carry out the Calibration and Validation work
- Users, who are (possibly) experts in neither the code nor modeling, but are applications specialists

Their activities cover the following four steps required to build credibility.

- Verification of Codes
- Certification (at a project-level)
- Calibration (tuning)
- Validation (over project-oriented range of parameters)

In their view (Rizzi and Vos, 1996), Verification [of Codes] and Certification are the responsibility of the first team, the code developers. The responsibility for Calibration (tuning) and Validation belongs to the second team, but the work of the second and third teams is interrelated because Calibration and Validation are specific to a particular design application and User community. (Recall the assertion in Chapter 2, that Validation, unlike Code Verification, is an ongoing activity.) The second team Calibrates the code and through Validation determines its range of applicability, which information is then transferred to the third team (end Users). Validation and Calibration are related. In Calibration, you improve the results of your Validation, [usually] at the expense of generality because you have modified or specialized the physical model to make it work for a certain type of generic flow. In Validation [for the more complex, systems-type flows] you identify all the generic flow types involved and then chose the appropriate model to make the predictions. The reasoning behind this is *inductive*, and therefore must be "Validated."

The question of code operation independent of the code builder is *not* an issue for Code Verification, in my opinion as stated in Chapter 2. However, it is very much an issue for Code QA (e.g., see Porter, 1996; Rizzi and Vos, 1996). In some situations, it is recommended that "an independent agency or group should oversee the process" (Porter, 1996).

11.9 PERSONNEL ROLES, CODE LEVELS, AND CODE SOURCES

It is necessary to distinguish between the *roles* of code builder (or code author) and the code User (or analyst), recognizing that in some projects, the two *roles* may be fulfilled by the same *person*, but that the code QA procedures are best formulated by maintaining the distinction between the functionalities. (The distinction is most evident in commercial codes.)

For some projects, it may be useful to have more than one level of classification, rather than a simple yes/no division. For the project described in Rechard et al (1991) we originally used 4 levels of QA classification, which we found appropriate for the situation wherein preliminary (yet formal) analyses were published over several years using software that was simultaneously undergoing changes due to algorithm development.

A – Full QA status
C – Partial QA status, possibly undergoing continued refinement
D – Obsolete software formerly in Class A or C
X – Entry level software in early stages of development or experimentation, with little QA requirement other than its announcement of availability

These will be elaborated in the later Section 11.24 on a template for a QA system.

Distinction was made between Scientific/Engineering Software (SES) (the type treated in this book) which model physical processes, and non-SES software, e.g. spread sheet programs, plotting programs, etc. Also, distinction was made between codes developed outside the project (e.g., commercial codes) and codes developed within the project.

11.10 DESIRABLE (BUT NOT REQUIRED) CODE CHARACTERISTICS

Not every desirable characteristic of code should be required by a formal QA system. To do so would lead to excessive rigidity.

11.10.1 Specific Coding Practices

The general criteria for scientific code quality are fairly well agreed upon by software professionals, and quality codes are readily recognized. However, the strict observances of specific practices of coding style and structure are not agreed upon, and standard practices and tools are still evolving. Further, there are elements of style which are unique to certain codes, including aspects that are simply personal and those that are specific to the intended applications.

Certainly code developers strive to attain objectives of readability, ease of maintenance, modularity, efficiency, etc. However, we see no significant advantage, and many operational disadvantages, to requiring specific coding practices in a QA program. There is no real need for uniformity in code writing, and any requirement for imposing stylistic practices on all code writers serves to distract them from the real task at hand, that of writing quality code, and thereby results in decreased productivity.

11.10.2 Language Standards

Likewise, there is no significant advantage, and great disadvantage, to requiring anything less than the most progressive software tools and language features available. Specifically, we reject any suggestion to limit the language features to ANSI Standard Fortran 77 or the Fortran 88. Industry-standard (or common) enhancements such as mixed-case variable names, use of the underscore in variable names, and extended name length very significantly add to the readability and mnemonic value and hence to the maintainability of codes, contrasted to the Fortran 77 requirement of six upper case alphanumeric characters.

11.10.3 Portability

Portability is a desirable but somewhat over-rated quality, in my opinion even for commercial codes. For complex modeling codes (such as groundwater hydrology simulation codes, radioactive decay modeling codes, etc.) the issue of code (general) maintainability is much more significant than the issue of (specific) transportability, which is the only advantage of a smallest-subset standard such as ANSI Standard Fortran 77. (Transportability is indeed a more significant issue for general purpose mathematical routines, e.g. linear equation solvers, special function generators, etc. For these, the ANSI Standard Fortran 77 requirement may be appropriate.) The most common failure in portability is in timing routines, which are not an accuracy issue, and which any systems manager can fix easily. There is no need for portability to be a trump card; it is a desirable feature, all other things being equal, but is not important in the context of modern languages, and when it results in imposition of outdated environments and languages like ANSI Standard Fortran 77, it constitutes and annoying obstruction to code development and maintenance. Note that DOE studies (Osborne, 1988) have shown that 50–70% of the life-cycle costs of software are attributable to code maintenance, including bug fixes, code enhancements, installation on new hardware systems, etc. These life-cycle costs are translatable from dollars to months; i.e. completion of an enhancement to a previously developed code will not only require less money but also less calendar time for a more readable code compared to an ANSI Standard code.

11.10.4 Robustness

A laudable goal of any software developer is to design a robust code (i.e., a code that can handle unique situations as well as more common general problems). Robustness (like portability) is a desirable but somewhat over-rated quality, in my opinion. This opinion reflects my preoccupations, first with accuracy, second with efficiency. I recognize that this is not the priority of most Users, especially Users of commercial codes and general purpose software. It is an unfortunate fact of life that most often there is an inherent (i.e., algorithmic) trade-off between robustness vs. accuracy and/or efficiency. Simple two-point upstream differencing of advection terms is robust but notoriously inaccurate. Likewise, a solver based on brute force direct Gaussian elimination is robust but terribly inefficient. For iterative solvers, substantial gains in efficiency (commonly a factor of 2, often an order of magnitude) can be achieved with some fine tuning of iteration parameters, which does not significantly change the answers and in fact make it easier to judge iterative convergence, yet are problem-dependent and therefore not robust across problem parameters. For design studies, multiple solutions of "nearby" problems can run into 100's. The possibility of very significant savings, in money and schedule, by sacrificing some robustness should not be discarded automatically. It is also true that the virtually bullet-proof robustness of direct Gaussian elimination can be a disadvantage when it gives a numerical solution to an ill-posed model formulation, e.g. some two-phase flow formulations. In such cases, the poor conditioning has at times caused iterative solvers to fail, and to provide clues to improved formulation of discretizations or basic governing equations.

The key to evaluating the importance of robustness is always to keep the purpose of the code in mind. Developing non-robust codes may not be a major issue if the code does what it is supposed to do for the project (the practical definition of quality). A code that is not robust may fail to produce a solution for a few parameters (e.g., the combination of an extreme parameter value and a particular solution technique), but this lack of robustness is not a major QA issue provided the analyst is aware that he or she has not solved the problem for those conditions. The troublesome parameter cases can then be attacked separately, perhaps with educated tweaking of purely numerical parameters (such as under-relaxation factors), or with code options for more robust but less efficient solvers.

11.10.5 Maintainability

Although these code characteristics cannot practically be mandated, they have a major impact on the maintainability of code. Roughly 50–70% of scientific software life-cycle costs can be attributed to code maintenance, including code enhancements (40%), bug fixes (15%), and installation with new hardware/software (15%). (Osborne, 1988). Consequently, defining the desired code structure, readability, portability, and robustness are important tasks of the code builder.

11.10.6 Summary on Code Characteristics

In summary, an enlightened QA system should encourage and enable quality programming developed in a flexible, professional, and *progressive* atmosphere, rather than mandate every desirable characteristic in a legalistic fashion.

11.11 CODE DOCUMENTATION

11.11.1 The Importance of Code Documentation

Code Documentation is of tremendous importance to QA. We agree with Mehta (1996): "The documentation is a very important quality factor in providing traceability and in helping qualify the User of the code." Note the implicit understanding that *User training* is part of QA for a project; see also Rizzi and Vos (1996), and Porter (1996). Weiss (1983) states that good "technical documentation increases the maintainability, modifiability and reliability."

User acceptance is also highly dependent on documentation. In my experience, code documentation is second to no other factor in User acceptance, not even ease of use. (Accuracy is not even in the race, for most Users.) The widespread acceptance of the USGS groundwater flow code MODFLOW (McDonald and Harbaugh, 1988) is apparently due to its admirable documentation, as well as its availability. (This is not to say that it is not accurate or efficient.) Although not flawless or 100% complete, the MODFLOW documentation is definitely methodical and extensive, and is totally devoid of the deliberate obscurantism often seen in scientific journals, using the minimum mathematical sophistication necessary to communicate the ideas to the reader.

The essential User documentation may cover only the input/output data, but full documentation, essential to maintenance, involves internal documentation as well. In each subprogram, internal documentation should describe all the parameters. It would be convenient if the internal documentation was complete and self-contained, but references to other internal or external documentation will be the more usual case.

11.11.2 Undocumented Options

Users often complain (e.g., Cosner, 1994, 1995) about the hundreds of options in codes and the fact that "many of these are undocumented features." In fact, there is a very good reason for not documenting all features. Documented features should be QA'ed, which takes time and money. The documentation itself takes time and money (see below). Often, undocumented features are artifacts of earlier code development. For example, one may originally build a code with an option in a multigrid solver, and after much experimentation, decide on a particular value. Rather than tear out the option (which takes time and money), it may be easier to leave it in. Also, there may arise a need for the option in the future, perhaps when a new class of problems is considered. If it is best for the average User not to mess with it, it may be better to leave it in as an undocumented option. In the RAVEN codes (Roache et al, 1990), we distinguished levels of options and how they would be accessed. The User was forewarned that undocumented options existed, and these were not accessible through conventional external data entry but only through internal data statements. They were intended and designed to be accessed only by the code builders.

11.11.3 The Cost of Code Documentation

Everyone likes good documentation. Unfortunately, code authors are often reluctant to make the effort, and management is usually unwilling to pay the price. The price is not a small perturbation on the budget. The cost (in money and time) of full documentation, including Verification and Confirmation tests with complete Input/Output specification (so that a User could duplicate the results), usually will be a significant fraction of the total cost of the code development itself. This unpleasant fact of life will not be welcomed by management.

In an article entitled "Why We Still Have So Little Technical Documentation," Weiss (1983) lists three "myths" subscribed to by most analysts and programmers.

Myth No. 1. — Documentation slows progress.
Myth No. 2. — Documentation is Historical.
Myth No. 3. — Documentation is for writers [i.e., is to be done by professional writers].

Weiss and similar writers are focusing on software systems, not just individual scientific codes. The assumption is that there will be multiple code authors for the system, design specifications, management and senior analysts directing junior analysts, professional writers to aid programmers/analysts who must do the core technical writing, QA audit teams, etc. In such an environment, it is clear that documentation will not slow progress, but rather that lack of good documentation will not only slow progress but sabotage progress.

The idea that documentation is not historical but should be done *before* system development may not be readily accepted by an algorithm developer or a graduate student. The processes of algorithm development and testing, conceptual model building (e.g., turbulence model development), code writing, and (larger scale) software systems development are all iterative processes, which Weiss and similar authors recognize. But algorithm development is more iterative than later stages, just because of the uncertain nature of research. Therefore, it makes less sense to emphasize early code documentation and other QA issues during algorithm development than in later stages. Likewise for conceptual model building.

11.11.4 Early Documentation

Weiss (1983) traces reasons for lack of good software documentation. One of the reasons given is the misconception that documentation is put off until late in the development cycle.

Code builders should consider the model of writing draft User documentation, even a draft User manual, *before* the code is written. In this approach, the manual becomes a design specification document. It helps to organize the overall design. It helps to focus early attention on the User interface, and on code module communications. In short, it works better than the scientist might think, and deserves some consideration even at the algorithm or conceptual model development stage. At the stage of full software system development, especially with multiple code authors, it is strongly recommended. Of course it requires some early work; "Pay now, or pay later."

11.12 CODE MODULE COMMUNICATION STRUCTURE

For readability, debugging, and maintenance, the best code structure is one that has no obscure paths of communication between code modules. This means that all data is passed through subprogram argument lists, and no unused data is passed.

Unfortunately, this code structure can cause severe inefficiencies in operation. For example, use of the "blank common" structure in Fortran can minimize gather-scatter operations in large arrays. In the experience of colleagues, a penalty factor of 2–3 can be encountered if the isolated communication path structure (all data passed between module through explicit argument lists) is used (Salari, 1995). For parallel computer languages, the isolated communication structure is not feasible.

11.13 CODE UPDATES

With multiple copies of codes in existence, and with later versions of codes under continual development by some personnel while earlier versions are being applied by other personnel, it is inevitable that inconsistencies arise. While efforts should be made to update earlier versions to reflect changes, we distinguish between functional changes (bug fixes, capability enhancements, etc.) and cosmetic/informational changes (e.g. changes in DO-loop indentations, additions to in-line comment cards, Input/Output alterations). Only functional changes should result in downgrading a QA certification or requiring re-certification. Bug fixes developed during development of later versions must be incorporated into earlier versions (if the fixes affect the application), but changes in indentations, additions of internal documentation, changes in I/O format, etc. need not be.

We would like to avoid legalism whenever possible, recognizing that some projects involving public policy and inevitable lawsuits must become legalistic. An extreme example occurs in some DOE projects that contain an entire section on the definition of "computer program." Such an ANSI-like section contributes to the legalistic tone that we are trying to avoid. Common sense indicates that anyone reading such a document had better know what a "computer program" is.

The engineering management of projects also requires attention to decidedly non-mathematical but nevertheless significant (and difficult!) issues of code traceability and retrievability, including authorship (for the original code, modifications, Verification exercises, documentations). Version Control can be especially difficult, as experienced by many individual code developers, e.g. when one cannot reproduce code results from one's own previous journal publications because one has "improved" the code in the interim, or because one has different versions of the code on the office computer, on the parallel computer system, on one's home PC, at the consulting customer's installation, etc. In the context of an engineering project with multiple authors and Users, Version Control can be a major issue, requiring a formal Software Management System that includes Archival Storage of source code, associated documentation, and (recommended) Confirmation cases. The very real needs of a formal Software Management

System will *necessarily* interfere with further code enhancements, from algorithmic improvements in accuracy or speed, to more descriptive graphics, even to bug fixes.

11.14 BUILT-IN AUTOMATIC ERROR DETECTION TESTS

Consider a CFD code that incorporates thermodynamics data with a limited range of applicability. Suppose that this range is exceeded in the course of a calculation. If the code was built to simply continue the calculations based on extrapolation of the data, this would not, strictly speaking, be a violation of Verification of the Code. The responsibility would lie with the code User. However, good coding practice would require the code to at least print out a warning message to the User. This is a major QA issue. Alternately, the code could be designed to simply terminate the run when the acceptable range is exceeded, or to offer the User the option of continuing with extrapolated data, or with the end-point data (i.e., zeroth extrapolation).

Oberkampf et al (1995) give several examples of recommended error detection whenever an aerodynamics CFD code deviates from the physical modeling assumptions.

- Incompressibility (local Mach Number exceeding some limit, e.g. 0.2)
- Thermodynamic Equilibrium
- Temperature Threshold for Chemically Reacting Flow
- Ranges on Thermodynamic Properties
- Ranges on Transport Properties

Many such checks will suggest themselves for various application areas.

Note a common feature of such error checks: they may require the User to provide more data than is required for the solution of the mathematically defined modeling problem. For example, consider the check on incompressibility. If the code is built to accept all the dimensional defining thermodynamic and other physical data for the problem, the code can calculate local Mach numbers, etc. But typical incompressible flow calculations do not require such information. For common fluid dynamics problems with assumed constant properties, a single dimensionless parameter (*Re*) suffices to define the problem, and a code may be built with only dimensionless parameters for input. In order for these codes to provide such error checking, the User will be required to provide additional dimensional data, e.g. a nominal speed of sound, and a characteristic dimensional length.

This requirement can become annoying to the User when the problem requires much of this sort of extraneous input data, since the User may prefer to take responsibility for the assumptions outside of the code. The correctness of the User's assumptions would then become a matter for Analysis QA rather than Code QA. As often happens, there may be a conflict between the needs of an engineering project and the desires of a scientifically oriented analyst, with the project management preferring clumsy input and rigid tests embedded in the code, and the analyst preferring mathematical elegance and simplicity.

Note a related issue affecting overall code quality; codes built on dimensional data inputs (1) require selection of units, and (2) are more susceptible to round-off error (machine error) problems than codes which are built on dimensionless data inputs (*Re*, *M*, etc.).

11.15 DESIGNING FOR CODE MAINTENANCE

One bit of folk wisdom accumulated by the code development community over decades is virtually universally acknowledged.

Codes are used much longer than the authors intended or expected.

This recognition mitigates against taking too many "shortcuts" in the code development process.

11.16 COMMERCIAL CODES AND THEIR USERS

Most Users of commercial codes are primarily concerned with issues other than accuracy. Also, the complexity of the option tree in a general-purpose commercial code makes code accuracy Verification of all option combinations difficult. Even if a code option set is Verified for numerical accuracy, it still remains to Verify the accuracy for a particular calculation, as described in Part II. Some commercial codes include a single grid error "estimator," better referred to as an error indicator; see Chapter 7. As noted therein, dependable error estimation with a practical definition of error (an error really of interest to an engineering calculation) is not yet available. It could be, with an automatic grid coarsening feature, but the typical programming environment is not conducive to inclusion of this feature. (The SciNapse system under development by SciComp, Inc. will address automated grid convergence testing; see Akers et al, 1997.) Although there are exception and signs of improvement, generally speaking, commercial code accuracy is in a sorry state (not in regard to flexibility or robustness, but accuracy).

The situation is difficult for anyone committed to the QA party line of responsiveness to the "customer." A commercial code user once told me that my concern with numerical accuracy was "Neanderthal," and that the "purpose of calculations is to motivate the cutting of metal." A director of a very successful consulting firm, specializing in simulation of enhanced oil recovery processes, stated that in some 20 years of experience he had *never* had a request from a customer for enhanced numerical accuracy. Even dealing with academic researchers, the experience of myself and colleagues has been disappointing. We have seen more concern over trivial issues of input data structure than numerical accuracy. Not that input data structures and ease of use are not important, but that the customer-users are too *easy* to please-they are happy if these problems are solved, with no concern for numerical accuracy. Commercial code builders cannot be faulted too much for not including dependable accuracy checks if customers are not concerned.

11.17 CODE TO CODE COMPARISONS

Given the suspicion of some regulators, it is inconsistent and surprising that they often like to use Code-to-Code comparisons as part of Code QA or Certification. Certainly, if a new code agrees with an old code that has already been shown to be of high quality, confidence is gained in the new code. In actual practice, such comparisons have often been misused, notably in the groundwater flow community. The practice is acceptable if one code is a trustworthy standard-not merely *old*, or commonly used, but accurate. Also, more limited use of Code-to-Code comparisons is often efficient, e.g. in testing the installation of a new solver, in which case the comparison is on efficiency only, and the Code-to-Code comparison is really used as a debugging device or an aid to confidence building. In general, Code-to-Code comparisons are suspect, and are no substitute for rigorous code accuracy Verification by comparison with an exact solution (see Chapter 3).

11.18 GENERAL SOFTWARE CERTIFICATION AND ISO 9000 STANDARDS

For a general overview of software certification practice and issues, not restricted to the solution of partial differential equations, see de Neuman (1989). See Porter (1996) and some excerpts in Chapter 2 for an overview of CFD Validation/Certification Activities by leading groups in the aerodynamics community.

The widespread acceptance of ISO 9000 standards will possibly impact scientific and engineering codes, especially commercial codes. In a brief informal note, Thibault (1994) has outlined the two ISO 9000 standards that are most likely to affect CFD software vendors. "ISO 9001 addresses the entire software life cycle and requires procedures for requirement specifications, design, implementation, testing and other related activities such as software configuration management, problem reporting, internal meetings and staff training. The less stringent ISO 9003 standard addresses only the testing phase of the life cycle. Both standards require an initial registration and audit, followed by yearly audits."

It seems clear, from this description and from any first- or second-hand contact with ISO 9000, that ISO 9000 is expensive and cumbersome, and not conducive to creative work, e.g. in algorithm development. Only time will tell whether it does more harm than good.

11.19 QA FOR LARGE PUBLIC POLICY PROJECTS

QA at its best is a flexible philosophy, at its worst a bureaucratic morass. It seems clear that the need for a comprehensive software QA system increases as the size of an engineering project increases, and as the public becomes more involved, e.g. in pressing environmental issues such as hazardous or nuclear waste disposal. The activity of "stakeholders" who may not be either technically qualified nor intellectually honest puts extreme pressures on QA management, especially in a litigious society such as the United States of America. Regulators may be more competent than the general public, but still technically unqualified, and more intrusive. (It is challenging to try to explain difficult algorithmic issues like conservation in an Approximate Factorization algorithm to people who are unfamiliar with conservation laws or Gaussian elimination.) This political situation leads to excessive rigidity and records-keeping in a QA system, beyond what the technical requirements might be. For anyone involved in such projects, the inevitable results will include reduced personal productivity and increased job stress.

My recommendations are

1. resist micro-management from unqualified regulators who are not supposed to be even macro-managing, and
2. discuss technical questions with members of the general public only if they understand high-school level science and mathematics. (Democracy will not work with an uneducated populace.)

11.20 QA OF ANALYSES

Since CFD and other simulation codes are so extensively used in modern engineering and science projects, it is perhaps worth emphasizing again that having the best code imaginable is not a guarantee of an adequate analysis. The analysis or design of an engineering project involves much more than the use of a (perhaps excellent) set of codes. Therefore, existence of a formal QA system for Codes or Software is not sufficient. If the QA approach is to be effective, there also must be formal QA procedures to review and evaluate the analyses themselves, and probably the resultant reports (e.g., Rechard et al, 1992).

11.21 QA / CERTIFICATION OF USERS AND REGULATORS

Porter (1996) and others have also considered the possibility of Certification not just of codes, but of *code Users*. Porter notes that "the utility of CFD in [a particular Air Force project] will be as dependent on the skill and experience of the analysts as on the codes themselves." To me, the most remarkable thing about this statement is that it had to be made at all. It should be patently obvious, but evidently is not.

It is my opinion based on some experience in a major project (WIPP PA Dept., 1992; Helton et al, 1995, 1996) involving sensitive public policy decisions, that *specific user training and Certification* (not just token Certification based on general credentials such as M. S. degree and 5 years of experience) is a much needed interpretation of the QA concept. Furthermore, it should be extended to *include regulators*.

11.22 ASSESSMENT OF CODES? OR USERS?

Strazisar and Denton (1995) presented a "CFD Code Assessment in Turbomachinery–A Progress Report." Intended to be (in the terminology used in this book) a Validation comparison of CFD codes, it became (in my assessment) a QA exercise on User training. The results are worth considering in some detail.

A Working Group in Turbomachinery, set up by the International Gas Turbine Institute, designed the assessment as a true *prediction* exercise, being of the opinion that comparisons with already published data-sets "do not provide an objective view of CFD capability." With the same tactful phrasing, the organizers stated "An objective view is not provided when a code is run by its developer." However, 8 of the 12 choosing to participate used their own codes. The test dataset on a transonic axial flow compressor rotor (what the organizers considered to be "almost the simplest example of a practical machine") was taken at NASA-Lewis. The data set was not distributed (internally or externally) until the 11 "blind" simulations were submitted; the 12th simulation was "non-blind" and performed to test feasibility of the project. After the original submission of results, the participants were free to perform additional postdiction simulations.

Tangential to the main thrust of the exercise, the organizers provided a valuable service to future algorithm/code developers by polling several major turbomachinery designers on their goals for CFD accuracy, shown below.

- Pressure rise ± 1–2%
- Temperature rise ± 1–2%
- Efficiency ± 0.5–1.0%
- Rotor exit flow angle ± 1–2°

However, the designers acknowledged beforehand that these levels were not presently realizable, but were thought to be "ultimately possible." The designers also noted two more encouraging aspects of design simulations.

- "Accurate prediction of *change* between two configurations is more important than prediction of *absolute levels*." [See discussion earlier in Chapter 8 on "Inadequacy of Single Grid Calculations for Parameter Trends" and in Chapter 9 on "Trends, Computational and Experimental."]
- "[A]ccurate prediction of the *shape* of radial distributions is more important than absolute accuracy." [This is an interesting variant on the "trends" question.]

Of the 12 submissions, 10 used the algebraic Baldwin-Lomax turbulence model and 2 used k-ε. Most (9 of 12) used H-type grids [which are known to be difficult to use accurately]. The number of grid points varied from 37,000 to 1,050,000; this factor of nearly 30 is not so suspicious as it may seem at first glance, since this represents only about a factor of 3 in each of the 3 coordinate directions. However, no mention (let alone details) of any Verification of Calculation is made in the article. Thus, it is *not* surprising to read the first conclusion of the organizers; "It has been surprisingly difficult to draw firm conclusions from the numerical predictions," thus echoing the first conclusions of the 1981 Stanford Turbulence "Olympics" held 14 years earlier (see discussion in Chapters 1, 9, 10). No numbers on accuracy are reported, only general assessments. The statement that "overall performance was reasonably well predicted by most methods" [some with postdiction] is difficult to reconcile with another statement, that "overall performance can be *greatly affected* [emphasis added] by the details of the turbulence modeling." Again echoing the 1981 Stanford Turbulence Olympics, "no conclusions can be drawn as to the best [turbulence] model."

The hopeless confusion is further evident in the statements about H-grids. "Unfortunately those using more complex grids always used more grid points and so it is not possible to identify a particular grid type as the source of any improvements." This statement seems to take it for granted that Verification of Calculations was not done adequately. If it were, there would be *no significant effect of grid type on the results*. What it would effect is *efficiency*, with other grid topologies (O-type grids and especially C-type grids) achieving the acceptable level of grid convergence with fewer grid points, if previous general experience is any guide. All joking aside (see the CFD parody in Appendix D), the mathematical solution is not supposed to depend on the coordinate system!

The ultimate confusion (and ultimate justification for the CFD parody in Appendix D) comes from these statements. "Three of the participants used the same method...[and]...code, and it is significant that they all obtained different results both for overall performance and for flow details. This occurred even though two of the simulations used a similar number of grid points. It must be concluded from this that the results are not only code dependent but also User dependent." Many comparison exercises have come to the same (presumably tongue-in-cheek) conclusion. It simply states the obvious: that different Users *who use different code parameters* will obtain different results. This conclusion hardly requires such an extensive comparative exercise.

The authors clearly understand the real issue, when they state: "It is clear that even tighter controls of the CFD code parameter space are necessary in future test case exercises." To put it another way, QA/Certification of Users and Analyses (as well as codes) is required.

11.23 QA PROCEDURES

This section on QA Procedures is adapted from our work in Rechard et al (1991, 1992).

A formal QA system involves not just goals but procedures. The purpose of these procedures is to address the quality of the software tools used by the analyst. Software QA procedures are necessary in most large projects because each analyst individually cannot adequately or efficiently assure software quality under reasonable time constraints. That is, the analyst cannot become a specialist in all of the subtle features in a code and Verify their proper implementation. The procedures establish a framework, forms, and personnel assignments for supplying software documentation, Verification reports, evidence of traceability and retrievability, etc. upon which the analyst can depend.

11.23.1 Properties of Software Addressed by QA Procedures

"Quality Assurance" of software means that the User has a reasonable degree of assurance that the software will correctly perform its stated capabilities when given the proper input. The software QA procedures accomplish this assurance by addressing the following properties: traceability, retrievability, documentation, and Verification/Validation/Confirmation.

Traceability and Retrievability

On long-term projects, the software User must be able to determine the version of software run on a particular analysis (traceability) and possibly retrieve an older version (retrievability). QA Procedures are established for traceability and retrievability within the Project so that Users can identify the software and version used in a calculation and retrieve copies of that software, if needed. Retrievability helps the User to explain earlier analyses of the project to reviewers.

Documentation

QA Procedures establish internal and external documentation software requirements so that the User has access to a description of the type of problem and equations solved in the software, the method of solution, Verification/Validation/Confirmation exercises, and the required input. This information helps the analyst to determine the suitability of a piece of software for a particular task. Good documentation also allows management to recognize areas for improving computational capabilities, minimizes duplication of effort, and permit the User to examine the evolution of the software (i.e., update history).

Verification/Validation/Confirmation Exercises

Verification /Validation/Confirmation exercises provide the User with confidence in the software capabilities. Additionally, the Confirmation exercises provide the User with checks on the User's understanding and application of the code. Although customarily relegated to external documentation, a summary of such exercises can also be included in code internal documentation. At the least, internal documentation (i.e., comment cards) can be provided to point the User to the external documents.

Properties Not Addressed by QA Procedures

Software maintainability involves issues of internal documentation, code structure, code readability, and undoubtedly others. Maintainability is an important software property, but QA standards are difficult to establish because of the diverse needs of software used in the project studies. Furthermore, QA standards for deciding whether software is qualified for solving a particular problem, in addition to the ongoing task of justifying such decisions, may be difficult to define in the QA system *per se*. (There seems little value in expanding the definition of software QA to include the entire management structure of a project.) Likewise, controls on the distribution of software outside the project could be included in the software QA system but are possibly best left out.

11.23.2 QA Procedures in Continuous Code Development

Over a long term project, requirements placed on an analyst constantly change; thus, the software must be constantly updated. Analysts exploring a complex phenomenon (e.g., contaminant transport) through the use of computer codes must frequently modify the software as a deeper understanding of the phenomenon evolves. For example, the software might be modified to model more closely the phenomenon or to improve computational capabilities, such as speed or size of the code. Indeed, if the code is being used by analysts, the changes can become almost continuous. Hence, the User needs to be able readily to convey needed enhancements (and bugs fixes) to those in charge of modifications. In addition, the User needs assurance that the software was changed by a qualified individual and the changes were Verified. (Code Verification is repeated for the modified code.)

Furthermore, because the development can be continuous, assuring the quality of the code likewise must be continuous. In theory, this is possible, but in practice, the characteristic of QA as a series of discrete steps means that software development can easily become mired in paperwork, such as providing formal external documentation (e.g., reports) and completing other required paperwork (e.g., filling out and responding to modification reports). A potential problem in this situation is that while waiting for the up-to-date *quality* tools to move through the slow evaluation and approval process, Users must work with out-of-date tools. When this happens, the QA procedures may become more of a means to police the software developer's or analyst's work, instead of acting as a framework to encourage true quality work.

Therefore, in a project environment involving continuous code development, QA procedures should strike a balance between (a) ensuring that the analysts have access to the most up-to-date capabilities and (b) providing the proper level of confidence that those capabilities are working correctly. The Users are in the best position to evaluate this balance, and this fact should be recognized by the QA procedures.

11.23.3 QA Audits

Audits are a necessary part of a formal QA program. Although audits are of secondary importance as far as the User is concerned, management does have needs from a Quality Assurance program. It is important that the plan and procedures are cost effective and produce the desired level of quality. Once these are resolved, management requires assurance that the plan and procedures are followed and any deficiencies are detected and corrected early. The QA System must define the audit responsibility and provide procedure, forms, resolution procedures, etc. This is QA aggravation at its worst, yet most essential, especially for large public-policy projects.

11.23.4 Data and Analysis QA

Quality Assurance for the project means that the results of analyses present a reliable and adequate view of the project. Decision makers (e.g., upper management, government agencies, general public) use the results in conjunction with feasibility, costs, and public concerns to determine whether the project complies with government regulations.

To ensure quality results, the QA system must exercise quality control over several areas, including Data, Software, Analysis, and Documentation. Together these procedures encourage quality work in *all* facets of the study, including gathering data, storing data, using the correct data, using the correct analysis tools, running accurate calculations, correctly interpreting results,

- Concept exploration
- Requirements specifications
- Design
- Implementation
- Test
- Installation and checkout
- Operation and maintenance
- Retirement

Table 11.23.5.1 Eight typical stages of software development. (After Rechard et al, 1991.)

and documenting the analysis so that it can be repeated if necessary. Review procedures are incorporated throughout the procedures so that potential problems can be discovered early in the process. The procedures are documented in separate reports (Rechard et al, 1991, 1992).

11.23.5 QA Procedures and Software Life-Cycle Requirements

Software experts (Sandia Software Guidelines, 1986; Miller, 1989) have identified the eight typical stages of software development listed in Table 11.23.5.1.

The question is, should QA procedures mandate that software developers document in reports each stage of development?

Such a requirement is appropriate for a large project involving analyses that are straight-forward and codes that are not expected to be in continuous development. For the types of analyses and codes primarily considered in this book, such a requirement would be counter-productive. The inherent iterative nature of project analysis, and specifics of the template for a QA system (to be given below) can ensure that the software development or modification process is a series of small steps within the first seven stages of software development. These steps will be repeated many times. Consequently, software discrepancies and deficiencies are discovered within the overall project analysis.

Significant benefits accrue to a self-imposed discipline of documenting a code's progress at any of these stages (notably at stage of requirements specifications, typically neglected by the scientific algorithm/code developer). However, when a formal QA system imposes this requirement, with the usual QA forms, formats, reviews, approvals, etc., the result is to discourage creative approaches, experimentation, and iterative improvements. Of course, these are precisely the activities that make the QA temperament type uncomfortable. If the QA temperament types win this argument, creative scientists and engineers become stressed and/or begin to look for new jobs.

	A	C	D	X
Status	Adjudicated	Candidate	Dormant	eXperimental
Source statements	Fixed	Fixed	Fixed	Not Fixed
Direct User accessibility	Yes	Yes	No[a]	No[b]
Traceability	Yes	Yes	c	No
Retrievability	Yes	Yes	c	No
Verification	Yes	In progress	c	No
Documentation				
Internal	Yes	Yes	c	No
Help File and Abstract	Yes	Yes	c	No
Formal Reports	Yes	In progress	c	No
Discrepancy Records	Yes	No	c	No

a Source code can be made available.
b Software available to individuals within project, but usually only from software developer.
c QA properties of last A or C classification.

Table 11.24.1.1 Relationship Between Software Class and Quality Assurance Requirements. (After Rechard et al, 1991.)

11.24 A TEMPLATE FOR A QA SYSTEM

QA and Certification systems should be tailored to specific projects. In spite of this, it is helpful to have some template of a system in mind, just to facilitate discussion of more general issues. The following description of a template is not suggested for any actual system, but may also serve as a starting point for a design of a particular system. It is based on an early version (Rechard et al, 1991, 1992) in which I was involved. (The system has since been supplanted in the project by a more rigid and elaborate one.) The general concepts still have value, even if titles and descriptors are changed.

11.24.1 Description of the Four Software Classes.

In this QA template, there are four software classes that represent the following stages of development.

A – Full QA status
C – Partial QA status, possibly undergoing continued refinement
D – Obsolete software formerly in Class A or C
X – Entry level, software in early stages of development or experimentation.

Although new software can be entered at any level, a Code Review Committee decides whether to classify software as Class A. The QA Coordinator decides whether software is entered into Class C, D, or X. Software should be classified at the highest possible QA level consistent with its importance or requirements of its tasks.

Classifying software is intended to offer the User the following assurances.

Class A – Class A software has been evaluated by the Code Review Committee. The software satisfies the QA requirements for traceability, retrievability, documentation, and Verification. The software is available to any interested User within the project.

Class C – Class C software is a candidate for Class A, but currently satisfies only the traceability and retrievability requirements (as evaluated by the QA Coordinator). The adequacy of documentation and Verification has not been formally evaluated. An up-to-date Help File is maintained, a Software Abstract has been written, and internal documentation exists. However, both Verification tests and external documentation are in progress. Class C software is often undergoing continued refinement to adapt the software's capabilities to changing requirements. The software is available to any interested User within the project.

Class D – Class D software is software that was formerly classified as Class A or C. Class D software is archived because it is no longer being used (usually because a more sophisticated version has superseded its use). Class D software can be retrieved from the archives, if necessary.

Class X – Class X software is in the early stages of development and has not been processed through any formal QA procedures. The primary reason for the Class X classification is to make the existence of the software known to potential Users. Usually, a User must obtain Class X software by contacting the software developer.

11.24.2 Software Categories

The project software is classified on two bases: *origin* and *type*.

Under *origin*, the two possibilities are *internal* or *external* software (i.e., developed within or outside the project organization).

Under *type*, the two possibilities are SES (for scientific/engineering software) or *non-SES* software. SES software models a physical process, often by the numerical solution of mathematical equations. It is also software that is used to establish the spatial variation of modeling parameters. A significant distinction of SES software is that the Validation process is subjective and often difficult because the software models a physical process. Non-SES software, such as translators and plot programs, does not have to be reviewed against physical data. For Classes X, C, and D, the QA review process is the same for all software categories. For Class A, SES software must be reviewed by the Code Review Committee; non-SES software is also reviewed by the Code Review Committee, though it may consist of only two individuals.

11.24.3 Participants in Software QA Procedures

The fundamental means of implementing both the specific requirements and the common sense needs of software Users (discussed below) is to assign a person to be a Sponsor to each code. This Sponsor is responsible for interacting directly with the software Users (the project Analysts) and ushering the assigned code through the QA process. However, because of the magnitude of the Sponsor's task, these procedures also define major supporting tasks for one other participant, the Code Consultant. Six additional participants also provide support for the Code Sponsor. A brief description of the participants and their responsibilities are given here.

Code Sponsor – Each code has a Sponsor, who is responsible for ushering the code through the project software QA procedures. The Sponsor may also serve as Consultant, when necessary.

Code Consultant – Some codes have a Code Consultant, who answers technical questions from Users on the mathematical and numerical basis of the code and its application.

QA Coordinator – The QA Coordinator administers the QA procedures by selecting Code Sponsors, Code Consultants, and Code Review Committee members, and by approving the contents of Software Abstracts.

Managerial QA Chief – The Managerial QA Chief is responsible for ensuring that the software QA procedures are established and implemented in accordance with the documents defining the QA process. Also, the Managerial QA Chief is responsible for conducting periodic audits, the selection process for review committees, establishing signature authority for change orders, approving forms for bug reports, designing conflict resolution procedures, etc.

Code Review Committee – Each code is assigned a Code Review Committee. The Code Review Committee evaluates software to determine whether it can be classified at the highest software QA level (Class A). The evaluation consists of determining whether the documentation (i.e., internal documentation, external formal reports, and external computer Help Files) and the code Verification studies are adequate. The committee is also responsible for reviewing (or adding) any limitations on the use of the software in the project studies. For scientific/ engineering software (SES) (Bingham, 1990; Miller, 1989) the committee consists of the Code Sponsor, Code Consultant, project modeling Task Leader, and peer reviewer outside the project Division. For non-SES, the "committee" usually consists of two individuals, the Code Sponsor and an independent peer reviewer, inside or outside the project Division.

Records Manager – The Records Manager maintains QA records on each code used in the project studies. The records include Software Abstracts, Code Review Committee reports, Verification reports, relevant Validation studies, and modification reports.

Task Leader for Project Modeling Task – The Task Leader is assigned to conduct specific modeling activities. The Task Leader provides design or acquisition requirements that are in accordance with the QA procedures, provides Code Sponsors and Consultants, if necessary, and participates on the Code Review Committee for the software used in his or her modeling activity.

Cognizant Analyst – The Cognizant Analyst applies the software to the project's problem. The analyst must ensure a high quality analysis by selecting the proper software, providing correct problem data, interpretation, and software Validation, and maintaining software quality by reporting suspected errors and suggested enhancements.

Cognizant Manager – We define "Cognizant Analyst" as the one who applies the code, and "Cognizant Manager" as the one who enforces the QA requirements. The Cognizant Manager is not responsible for revising of computer programs, but for managing the revisions of computer programs.

Also, Cognizant Manager is responsible for maintaining a written procedure and log for QA compliance audits. In some cases, "Cognizant Analyst," "Code Developer" and "Code Sponsor" were the same person. "Code Sponsor" is required as a separate role to cover the common situation of a project depending upon some codes developed by Code Builders outside the project team; it is still a good idea to have a "point of contact" or responsible person for that code.

Although the relationship of the individual participants in QA may appear straightforward, the actual relationship of the software QA task with project modeling tasks may be more complicated. The project or company structure may not be organized by software-development categories (such as code-definition, programming, and installation), but rather by task assignments in specific project problem areas. Hence, the QA responsibilities will not be assigned broadly to groups but to individuals, who may be part of other groups.

11.24.4 Further Comments on Division of Responsibilities

Viewpoint

The importance of the QA issues addressed by these procedures differs among the classes and categories of software. Assuring the quality of software involves common sense issues. Because issues are less likely to command unreasonable attention if they are shown in context, the issues are presented from the viewpoint of the software User, the project Analyst.

Properties Not Controlled by Software QA Procedures

Software maintainability (i.e., internal documentation, code structure, and code readability) is encouraged but not controlled by these template QA procedures. QA standards have not been established nor are they likely to be forthcoming. Furthermore, QA standards for deciding whether software is qualified for solving a particular problem, in addition to the ongoing task of Validating this decision, are not defined. The Task Leader for each modeling task is assigned the responsibility for addressing the issues of software Validation as well as maintainability.

Controls on the distribution of software outside the project are not established by these template QA procedures. (Most organizations have established controls.)

Responsibility with Continuous Code Development

The fundamental means of addressing the User needs and difficulties engendered by continuous code development is to assign a Sponsor to each code, whether the software was developed internally or externally. All modifications (enhancements or bug fixes) are under the purview of the Code Sponsor. After modifying the software, the Code Sponsor is responsible for updating a computerized Help File and making the new version readily available to software Users within the project organization. The Code Sponsor also maintains the most recent version of the code.

Although potentially a large burden for one individual, the fact that most of the software development and necessary enhancements are problem driven makes fast responses to User needs of paramount importance. Empowering one individual to make the changes and perform the necessary QA functions improves the response.

Responsibility for Traceability and Retrievability

On long-term projects, the software User must be able to determine the version of software run on a particular analysis (traceability) and possibly retrieve an older version (retrievability). The Code Sponsor is responsible for placing on all output the software name, the version identification, and the date to ensure traceability. The Code Sponsor is also responsible for maintaining the previous version of the software (short-term retrievability). For retrievability over the long term, the QA coordinator is responsible for developing a reference system for tracking current versions of software and related documents and for archiving older versions and documents. The intent is to archive software used in a major project study so that an archived version is tied to a particular project document. Class C software, which may change because of continual refinement, is archived at the request of Users. Class A software, which changes much less frequently and is usually tied to a major project study, is always archived. The QA coordinator is also responsible for distributing software to interested individuals or agencies outside the organization, if appropriate. No configuration control for externally distributed codes is required.

Responsibility for Documentation

For all Class C and A software, the Code Sponsor is responsible for producing a Software Abstract for distribution to a Central Files System (accessible to everyone involved in the project) as well as a computerized Help File containing software execution instructions, input commands and instructions, and update history information. The Help File is available to all participants who have access to the project computers.

For software that models complex physical phenomena, documentation must also include the theory (including the type of problem, equations solved, and the method of solution) underlying the code so that the analyst can properly use the software. However, because it is impossible to expect that every analyst will become a specialist in the intricacies of codes that model complex physical processes, Code Consultants are also assigned to bridge the gap between theory documentation and the User.

Responsibility for Documentation Considering Software Life-Cycle Requirements

As noted earlier, experienced software developers (Sandia, 1986; Miller, 1989) have identified eight typical stages of software development. The question is, should QA procedures mandate that software developers document in reports each stage of development? In this suggested template for a QA system, such reports are not required.

Responsibility for Coding Style, Structure, Portability, Robustness, Maintainability

Indirectly, code maintenance issues, such as internal documentation, code readability, code structure, and code portability to future computers and operating systems, can become important QA properties to the software User in some projects. Whether or not to address these issues in a QA system would be highly specific to a project. While the easy response (and the typical QA personality response) is to insist on it, I believe that most engineering and science projects would suffer from the requirement. For most projects, code developments and enhancements are *not* aimed at general situations but instead are performed to solve specific problems requested or needed by the project. In other words, the Code Sponsor is not producing or supporting general purpose software or software that must be applied outside the project. Also, for precise calculations of complex physical phenomena, nonlinear behavior must often be included, a characteristic that can raise havoc when a developer is trying to build a robust code. Therefore, in these template QA procedures, the laudable goals and desirable features such as coding style structure, language, code robustness, software distribution control, and software portability are addressed only as needed by the Task Leader for the code development and modeling task.

Responsibility for Audits

The QA System must define the audit responsibility and provide procedure, forms, resolution procedures, etc. This is QA at its worst, yet most essential, especially for large public-policy projects. The audit function is performed by the designated Managerial QA Chief who has the responsibility for evaluating the level of compliance.

Responsibility for Verification, Validation, Certification

The responsibility for Verification of a Code basically and ordinarily rests with the code builder. However, if the code builder is not available (e.g., a commercial code developed external to the project organization is being used) then that responsibility goes to the Code Consultant. (If the code was developed internally, the code builder, or a member of the team that built the code, may be the Code Consultant.) In this template QA system, the point of contact for the User will be the Code Sponsor, who may also be the Code Consultant. Especially in a large project, it may be clearly advantageous to somewhat isolate the code builder (who is probably more ego-involved with newer work) from the User needs. In this template QA system, the interface is provided by the Code Sponsor.

In any engineering or scientific project, the person performing the analysis (the Project Analyst) has the responsibility to perform Verification of the calculations. Also, the Project Analyst has the responsibility to select the code used, and therefore to ensure that the code solves the correct problem, i.e. to accomplish Validation. This would indicate that, in this template of QA procedures, the role of Project Analyst be given the responsibility for Validation. However, in a large project, it will often be of some advantage to use the same code (or code families) for different analyses. Also, Certification (as understood in Chapter 2) involves Validation for a project-specific range of parameters, often broader than what individual analysts would consider for their more narrowly defined problems. Thus, in these template QA procedures, responsibility for Certification rests with the Task Leader. Of course, "responsibility" does not mean that the Task Leader operates in a vacuum, or cannot delegate the responsibility. Also, Validating and Certifying that a PDE code acceptably represents a complex physical process is not straightforward, as discussed in Chapters 2, 9,10. Thus, although Validation and Certification are primarily under the purview of the project Analyst and the Task Leader, code modifications and enhancements will often be required. Often the iterative process of Validation, code modification, Code Verification, further Validation, etc. will be virtually a continuous process, involving Task Leader, Project Analyst, Code Sponsor, and Code Consultant/Builder.

11.25 CONCLUDING REMARKS ON QA

Like any bureaucracy, a formal QA system can be frustrating. Like most bureaucracies, it is a necessary evil for large projects. Provided that participants exhibit some patience, tolerance, and common sense, a formal Quality Assurance system can actually contribute to true "quality work," especially for large engineering projects. It can also provide creative scientists, mathematicians, and engineers with a different kind of challenge, and can provide its own sense of satisfaction.

REFERENCES FOR CHAPTER 11

Akers, R. L., Kant, E., Randall, C. J., Steinberg, S. and Young, R. L. (1997), "SciNapse: A Problem-Solving Environment for Partial Differential Equations," *IEEE Computational Science and Engineering*, Vol. 4, No. 3, July–September 1997, pp. 32–42.

Barber, T. J. (1996), "The Role of Code Validation and Certification in the Design Environment," AIAA Paper 96–2033, 27th AIAA Fluid Dynamics Conference, New Orleans, Louisiana, 17–20 June 1996.

Bingham, F. W. (1990), *Software Quality Assurance Plan*, SLTR 89–2001, Nuclear Waste Repository Technology Department, Sandia National Laboratories, Albuquerque, New Mexico.

Biot, M. A. (1956), "Applied Mathematics: And Art and a Science," *Journal of the Aeronautical Sciences*, May 1956, pp. 406–410.

Blottner, F. G. (1990), "Accurate Navier-Stokes Results for the Hypersonic Flow over a Spherical Nosetip," *AIAA Journal of Spacecraft and Rockets*, Vol. 27, No. 2, March–April 1990, pp. 113–122.

Cosner, R. R. (1994), "Issues in Aerospace Application of CFD Analysis," AIAA Paper 94–0464, 32nd Aerospace Sciences Meeting, Reno, Nevada, January 10–13, 1994.

Cosner, R. R. (1995), "CFD Validation Requirement for Technology Transition," AIAA Paper 95–2227, 26th AIAA Fluid Dynamics Conference, San Diego, California, June 19–22, 1995.

de Neuman, B. (1989), Ed., *Software Certification*, Elsevier Applied Science Publishers Ltd., Barking, United Kingdom.

Hatton, L. (1997), "The T Experiments: Errors in Scientific Software," *Computational Science and Engineering*, Vol. 4, No. 2, April–June 1997, pp. 27–38.

Helton, J. C., Anderson, D. R., Roache, P. J., et al (1995), "Effect of Alternative Conceptual Models in a Preliminary Performance Assessment for the Waste Isolation Pilot Plant," *Nuclear Engineering and Design*, Vol. 154, 1995, pp. 251–344.

Helton, J. C., Anderson, D. R., Roache, P. J., et al (1996), "Uncertainty and Sensitivity Analysis Results Obtained in the 1992 Performance Assessment for the Waste Isolation Pilot Plant," *Reliability Engineering and System Safety*, Vol. 51, 1996, pp. 251–53–100.

ISO (1991), *Quality Management and Quality Assurance Standards–Part 3: Guidelines for the Application of ISO 9001 to the Development, Supply, and Maintenance of Software*, First Edition, International Organization for Standardization, ISO 9000–3, Geneva, Switzerland, 1991, reprinted 1993.

Keirsey, D. and Bates, M. (1984), *Please Understand Me: Character and Temperament Types*, Prometheus Book Company, Del Mar, California.

McDonald, M. G. and Harbaugh, A. W. (1988), *A Modular Three-Dimensional Finite-Difference Ground-Water Flow Model*, U. S. Geological Survey Book 6, Modeling Techniques. Available from Scientific Software Group, P. O. Box 23041, Washington, D. C. 20226–3041.

Melnik, R. E., Siclari, M. J., Marconi, F., Barber, T. and Verhoff, A. (1995), "An Overview of a Recent Industry Effort at CFD Code Validation," AIAA Paper 95–2229, 26th AIAA Fluid Dynamics Conference, San Diego, California, June 19–23, 1995.

Miller, W. (1989), *Software Quality Assurance Requirements, Revision B, Department Operating Procedures*, Yucca Mountain Project Department, Sandia National Laboratories, Albuquerque, New Mexico.

Myers, I. (1962), *Manual: The Myers-Briggs Type Indicator*, Consulting Psychologists Press, Palo Alto, California.

Oberkampf, W. L., Blottner, F. G., and Aeschliman, D. P. (1995), "Methodology for Computational Fluid Dynamics Code Verification/Validation," AIAA Paper 95–2226, 26th AIAA Fluid Dynamics Conference, San Diego, California, 19–22 June 1995.

Osborne, W. M. (1988), "Developers Must Design In Maintainability," *IEEE Software*, May 1988.

Porter, J. L. (1996), "A Summary/Overview of Selected Computational Fluid Dynamics (CFD) Code Validation/Calibration Activities," AIAA Paper 96–2053, 27th AIAA Fluid Dynamics Conference, New Orleans, Louisiana, June 17–20, 1996.

Rechard, R. P., Roache, P. J., Blaine, R, L., Gilkey, A. P., and Rudeen, D. K. (1991), *Quality Assurance Procedures for Computer Software Supporting Performance Assessments of the Waste Isolation Pilot Plant*, SAND90–1240, Sandia National Laboratories, Albuquerque, New Mexico, April 1991.

Rechard, R. P., Rudeen, D. K., and Roache, P. J. (1992), *Quality Assurance Procedures for Analyses and Report Reviews Supporting Performance Assessments of the Waste Isolation Pilot Plant*, SAND91–0428, Sandia National Laboratories, Albuquerque, New Mexico, October 1992.

Rizzi, A. and Vos, J. (1996), "Towards Establishing Credibility in CFD Simulations," AIAA Paper 96–2029, 27th AIAA Fluid Dynamics Conference, New Orleans, Louisiana, June 17–20, 1996.

Roache, P. J., Mikatarian, R. R., Chandler, L. J., Steinberg, S., Salari, K., and Sery, P. (1990), "Adaptive Grids for RAVEN: Vol. I of IV, Project Summary," WL–TN–90–10, Air Force Systems Command, Kirtland AFB, NM, June 1990.

Salari, K. (1995), personal communication.

Sandia (1986), *Sandia Software Guidelines*, Vol. 3, SAND85–2346, Sandia National Laboratories, Albuquerque, New Mexico.

Steinberg, S. (1997), personal communication.

Strazisar, A. J. and Denton, J. D. (1995), "CFD Code Assessment in Turbomachinery–A Progress Report," *ASME Global Gas Turbine News*, May/June 1995, pp. 12–14.

Thibault, P. (1994), "Code Validation: A First Step Towards Quality Assurance for CFD Software," *CFD Bulletin*, CFD Society of Canada, Winter 1994, p. 6.

Weiss, E. H. (1983), "Why We Still Have So Little Technical Documentation," *Infosystems*, June 1983, pp. 88–89.

WIPP PA Dept. (1992), *Annual Performance Assessment for the Waste Isolation Pilot Plant, Dec. 1992, Vol. 1, Preliminary Comparison with 40 CFR 191, Subpart B, Vol. 2: Technical Basis*, Performance Assessment Department, Sandia National Laboratories, Albuquerque, New Mexico.

12

CONCLUSIONS

12.1 THE OVERALL PROCESS FOR QUANTIFICATION OF UNCERTAINTY

This book has treated the general area of Quantification of Uncertainty in Computational Science and Engineering and related disciplines that utilize the numerical solution of partial differential equations. Included are the activities of Verification of numerical accuracy of Codes, Verification of Calculations, Validation of Codes (or the conceptual and mathematical models on which the code is built), and the broader areas of Code Certification and QA. Certainly, more development work remains to be done, especially in removing burden of Verification of Calculations from Users of commercial codes, and certainly Validation is and will remain an ongoing activity. But there is no excuse for not doing a convincing, credible treatment of the numerics. We *now* have algorithms, codes, methodology, and computer power to do a good job on the numerical accuracy. Note that the accuracy issues associated with "nonlinear dynamics of numerical methods" (see Chapter 8) are easily handled by straightforward techniques, and the "basins of attractions" phenomena relate only to code iterative convergence and robustness. While certainly important to Code Users, these are not *accuracy* issues.

Techniques are presently available to generate an exact solution (the Method of Manufactured Solutions, Chapter 3) for virtually any PDE code including commercial codes. This method (or another equally complete method) should be used first to rigorously Verify the numerical accuracy of a Code, before Verifying the individual calculations. The Verified code order (say 2nd order) can then be used to verify an actual calculation over 2 grids, or, preferably, the observed order for an actual calculation can be determined experimentally by calculations over a minimum of three grids.

Also, good practice requires that one investigate changing the time step even for steady state and for adaptive time step methods if nonlinear time stepping algorithms like 4th order nonlinear Runge-Kutta have been used.

Once the mathematical veracity of the code and the calculation have been established (i.e., Verification), then one can proceed to attempt Validation of the code (or better, the conceptual model embodied in the code) by comparison to experiments. The Validation is ad hoc, particular to the physical experiment and a (usually small, but hopefully usable) parameter range around the particular experiment. The voices of extensive experience clearly warn the practitioner (or team): do not attempt the Validation process until you have confidence in the Verifications.

The various approaches to Verification of Calculations can be compared on the basis of their reliability, flexibility, and work required. Single grid error estimators are perhaps most efficient, and are appealing because they do not require additional grid generation. However, they do require additional code development and numerical experimentation to correlate (calibrate) their predictions with more straightforward grid convergence studies. This work must be repeated (re-calibrated) for any change in the mathematical model, which becomes a serious issue in code maintenance. Nevertheless, these methods are attractive for large suites of calculations such as those required in Monte Carlo studies and optimization studies. This re-calibration of the error estimator is not required with grid convergence studies and error estimation (or banding) based on Richardson Extrapolation. Also, generalized Richardson Extrapolation methods such as the Grid Convergence Index are the most reliable method, the most quantifiable, are specific to whatever practical error measure is of interest, and are adaptable to various order of accuracy; they can be used to Verify and extract the observed order of convergence for a calculation. Methods using separate PDE's to transport the errors require the most work, and are fragile; they are interesting but not recommended.

The convincing Verification of a Calculation sometimes involves subtleties, but generally does not involve difficult concepts or mathematics. It is somewhat tedious and requires attention to detail. Melnik's advice is realistic: "Think like an accountant". I would add, "Go do your homework".

The level of confidence required depends on the project, i.e. on the intended use of the results. As we noted in the 1986 ASME Journal of Fluids Engineering Policy Statement (see Chapter 1), many industrial calculations will probably continue to be single grid calculations without an error estimate, with adequacy of the grid resolution estimated by intuition and experience. But the "practical engineer" must recognize that CFD practitioners with an extensive experience base usually cannot infer grid resolution requirements from ostensibly "nearby" calculations. This is a task for the code user, not the code builder. It is just as difficult when a commercial code is used, no matter what the advertising promises may imply. Furthermore, although merely predicting trends produced by changes in parameters will generally require less grid resolution than predicting absolute values, the user cannot be casual; under-resolved calculations cannot be expected to always produce correct trends, even if the solutions are qualitatively correct. There exist counter-examples showing that grossly or even moderately under-resolved solutions do not show the correct trends. Also, such trends or sensitivity studies may require tighter iterative convergence criteria than absolute values do.

Compared to just obtaining a single grid solution, it requires (very) roughly twice as much work to Verify a new calculation, if we are using a previously Verified code and the same parameter range, and if the Verification is positive on the first try. If the new problem exhibits new features (compared to the previously Verified problem), perhaps three times the effort will be required

12.2 INTERNET ARCHIVE

An Internet archive of a bibliography for CFD Verification and Validation is maintained by Prof. R. Barron of the University of Windsor, Ontario, Canada, and can be accessed at the following address.

http://www.lpac.ac.uk/SEL-HPC/Articles/ValCFD.html

12.3 FULFILLING THE PROMISE OF COMPUTATIONAL POWER

With the often noted tremendous increases in computer speed and memory, and with the less often acknowledged but equally powerful increases in algorithmic accuracy and efficiency, a natural question suggests itself. What are we doing with the new computer power? with the new GUI and other set-up advances? with the new algorithms? What *should* we do? Shall we continue to produce more pseudo-solutions, or shall we finally solve the problems correctly? I suggest that, when all things are considered, including the inevitable fruits of quick-and-dirty calculations - anxieties and guesswork and misdirections and distrust and tarnished reputations, both for individuals and for the computational community in general–that it is simpler and easier to "just do it!" *Get the right answer.*

The statement of Ferziger (previously quoted in Chapter 3) bears repeating.

"*... the frequently heard argument 'any solution is better than none' can be dangerous in the extreme. The greatest disaster one can encounter in computation is not instability or lack of convergence but results that are simultaneously good enough to be believable but bad enough to cause trouble.*"

APPENDIX A. NEED FOR CONTROL OF NUMERICAL ACCURACY

The following is a reproduction of Roache, P. J. (1990), "Need for Control of Numerical Accuracy," AIAA *Journal of Spacecraft and Rockets*, Vol. 27, No. 2, March–April 1990, pp. 98–102. See also AIAA Paper No. 89–1669.

Need for Control of Numerical Accuracy*

Patrick J. Roache
Ecodynamics Research Associates, Albuquerque, New Mexico 87198

The need for the control of numerical accuracy in computational fluid dynamics (CFD) code solutions is reviewed in the light of current journal practice and experience with implementation of an editorial policy on the same subject published by the *Journal of Fluids Engineering*. Various actual objections to that policy are listed and responses are given. The general successes and particular difficulties experienced in the implementation of the policy are noted. The broader

* Presented as Paper 89-1669 at the 24th AIAA Thermophysics Conference, Buffalo, NY, June 12-14, 1989; Received July 8, 1989.

* President. Senior Member AIAA.

question of code verification, validation, and certification is considered. It is suggested that professional societies such as the AIAA and American Society of Mechanical Engineers may ultimately become involved in the task of certification of available CFD commercial codes.

I. INTRODUCTION

This paper, on the general philosophy and need for numerical accuracy control in computational fluid dynamics (CFD) codes, is based on experience with the implementation of an editorial policy statement by the American Society of Mechanical Engineers (ASME), published in the *Journal of Fluids Engineering* (*JFE*) [1]. The policy statement was conceived following the creation of the position of Associate Editor for Numerical Methods in the *JFE,* which formally recognized the special needs of this discipline. The ASME policy and supporting statements are reproduced in the Appendix. The rationale and needs for the policy statement are explained in the announcement. The *JFE* had previously published and had many years of experience with a similar requirement for uncertainty analysis in experimental papers. Following our early experience with this new policy, a similar policy was also adopted by the ASME's *Journal of Heat Transfer.* The general subject of control of numerical accuracy has become something of a "hot topic," with special reference to the National Aero-Space Plane, a session at the AIAA 1989 Thermophysics Conference, ASME sessions at the 1988 and 1989 Winter Annual Meetings, and by the Texas Institute for Computational Mechanics workshop on the slightly broader topic of reliability in computational mechanics in October 1989.

II. RESISTANCE AND OBJECTIONS

Although Roache et al [1] thought that the policy statement and the editorial requirement were quite mild, it was not universally welcomed. Objections were offered by some of the other editorial board members and by other members of the professional community from whom I solicited comments in the months following the publication of the statement. Some of these objections, all of which are actual (i.e., not straw-man objections), are listed below, together with my responses.

Objection 1

It is too expensive of computer time to do mesh doubling calculations in order to ascertain grid convergence.

There are, of course, other ways to ascertain grid convergence than the straightforward method of grid doubling. In fact, this is probably the most reliable method available, but there are other approaches, as briefly touched upon in the original policy statement. If the cost of computer resources is not a problem to the researcher, this is certainly the easiest approach to take. However, ıf computer resources are a problem, there are other methods that are not intensive users of computer time. It seems that the greater objection for doing grid convergence studies is the fact that it requires a bit of conscientious work on the part of researchers.

Also, if it is argued that it is simply too expensive to do any kind of control of numerical accuracy, then I would argue that the author simply cannot be in this CFD business. After all, if you do not have a wind tunnel you cannot do experimental testing. My impression of the situation is actually worse than this. Journal articles in the late 1960s and early 1970s commonly predicted high resolution accuracy runs when the next generation of computers became available. But for the most part they have not been used that way. Generally the tremendous advance of computing power has been used to produce more mediocre papers rather than fewer reliable ones.

Objection 2

Some exception to the policy should be made for expensive calculations, particularly three-dimensional turbulent studies.

I do not think that the overall cost of the computations should be a consideration. It seems clear that the incremental cost of performing a grid convergence test should be normalized by the cost of the base case. In this sense it is cheaper to validate the grid convergence on a three-dimensional problem than on a two-dimensional problem, presuming that the incremental work involves an extra coarse grid computation. This again relates to the first point, which states that it is not necessary to do a grid doubling in order to ascertain some index of numerical accuracy. A grid halving is also appropriate. Of course, the advantage lies in doing a grid doubling test because the error bounds will be sharper.

Objection 3

Turbulence modeling, rather than the numerical solution of the partial differential equations, is the real determinator of accuracy.

My response is, yes and no. Accuracy is a question to be addressed even for laminar flow calculations, wherein the constitutive equations are not in doubt. The discretization error does not disappear just because one uses a turbulence model! Our point in the *JFE* policy statement, and the first criticism which our evaluation committee made at the 1980/81 Stanford meeting on complex turbulence flows [2], is that one cannot evaluate different turbulence models unless one first satisfies grid convergence. There are yet more considerations in the overall accuracy question; including, for example, the attainment of a true steady state in the computations (yet another "convergence"), inner-loop convergence for incompressible flows, equation of state accuracy, low Mach number approximation, geometry representation, accuracy of viscosity and conductivity coefficients, constant Prandtl number assumptions, chemical reaction rates, and so forth. And, of course, coding errors! These all affect accuracy but do not remove or over-ride the requirement for grid convergence testing. I am of the opinion that these modeling questions, including turbulence modeling, should be kept separate from the question of the numerical accuracy of the solution to those mathematical models.

However, there is in fact another special problem with turbulence modeling in that both the fine and coarse grids have to get some points into the viscous sublayer. I do not know any easy way around this or any other problem that generates such a range of significant length scales which must be adequately resolved. However, thorough work can be done. See, for example, Ref. 3.

Objection 4

The policy statement does not go far enough.

This is certainly true. We consider the *JFE* policy statement to be a minimalist statement. The idea is to make it clear that the authors have to address the topic of numerical accuracy control, and to give the editors and the referees the support they need in demanding some effort in this regard. For example, in the policy statement, no mention is made of the control of errors due to far-field boundary conditions. Cheng [4] pointed out two decades ago that this computational modeling error does not improve as the grid is refined. The only way to test for its effect is to move the position of the far-field boundary. Likewise, the simple reporting of, for example, a 5 or 10% difference in some function of the solution between two grids is not really a rigorous indicator that the solution has been obtained to 5 or 10% accuracy. Although claims have been made that the finite-element methodology offers a more rigorous approach to error estimation, I have not been convinced that anything useful and dependable has been produced for nonlinear systems of partial differential equations. (The theoretical work that has been done is interesting to some extent, but it is usually limited to a very simple model equation.) Schonauer [5] has devoted much time and effort to producing accurate error estimators for Navier-Stokes codes. In fact, his goal is not merely to produce accurate solutions but to accurately predict the error bounds on those solutions. I suppose a really tough journal policy or certification policy would require such effort before accepting a solution, but the journal policy that we adopted was much more lax than this. The only intent was to avoid the syndrome of producing a single calculation on a single grid (the all too common "take it or leave it" attitude).

It is important to note here that grid convergence is really only one aspect of numerical accuracy control; it is, however, a necessary component. For example, a statement of results on grid convergence studies really has nothing to do with coding errors (unless an exact solution is known, as will be discussed subsequently).

Mehta [6] is concerned with code validation, verification, and certification. Note that a *code* is certified, but a particular calculation still needs to be examined for the control of numerical accuracy. A wrong code can "converge" and, conversely, a correct (i.e., certified) code can be applied to a particular problem with an inadequate resolution. Therefore the question of grid convergence testing really presumes ahead of time that we have a correct code. The only question then remaining is whether or not the code has been applied with adequate resolution to obtain an accurate answer to the problem at hand.

Common sense and experience certainly indicate that this presumption of code correctness is not always justified. There are now tools available to perform this kind of verification quite convincingly, in my opinion. In Ref. 7, we showed how to verify a FORTRAN code that was produced entirely by a symbolic manipulation code. The idea is to generate a selected analytic solution to a problem by introducing forcing terms and then to monitor the convergence of the CFD code to that solution as the grid size is reduced. This procedure verifies the coding accuracy, the grid transformation equations, and even the order of accuracy of numerical method. In Ref. 7, it was performed only for the Poisson equation in a transformed three-dimensional coordinate system and for the most common elliptic grid generation equations. Shih et al [8] have applied this concept to the full incompressible Navier-Stokes equations in two-dimensions. I consider this paper to be one of fundamental importance in the area of code verification. It would seem that the principle can readily be extended to compressible and three-dimensional flows using symbolic manipulation programs such as MACSYMA. Turbulence models also appear to be amenable to this approach of accuracy checking-noting of course that we are not talking about the adequacy of the turbulent representation of the real physics but rather the correct coding and the numerical solution of the turbulence model. Shocks, however, seem to be a more difficult topic to approach in this manner.

Note that there is an earlier history of using very simple solutions to verify models, but false indicators of accuracy sometimes have been obtained. The key in such an exercise is to choose a solution with enough structure in it to exercise all of the terms in the equations and all of the leading error terms in the discretization. If, for example, one selects a Couette flow solution, one will be lead to the erroneous conclusion that a first-order accurate method is perfectly adequate even with a coarse-grid resolution. The power of the Shih et al method [8] is that the particular solution they have constructed behaves like a real fluid dynamics problem with boundary layers and significant solution structure.

Objection 5

First-order methods and hybrid methods are not more difficult to judge for convergence than second-order methods, contrary to the discussion in the policy statement.

I do not understand this objection. It is an elementary behavior of numerical methods that higher order accurate methods, as they approach convergence, do so more quickly. For example, a fourth-order Runge-Kutta method applied to an ordinary differential equation is easier to judge as being converged than is a second-order method. If a paper reports that a successive grid doubling of a solution of a linearized advection-diffusion equation produces only a 5% change in the answers, one could expect with some confidence that the fine grid solution is indeed within 5% of the true solution, if the method used is fourth-order accurate. If the method used is second-order accurate, it is more problematical; if the method used is only first order accurate, I would be skeptical of the 5% limit on the "accuracy."

{*Addition to the original JSR paper*. In view of my later analysis in the paper on the Grid Convergence Index (Roache, 1994 cited in Chapter 5) I am now more optimistic, especially for a grid doubling with a second order method. However, the basic point, that reliable grid convergence is more difficult to ascertain with lower order methods, is still valid.}

Objection 6

Agreement of the calculations with experiment is enough justification for the solution accuracy, without any need for doing systematic grid convergence testing.

This objection is quite attractive at first glance. In practice, it does not seem to work very often because the agreement with the experiment is usually not universal. This is especially obvious (as stated in the original policy statement) with turbulence modeling. Whether or not the turbulence parameters have been "tuned" to a particular problem, we still have a "package deal" of discretization errors and turbulence modeling errors. The discretization errors should be separated from the turbulence modeling errors. If constitutive equations are not an issue, and if the agreement with a very good experiment is complete, I suppose I would have to relent. But I would still maintain that a *better* paper would result if two grids or some other grid convergence testing were used. For one thing, it would give some idea of how difficult the problem is numerically. (For example, perhaps the results presented were on 100×100 and a 30×30 grids might have been sufficient.) Also, such an exercise may help to verify the code, and would also verify the order of convergence of the code. As Blottner [9] demonstrated two decades ago, plausible second-order, boundary-layer approximations in fact do not always behave in a second-order manner, and this rate of convergence can be established by a systematic grid refinement testing.

Objection 7

I do not have any "truncation error" in my solution since I use finite-element methods (FEM).

Sure you do! (Whether or not the Taylor series is used in the *derivation* of a discrete method, it can still be used in an *analysis* of the method.) But it would have been better in the *JFE* policy statement if we had used the more general term of "discretization error." (Also, "truncation error" strictly speaking is not applicable in the presence of discontinuities.) I would also note that the commonly referred to FEM "error evaluation" practice of substituting the basis function and the solution values into the original partial differential equation is perhaps a valid index of discretization error, but it is not absolutely the error evaluation that we ultimately want. That is, this procedure does not tell us the difference between our discrete solution and the exact solution.

Objection 8

It should not be necessary to legislate such a requirement for doing conscientious work.

It should not have been necessary, but it was. The need is indicative of the problem; that is always the case for legislation, rules, formal ethics committees, and so forth. If a 25th century historian were to read through Albuquerque newspapers from 1989 he would find Sunwest Bank advertising that it has been "Safe, Strong, and Sound for 65 Years." The historian would rightly conclude that the U. S. banking industry in 1989 was in trouble. He would also read a lot about congressional ethics committees and rightly conclude that political ethics were somewhat lacking. And if he read our JFE policy statement on the control of numerical accuracy, he would rightly conclude that there was a bit of sloppy work going on. As Professor George Raithby wrote to me after Ref. 1 was published, "The need to legislate that authors check accuracy is sad, but the enforced discipline will benefit the field."

III. DIFFICULTIES IN APPLICATIONS

Generally speaking, the response from the reviewers has been positive. Also, the authors have been willing to perform the extra work required. The existence of the policy statement has helped since both the reviewers and the editors can refer to the policy statement and thus avoid rehashing the arguments with the authors. However, there were some particular difficulties.

1. The policy statement was phrased in such a way as to allow the editor to reject papers outright (without bothering reviewers) for failure to address the numerical accuracy issue. In practice, this seemed too severe and would have led to excessive publication lag since it would have put the numerical accuracy review/response in series with the other communications. Consequently, I often did initiate the review process even on papers that clearly did not abide by the policy. This also made it easier to be firm with the authors later, since I had the additional moral support of the reviewer's comments.
2. Some authors chose to abide by the letter of the law in the policy statement with generic statements that could be inserted into any paper, such as "the results from different grid resolutions were compared, and a 13×13 grid was shown to be adequate." Although our policy statement was not very demanding, I did reject such "trust me" statements and insisted on some quantifiable grid convergence results. As someone put it, "Convergence lies in the eyes of the beholder." I agree, but that is exactly the point of requiring some quantifiable criterion; it allows the reader to decide for himself. It would be naive in the extreme to think that there is any consensus agreement on what constitutes adequate convergence testing. We were not requiring a priori any particular method or any particular quantitative measure of convergence, but it is necessary to supply the reader with some *numbers*. Note that compliance with this requirement does not necessarily guarantee acceptance of the paper; it is still up to the reviewers and editors to exercise their own professional judgment on the adequacy of the quantified convergence criteria.
3. Some authors would prefer to refer to other papers on similar problems for the grid convergence test. This approach is legitimate in theory, and sometimes in practice, but for reasons noted in the previous paragraph, it is difficult and problematical. It is certainly preferable to have every paper self-contained in this regard.

Some of the papers submitted and accepted did *not* display very convincing convergence results. This again alludes to the difference of opinions mentioned previously, and it is a difficult subject. During my tenure as associate editor for *JFE, I* was quite liberal in accepting papers as long as the grid convergence numbers were provided so that the reader could judge for him- or herself. There are good and bad, sensitive and insensitive indicators of convergence. As noted earlier, it is possible to be fooled by only looking at a neighboring problem. An obvious example is that of Poiseuille flow. If one tries to answer the question of "how many points are required to be in a boundary layer on an aerodynamics problem" by considering the neighboring problem of two-dimensional Poiseuille flow, one quickly comes to the conclusion that a single point in the boundary layer gives perfect accuracy! The obvious reason is that the solution to two-dimensional Poiseuille flow is a parabola, and therefore any second-order finite-difference method will give the exact answer.

Likewise, one can erroneously conclude that a uniform mesh is optimal [10]. A not-so-obvious bad indicator of grid convergence is the vortex wake solution. Many authors have "pointed with pride" to their quite accurate predictions (compared to experiment) of the Strouhal number for the near wake of a circular cylinder. It is apparently not very widely recognized that this is an easy calculation, because the frequency of the von Kármán vortex wake is essentially an inviscid phenomena, and therefore the accuracy of its prediction is not a good indicator of the accuracy of predicted boundary layer heat transfer, friction drag, or wake decay.

IV. EXAMPLES OF WHAT CAN BE DONE

I would like to point out a few exemplary publications addressing the question of the control of numerical accuracy. I have already mentioned the turbulence calculations of Shirazi and Truman [3], Schonauer's ambitious approach [5], and the work of Shih et al [8] in which both the accuracy of the code (i.e., freedom from coding errors) and its second-order convergence for a particular problem are convincingly demonstrated. Another paper, by Kuruvila and Anderson," is nicely illustrative of the difficulties and pitfalls of doing convergence studies with artificial dissipation terms in the equations. The paper by Fujii et al [12] was illuminating in demonstrating the importance of grid resolution for Euler equation solutions to the leading edge separation problem. Thareja et al [13] present solution-adaptive, nonstructured finite-element solutions of the supersonic blunt body problem, which include, in the methodology, the control of numerical accuracy. Dietrich et al [14] present systematic grid truncation error testing of ocean circulation codes with four different methods. This procedure is so rare in the geophysical community as to be virtually nonexistent. The paper by Durst and Pereira [15] demonstrates that the entire procedure is readily applicable to time dependent problems as well as steady state problems. Blottner's paper [16] demonstrates his thorough tests for hypersonic nose tips. Nguyen and Maclaine-Cross [17] use Richardson extrapolation to zero mesh size to produce reliable curve fits to incremental pressure drop number in heat exchangers from full Navier-Stokes solutions. Finally, any such list, no matter how fragmentary, must include the classic study of the natural convection benchmark problem by de Vahl Davis [18].

V. CONCLUSIONS AND RECOMMENDATIONS

The control of numerical accuracy as addressed in the *Journal of Fluids Engineering* policy statement is a necessary but not sufficient component of the broader problem of code validation and verification. It is important to realize, however, that a code can be validated and verified and may be indeed certifiable as "error free," but this does not obviate the need for systematic grid convergence testing (or other systematic, quantitative error estimation) in any new application.

How far should we go? In research journals, I think a broad application of some numerical accuracy control philosophy would do much to improve the quality of published work. (Also, by putting a burden of more work on researchers, it would probably cut down on the number of papers. I, for one, would shed no tears.) In the long run, I would also be in favor of certification of CFD commercial codes. The American National Standards Institute and the American Society of Testing Materials are already involved in such certification activities in many areas, including FORTRAN compilers. I have some doubts, however, regarding the ability of such organizations to handle certification for complex CFD codes. It is probably worthwhile to consider now that AIAA and ASME may be active in the area of CFD code certification (and perhaps *de*certification) in the future. I hasten to add that this is *not* the kind of work that I would personally enjoy or am qualified to do. But, like it or not, I can envision seeing, in my lifetime, CFD code certification by professional societies or joint committees thereof. Note that CFD code certification was first suggested by Mehta [19].

I do not particularly like the idea of CFD code certification, and perhaps many readers do not either. But *users* do, and if we do not involve ourselves, I think we will find ourselves frozen out. Clearly, users of CFD codes (not fellow CFD researchers, but real engineers interested in building something) want something that they can use with ease and confidence. Without a certification program, I am confident that simple "market forces" will not enforce high standards, but in fact will strongly favor pseudo-robust codes. By "pseudo-robust" I mean a code that appears to give a reasonable answer to most any problem. A truly robust code, which really does give a reasonably accurate answer to most any problem, is most difficult to achieve as we all know. From the user's point of view, certification is extremely important. In my own experience, it is a major consideration in codes for ground water hydrology studies. Older codes, with perhaps a 20-year history and correspondingly archaic FORTRAN styles, are inefficient, clumsy to use, unreadable, and therefore virtually un-modifiable; nevertheless, they are certified and they are the standards of performance for new codes.

Available commercial codes certified by some agency or professional society must include attempts to guard against misuse. It is of course impossible to fulfill this goal completely. In fact, the experience of many of our professional colleagues suggest that it is well nigh hopeless. (I asked a professional colleague at a national laboratory if his code was publicly available, and he said no, but he would give it to me, because he knew or had a reasonable expectation that I would not misuse it.) CFD codes are much more difficult than linear algebra packages or statistical packages, but there are plenty of horror stories regarding the misuse of even these more straightforward codes. The reader may be of the opinion that we should not even turn loose on the world a general-purpose CFD code, but rather leave it in the hands of us "experts," at least until we can agree amongst ourselves. Unfortunately this agreement is not forthcoming, and in the meantime, commercial codes are already available (none of which is very impressive, in my opinion).

One thing that we (professionally) could require is that any code, in order to be certified, have built into it automatic error estimators, *not* as user options, but as hard-wired additional output. Like the experimental error bars on data, the error estimates can be ignored or incorporated by the practicing engineer according to his judgment, but he must be apprised of the estimates. There is a natural human desire for easy and unambiguous answers; it is our professional obligation to refuse to give them, in spite of "market forces," which I do not hold to be sacrosanct.

APPENDIX: EDITORIAL POLICY STATEMENT ON THE CONTROL OF NUMERICAL ACCURACY†

A professional problem exists in the computational fluid dynamics community and also in the broader area of computational physics. Namely, there is a need for higher standards on the control of numerical accuracy.

The numerical fluid dynamics community is aware of this problem but, although individual researchers strive to control accuracy, the issue has not to our knowledge been addressed collectively and formally by any professional society or journal editorial board. The problem is certainly not unique to the *JFE* and came into even sharper focus at the 1980–81 AFOSR-HTTM-Stanford Conference on Complex Turbulent Flows. It was a conclusion of that conference's Evaluation Committee‡ that, in most of the submissions to the conference, it was impossible to evaluate and compare the accuracy of different turbulence models, since one could not distinguish physical modeling errors from numerical errors related to the algorithm and grid. This is especially the case for first order accurate methods and hybrid methods.

The practice of publishing comparisons based on coarse grid solutions, without systematic truncation error testing, may have been acceptable in the past. Certainly 10–15 years ago any calculation was of interest, and much of the exploratory work deserved publication, as many researchers lacked the computational power or funds to do a thorough and systematic error estimation. We are of the opinion that this practice, however understandable in the past, is outmoded and that, with powerful computers becoming more common, standards should be raised. Consequently, this journal hereby announces the following policy:

The Journal of Fluids Engineering will not accept for publication any paper reporting the numerical solution of a fluids engineering problem that fails to address the task of systematic truncation error testing and accuracy estimation.

Although the formal announcement of this journal policy is new, it has been the practice of many of our conscientious reviewers. Thus the present announcement is not a change in policy so much as a clarification and standardization.

† Originally published by the American Society of Mechanical Engineers as "Editorial Policy Statement on the Control of Numerical Accuracy," by P. J. Roache, K. N. Ghia, and F. M. White, in the *ASME Journal of Fluids Engineering*, Vol. 108, No. 1, 1986; reprinted with permission.

‡ Emmons, H. W. (Chairman), "Evaluation Committee Report," pp. 979-986 in *Proc. 1980-81 AFOSR-HTTM-Stanford Conference on Complex Turbulent Flows*, Vol. 11, "Taxonomies, Reporter's Summaries, Evaluation, and Conclusions," Thermosciences Division, Mechanical Engineering Department, Stanford University.

Methods are available to accomplish this task, such as Richardson Extrapolation (when applicable), calculations with a high- and low-order method on the same grid, and straightforward repeat calculations with finer or coarser grids. As in the case of experimental uncertainty analysis, "...any appropriate analysis is far better than none as long as the procedure is explained."[§] Whatever the authors use will be considered in the review process, but we must make it clear that *a single calculation in a fixed grid will not be acceptable,* since it is impossible to infer an accuracy estimate from such a calculation. Also, the editors will not consider a reasonable agreement with experimental data to be sufficient proof of accuracy, especially if any adjustable parameters are involved, as in turbulence modeling.

We recognize that it can be costly to do a thorough study, and that many practical engineering calculations will continue to be performed on a single fixed grid. However, this practice is insufficient for publication in an archival journal.

ACKNOWLEDGMENTS

This work was partially supported by contracts from the Air Force Weapons Laboratory (Dr. N. L. Rapagnani, contract monitor), the Naval Underwater Systems Center (S. Jordan, contract monitor), Sandia National Laboratories in Albuquerque (Dr. M. Marietta, contract monitor), and by a grant from the National Science Foundation.

REFERENCES

1. Roache, P. J., Ghia, K. N., and White, F. M., "Editorial Policy Statement on the Control of Numerical Accuracy," *ASME Journal of Fluids Engineering,* Vol. 108, No. 1, 1986, p. 2.

2. Emmons, H. W. (Chairman), "Evaluation Committee Report," *Proceedings of the 1980–81 AFOSR–HTTM–Stanford Conference on Complex Turbulent Flows, Taxonomies, Reporter's Summaries, Evaluation, and Conclusions, Vol.* 11, Thermosciences Division, Mechanical Engineering Department, Stanford University, Stanford, CA, 1981, pp. 979–986.

3. Shirazi, S. A., and Truman, C. R., "Evaluation of Algebraic Turbulence Models for PNS Predictions of Supersonic Flow Past a Sphere-Cone," *AIAA Journal,* Vol. 27, No. 5, 1989, pp. 560–568.

4. Cheng, S. I., "Numerical Integration of Navier-Stokes Equations," *AIAA Journal,* Vol. 8, Dec. 1970, pp. 2115–2122.

5. Schonauer, W., private communication; see also, Schonauer, W., Raith, K., and Glotz, G., "The Principle of the Difference of Difference Quotients as a Key to the Self-Adaptive

---

[§] Kline, S. J., "The Purposes of Uncertainty Analysis," *Journal of Fluids Engineering,* Vol. 107, June 1985, pp. 153-164.

Solution of Nonlinear Partial Differential Equations," *Computer Methods in Applied Mechanics and Engineering,* Vol. 28, No. 3, 1981, pp. 327–359.

6. Mehta, U., "Computational Requirements for Hypersonic-Flight Performance Estimates," AIAA Paper 89–1670, June 1989.

7. Steinberg, S., and Roache, P. J., "Symbolic Manipulation and Computational Fluid Dynamics," *Journal of Computational Physics,* Vol. 57, No. 30, 1985, pp. 251–283.

8. Shih, T. M., Tan, C. H., and Hwang, B. C., "Effects of Grid Staggering on Numerical Schemes," *International Journal for Numerical Methods in Fluids,* Vol. 9, No. 2, 1989, pp. 193–212.

9. Blottner, F. G., "Finite Difference Methods of Solution of the Boundary Layer Equations," *AIAA Journal,* Vol. 8, No. 2, 1970, pp. 193–199; see also, "Finite-Difference Methods for Solving the Boundary Layer Equations with Second-Order Accuracy," *Proceedings of the Second International Conference on Numerical Methods in Fluid Dynamics,* Lecture Notes in Physics, Vol. 8, edited by M. Holt, Springer-Verlag, New York, 1971.

10. Blottner, F. G., and Roache, P. J., "Nonuniform Mesh Systems," *Journal of Computational Physics,* Vol. 8, No. 3, 1971, pp. 498–499.

11. Kuruvila, G., and Anderson, J. D., "A Study of the Effects of Numerical Dissipation on the Calculation of Supersonic Separated Flows," AIAA Paper 85–0301, Jan. 1985.

12. Fujii, K., Gavali, S., and Holst, T. L., "Evaluation of Navier-Stokes and Euler Solutions for Leading-Edge Separation Vortices," *International Journal for Numerical Methods in Fluids,* Vol. 8, 1988, pp. 1319–1329.

13. Thareja, R. R., Stewart, J. R., Hassan, O., Morgan, K., and Peraire, J., "A Point Implicit Unstructured Grid Solver for the Euler and Navier-Stokes Equations," *International Journal for Numerical Methods in Fluids,* Vol. 9, 1989, pp. 405–425.

14. Dietrich, D. E., Roache, P. J., and Marietta, M. G., "Convergence Studies with the Sandia Ocean Modeling System," *International Journal for Numerical Methods in Fluids* (to be published).

15. Durst, F., and Pereira, J. C. F., "Time-Dependent, Backward Facing Step Flow," *ASME Journal of Fluid Engineering,* Vol. 110, No. 3, 1988, pp. 289–296.

16. Blottner, F. G., "Accurate Navier-Stokes Results for the Hypersonic Flow over a Spherical Nosetip," AIAA Paper 89–071, June 1989.

17. Nguyen, T. V., and Maclaine-Cross, 1. L., "Incremental Pressure Drop Number in Parallel-Plate Heat Exchangers," *ASME Journal of Fluids Engineering,* Vol. I 10, No. 1, 1988, pp. 93–96.

18. de Vahl Davis, G., "Natural Convection of Air in a Square Cavity: A Bench Mark Numerical Solution," *International Journal for Numerical Methods in Fluids,* Vol. 3, No. 3, 1983, pp. 249–264.

19. Mehta, U., "Flight Performance Estimation Utilizing Computational Fluid Dynamics," *Proceedings of the Fifth National Aero-Space Plane Technology Symposium,* NASA Langley Research Center, Hampton, VA, Oct. 1988.

APPENDIX B. OTHER JOURNAL POLICY STATEMENTS ON CONTROL OF NUMERICAL ACCURACY

Chapter 1 (see also Appendix A) contains a reprint of the first (1986) journal policy statement on the subject of control of numerical accuracy, from the *ASME Journal of Fluids Engineering*. This appendix contains four others, from the *ASME Journal of Heat Transfer*, the *International Journal for Numerical Methods in Fluids*, the *AIAA Journal*, and a later (1993) more specific version from the *ASME Journal of Fluids Engineering*.

JOURNAL OF HEAT TRANSFER EDITORIAL POLICY STATEMENT ON NUMERICAL ACCURACY

Editorial Board (1994), "Journal of Heat Transfer Editorial Policy Statement on Numerical Accuracy," *ASME Journal of Heat Transfer*, Vol. 116, November 1994, pp. 797–798.

The purpose of this statement is to establish and clarify acceptable standards for the publication of numerical results, and supplements the previously established JOURNAL OF HEAT TRANSFER Policy on Reporting Uncertainties in Experimental Measurements and Results [1]. In developing the current policy, the policy statement on control of numerical accuracy adopted by the *ASME Journal of Fluids Engineering* [2] was consulted, as well as similar policy statements currently under consideration by the AIAA journals and the *International Journal for Numerical Methods in Fluids*. We appreciate the willingness of those involved with the development of these latter policies to share their thinking with us and to allow our adoption of specific elements of their policy statements.

Succinctly stated, the editorial policy of the Journal of Heat Transfer regarding the review and publication of numerical studies is as follows:

The Journal of Heat Transfer will not accept for review or publication any manuscript reporting the numerical solution of a heat transfer problem that fails to establish adequately the accuracy of the computed results.

The implementation of this policy will be at the discretion of the editor and associate editors in association with the reviewers, and will be guided, in essence, by the considerations set forth below and by the view attributed to Kline [3] in the context of experimental uncertainty analysis, that "...any appropriate analysis is far better than none as long as the procedure is explained."

To be specific, all manuscripts submitted for review that include numerical simulations must contain the following essential elements:

- A problem statement that is of sufficient clarity and completeness to allow the reproduction of the results by informed readers. As a minimum, this should include a statement of the governing equations solved, all relevant boundary and initial conditions, and values of associated physical and numerical parameters. Values for any adjustable or arbitrary parameters employed to obtain the solution must be explicitly given.
- A description of the solution technique employed. If a standard method is used. the description can be via reference to appropriate prior publications. If a new method is introduced, the description must be sufficiently complete to allow implementation of the numerical scheme and replication of relevant results by informed readers.
- The numerical solution must be supplemented with acceptable accuracy estimates for both the method employed and the results presented. A single calculation using a fixed discretization will not be acceptable, since no error estimate can be inferred from such a calculation.
- Authors may use any appropriate method for the estimation of errors. One, or more, of the following approaches may be useful in this regard:
- Comparison of numerical results with those from a sufficiently similar model problem available in the literature, possessing a known exact or highly accurate approximate analytical solution; or with an established, high-accuracy, fine-grid, numerical benchmark solution of the same, or closely similar, problem.
- A precisely defined and documented and refinement or grid coarsening study. Marginal refinement showing a qualitative convergence trend is not acceptable. Other numerical and arbitrary parameters such as time step, convergence criterion, and boundaries of the computational domain should also be varied to ensure that the results are independent of these quantities.
- Comparison with reliable experimental results that possess an associated established uncertainty. Noting "reasonable agreement" with experimental data is not, in general,

sufficient justification for acceptance of numerical results, especially when adjustable parameters are involved. Numerical and experimental results may be plotted or tabulated to indicate the level of agreement.

In the approaches described above references to grid refinement are intended to be interpreted in a general sense to include numerical methods that are not explicitly dependent on a computational grid in the manner typically associated with, say, finite difference methods. For example, grid-free methods would require modification of the number, or size, of discrete elements of the computation to establish error estimates. An additional comment is in order regarding comparisons of numerical results with experimental data. It is recognized that, when the results of a numerical simulation are compared with experimental data, it may not be possible to separate errors in modeling from those associated with the numerical method. In these situations, a separate estimate of the numerical error should be established. In rare situations where none of the above approaches can be satisfactorily employed to establish the accuracy of the results, other methods of error estimation acceptable to the editor, associate editors, and reviewers may be appropriate. For an illustration of many of the elements involved in establishing the accuracy of a numerical solution, the study of natural convection reported by de Vahl Davis [4], or the method discussed by Roache [5] for the uniform reporting of grid refinement studies, may be consulted.

By implementation of this policy, it is the intent of the editorial board to establish guideline requirements for the publication of numerical results and to enhance the quality of publications involving numerical simulations. It is not our intent to effect a significant increase in the length of papers published in the journal, or to impose excessive requirements on prospective authors. Rather, we hope to elicit a "good faith" effort from authors to establish the accuracy of their numerical simulations.

The Editorial Board

1. *ASME Journal of Heat Transfer*. Vol. 115, February 1993, pp. 5–6

2. *ASME Journal of Fluids Engineering*, Vol. 115, September 1993, pp. 339–340

3. Kline, S. J., 1985, "The Purposes of Uncertainty Analysis," *ASME Journal of Fluids Engineering*, Vol. 107, pp. 153–164.

4. de Vahl Davis, G., 1983, "Natural Convection of Air in a Square Cavity: A Bench Mark Numerical Solution," *Int. J. Numer. Methods Fluids,* Vol. 3, pp. 249–264.

5. Roache, P. J., 1993, "A Method for Uniform Reporting of Grid Refinement Studies," *Quantification of Uncertainty in Computational Fluid Dynamics*, ASME FED–Vol. 158, pp. 109–120.

INTERNATIONAL JOURNAL FOR NUMERICAL METHODS IN FLUIDS–EDITORIAL

Gresho, P. M. and Taylor, C. (1994), "Editorial," *International Journal for Numerical Methods in Fluids*, Vol. 19, 1994, p. iii.

In order to increase the Journal's value to its readership, we will raise the quality of published papers by requiring the authors to include two key features in submitted papers beginning 1 January 1995.

1. The problem statement must be sufficiently clear and complete so that the results presented could be essentially reproduced by an interested reader. This will require (at least) a statement of the governing equations, all relevant boundary conditions, initial conditions, and values of associated physical parameters. If a new method is being introduced, its description must be sufficiently clear that others could implement it and replicate the presented results.
2. The numerical solution must be supplemented with an acceptable accuracy estimation of both the method employed (general) and the results presented (specific).

The accuracy estimate may be based on one or more of the following:

- Comparison with the results of using the method on a *sufficiently similar* model problem with a known exact or highly accurate approximate analytical solution (such as a multi-term series expansion, for example) or an *established,* high-accuracy, very-fine-grid numerical solution (benchmark) of the same (or similar) problem.
- A serious (meaningful) grid-refinement study, possibly coupled with Richardson Extrapolation when appropriate. Marginal refinement showing a qualitative convergence trend is not acceptable. For 3D problems, it may sometimes be the case that only grid *coarsening* is viable for error estimation.
- For methods using either inherent or explicitly added artificial viscosity (or diffusivity), an estimate of the relative magnitude of these terms or the corresponding artificial grid Reynolds (or Péclet) numbers. This is *especially* important for methods based on switching or blending strategies using combinations of first- and higher-order methods.
- Any other method acceptable to the Editors via their referees.

The error estimate should not, in general, be based simply on "reasonable agreement" with experimental data, especially if any adjustable empirical parameters are involved.

Realizing that computer facilities and budgets are often not "adjustable parameters" for the author, the Editors and their referees will entertain arguments as to why quantitative error estimates are *not* provided—which arguments should be stated in the paper, not in a plea-bargain letter of transmittal.

Determining how well authors meet/achieve the above criteria will be at the discretion of the Editors and will be based, in part, on advice from the referees.

P. M. Gresho
C. Taylor

AIAA EDITORIAL POLICY STATEMENT ON NUMERICAL ACCURACY AND EXPERIMENTAL UNCERTAINTY

Anon. (1994), Editorial Policy Statement on Numerical Accuracy and Experimental Uncertainty, *AIAA Journal*, Vol. 32, No. 1, January 1994, p. 3.

The purpose of this statement is to reiterate the desire to have high quality investigations with properly documented results published in the AIAA journals, and to clarify acceptable standards for presentation of numerical and experimental results. Recently there has been considerable concern with the quality of published numerical solutions. Also the practice of including error bars on experimental results is often lacking. In response to these problems, a succinct policy statement on these items is as follows:

The AIAA journals will not accept for publication any paper reporting (1) numerical solutions of an engineering problem that fails adequately to address accuracy of the computed results or (2) experimental results unless the accuracy of the data is adequately presented

The implementation of this policy will be at the discretion of the Editors and Associate Editors of the journals.

The accuracy of the computed results is concerned with how well the specified governing equations in the paper have been solved numerically. The appropriateness of the governing equations for modeling the physical phenomena and comparison with experimental data is not part of this evaluation. Accuracy of the numerical results can be judged from grid refinement studies, variation of numerical parameters that influence the results, comparison with exact solutions, and any other technique the author selects. The validity of the accuracy estimation will be judged by the reviewers of the paper. An estimate of accuracy of the numerical results must be presented when comparisons with other numerical and experimental results are given, and when new results of the author will likely become data for future comparisons. Since accuracy of various computed results obtained from a numerical solution can vary significantly, the accuracy of the result being used must be stated. Accuracy of results from a validated code must still be established to show that proper input parameters have been used with the code.

Estimates of experimental uncertainty are required for all plotted or tabulated data obtained by authors. If data from other workers are used, they require no uncertainty. Unless otherwise stated and properly referenced, it is assumed that the uncertainty of authors' output data is estimated by the small-sample method [l] with assumed odds 20:1. All reported data must show uncertainty estimates if used in text or tables; for example, $T = 642 \pm 8$ K. All figures reporting new data should contain uncertainty estimates either on the figure with error bars in both coordinate directions or in the caption; for example, uncertainty in T = ±8 K at 20:1 odds. Investigations with limited data should present tabulated results in the paper while extensive data should be available elsewhere in tabulated form for use by other workers.

Finally, the accepted documentation procedures for a technical investigation must be used. For computational papers, the author must provide an adequate description of the numerical solution procedure, if not documented elsewhere. In addition, the complete governing equations must be specified with sufficient detail along with the input parameters to the code so that a reader could reproduce the results of the paper. For papers concerned with experimental test, thorough documentation of the experimental conditions, instrumentation, and data reduction techniques is required.

Reference

1. Kline, S. J., and McClintock, F. A., "Describing Uncertainties in Simple-Sample Experiments," *Mechanical Engineering*, Jan 1953, pp. 3–8.

JOURNAL OF FLUIDS ENGINEERING–EDITORIAL AND POLICY STATEMENT

Freitas, C. J. (1993), "Editorial Policy Statement on the Control of Numerical Accuracy," *ASME Journal of Fluids Engineering*, Vol. 115, No. 2, June 1993, p. 339.

The continued advancements in high performance computing, both in hardware and software, and the cost competitive advantage of numerical simulations over laboratory experiments have made computational fluid dynamics an integral tool in the study of science and engineering problems. However, unlike experimental methods for which a large body of knowledge and techniques exist for the evaluation of experimental error and uncertainty (and are widely accepted), equivalent techniques for the evaluation of numerical error and uncertainty are less well developed and accepted. A recent article in the *Journal of Fluids Engineering* (Celik, 1993) highlighted this point and identified possible reasons for both the lack of interest in this topic and lack of accepted methods for the evaluation of uncertainty by the computational science community. Celik went on to identify three major topics of relevance to the issue of numerical accuracy and uncertainty. These are: (i) the separation of numerical errors from modeling errors, (ii) the identification, estimation, and reduction of numerical errors; and (iii) the assessment of codes and computational schemes with respect to numerical uncertainty through benchmarking. As Celik indicated, there is the need to improve the quality of the large number of papers being published today in computational fluid dynamics, and thus there is an urgent need for implementing a policy regarding numerical uncertainty analysis or quantitative error estimation.

The Fluids Engineering Division (FED) of the ASME, through the Coordinating Group on Computational Fluid Dynamics (CGCFD), has taken on the task of leading the computational fluid dynamics community and focusing their attention on the formulation of reasonable measures for numerical accuracy. The CGCFD has by charter the responsibility to promote discussion and interest in research into all areas of computational fluid dynamics and principal among these are methods for the evaluation of numerical accuracy. The CGCFD has performed its function by conducting symposia, forums, and panel discussions addressing this complex topic. The objective here is to delineate standard practices by which computational studies may be performed and the standards by which archival publications will be gauged. As a result of these meetings and discussions a new level of standards for the evaluation of journal publications has been promulgated for the *Journal of Fluids Engineering*.

The *Journal of Fluids Engineering* has had the policy that it: *will not accept for publication any paper reporting the numerical solution of a fluids engineering problem that fails to address the task of systematic truncation error testing and accuracy estimation*. This policy statement, originally presented in Roache, Ghia, and White (1986), was the first of its kind and provided a very general standard for evaluating journal publications. In the seven years since this policy's implementation, significant advances in computer hardware and computational software have occurred such that CFD is no longer in its infancy, but is a full-fledged tool in engineering and scientific problem solving. As Roache, Ghia, and White (1986) point out, fifteen years before the publication of their editorial, any successful calculation was of interest, and much of this exploratory work deserved publication. But even in their time it was recognized that this practice was outmoded. Therefore, it seems only logical and scientifically correct that the CFD community meet higher standards for evaluation of accuracy, standards on par with those required of the experimental community. What follows below then, are refinements and enhancements to the Journal's original statement on numerical accuracy, which attempt to elucidate the criteria by which Journal papers will be judged.

Finally, it is not the intent of this new policy statement to eliminate a class of simulations which some have referred to as "practical engineering project simulations." The justification by these individuals for performing a single grid simulation has been that budget constraints, schedule constraints, or computer resource constraints prevent a systematic analysis of accuracy from being performed. It is assumed that in performing CFD analyses for "practical engineering projects," for which experimental data is usually not available, that one must perform, in the natural course of the project, an evaluation of the accuracy of the simulation results in order to determine the validity of these particular calculations. Without such an effort there is no clear justification for presenting a simulation as representative of the physical phenomena. Therefore, it would seem only natural, even in the solution of practical engineering problems, that the items addressed here, be used to validate a simulation.

The author wishes to acknowledge the comments and suggestions to this new policy statement from Prof. B. P. Leonard (University of Akron), Dr. P. J. Roache (Ecodynamics), Dr. Ramesh Agarwal (McDonnell Douglas), Prof. Oktay Baysal (Old Dominion University), and Prof Demetri Telionis (Virginia Polytechnic Institute and State University).

The guidelines on the following page have been approved by the FED Coordinating Group on Computational Fluid Mechanics and the Editorial Board of the Journal.

Dr. Christopher J. Freitas
Chairman of the Coordinating Group
on Computational Fluid Dynamics

POLICY STATEMENT ON THE CONTROL OF NUMERICAL ACCURACY

Although no standard method for evaluating numerical uncertainty is currently accepted by the CFD community, there are numerous methods and techniques available to the user to accomplish this task. The following is a list of guidelines enumerating the criteria to be considered for archival publication of computational results in the *Journal of Fluids Engineering*

1. Authors must be precise in describing the numerical method used; this includes an assessment of the formal order of accuracy of the truncation error introduced by individual terms in the governing equations, such as diffusive terms, source terms, and most importantly, the convective terms. It is not enough to state, for example, that the method is based on a "conservative finite-volume formulation," giving then a reference to a general CFD textbook.

2. The numerical method used must be at least formally second-order accurate in space (based on a Taylor series expansion) for nodes in the interior of the computational grid. The computational expense of second, third, and higher order methods are more expensive (per grid point) than first order schemes, but the computational efficiency of these higher order methods (accuracy per overall cost) is much greater. And, it has been demonstrated many times that, for first order methods, the effect of numerical diffusion on the solution accuracy is devastating.

3. Methods using a blending or switching strategy between first and second order methods (in particular, the well-known "hybrid," "power-law," and related exponential schemes) will be viewed as first-order methods, unless it can be demonstrated that their inherent numerical diffusion does not swamp or replace important modeled physical diffusion terms. A similar policy applies to methods invoking significant amounts of explicitly added artificial viscosity or diffusivity.

4. Solutions over a range of significantly different grid resolutions should be presented to demonstrate grid-independent or grid-convergent results. This criterion specifically addresses the use of improved grid resolution to systematically evaluate truncation error and accuracy. The use of error estimates based on methods such as Richardson Extrapolation or those techniques now used in adaptive grid methods, may also be used to demonstrate solution accuracy.

5. Stopping criteria for iterative calculations need to be precisely explained. Estimates must be given for the corresponding convergence error.

6. In time-dependent solutions, temporal accuracy must be demonstrated so that the spurious effects of phase error are shown to be limited. In particular, it should be demonstrated that unphysical oscillations due to numerical dispersion are significantly smaller in amplitude than captured short-wavelength (in time) features of the flow.

7. Clear statements defining the methods used to implement boundary and initial conditions must be presented. Typically, the overall accuracy of a simulation is strongly affected by the implementation and order of the boundary conditions. When appropriate, particular attention should be paid to the treatment of inflow and outflow boundary conditions.

8. In the presentation of an existing algorithm or code, all pertinent references or other publications must be cited in the paper, thus aiding the reader in evaluating the code and its method without the need to redefine details of the methods in the current paper. However, basic features of the code must be outlined according to Item 1, above.

9. Comparison to appropriate analytical or well-established numerical benchmark solutions may be used to demonstrate accuracy for another class of problems. However, in general this does not demonstrate accuracy for another class of problems, especially if any adjustable parameters are involved, as in turbulence modeling.

10. Comparison with reliable experimental results is appropriate, provided experimental uncertainty is established. However, "reasonable agreement" with experimental data alone will not be enough to justify a given single-grid calculation, especially if adjustable parameters are involved.

These ten items lay down a set of criteria by which the editors and reviewers of this Journal will judge the archival quality of publications dealing with computational studies for the *Journal of Fluids Engineering*. We recognize that the effort to perform a thorough study of numerical accuracy may be great and that many practical engineering calculations will continue to be performed by first order methods, on a single fixed grid. However, such analyses would not be appropriate for presentation in this archival journal. With the gains in performance of low-end workstations, it is now reasonable to require papers on solutions by CFD to meet these fundamental criteria for archiving of a publication.

With the details of these ten criteria now presented, a shortened statement will appear in each volume of the journal. This statement will appear as follows:

The Journal of Fluids Engineering will not consider any paper reporting the numerical solution of a fluids engineering problem that fails to address the task of systematic truncation error testing and accuracy estimation. Authors should address the following criteria for assessing numerical uncertainty.

1. The basic features of the method including formal truncation error of individual terms in the governing numerical equations must be described.

2. Methods must be at least second order accurate in space.
3. Inherent or explicit artificial viscosity (or diffusivity) must be assessed and minimized.
4. Grid independence or convergence must be established.
5. When appropriate, iterative convergence must be addressed.
6. In transient calculations, phase error must be assessed and minimized.
7. The accuracy and implementation of boundary and initial conditions must be fully explained.
8. An existing code must be fully cited in easily available references.
9. Benchmark solutions may be used for validation for a specific class of problems.
10. Reliable experimental results may be used to validate a solution.

References

Celik, I., 1993, "Numerical Uncertainty in Fluid Flow Calculations: Needs for Future Research, *ASME Journal Of Fluids Engineering*, Vol. 115, June 1993, pp. 194–195.

Roache, P. J., Ghia, K. N., and White, F. M., 1986, "Editorial Policy Statement on the Control of Numerical Accuracy," *ASME Journal Of Fluids Engineering*, Vol. 108, Mar. 1986, pp. 2.

APPENDIX C. COMMENT ON ORESKES et al

The following is the manuscript of a Letter to the Editor of *Science* magazine, rejected for publication in 1994.

COMMENT ON "VERIFICATION, VALIDATION, AND CONFIRMATION OF NUMERICAL MODELS IN THE EARTH SCIENCES"

It is hard to decide where to begin criticizing this article (Oreskes et al, 1994) since it fails on so many levels: as science, as philosophy of science, as logic, as semantics, as engineering, as public policy debate, as common sense. But therein lies a beginning: the article fails to distinguish between these arenas, particularly between engineering and philosophy of science.

The provocative focus is on engineering projects such as the proposed Yucca Mountain radioactive waste disposal site. This is not "science." Like all engineering, it is built on scientific principles, but only at an adequate level (a term that is evidently offensive to the authors). Engineers (and applied scientists, applied mathematicians, lawyers, accountants, sociologists, etc.) are trying to get rid of a public health problem, but the authors are agonizing over the knowability of the physical world, the nature of scientific truth, dictionary definitions of "valid," the possibility of certainty, etc. When the subject is the viability of an engineering project, it is difficult to take seriously people who make the following statements.

"Therefore, no general empirical proposition about the natural world can ever be certain."

"...the establishment that a model accurately represents the 'actual processes occurring in a real system' is not even a theoretical possibility."

"We are left with the conclusion that we can never verify a scientific hypothesis of any kind."

They even seriously consider (in footnote 36) the question of whether we can in fact prove a hypothesis *false*!

Well then, whatever *shall* we do? We obviously cannot put a man on the moon, or use numerical models in oil exploration. But closer to their focus of Earth Sciences vis-à-vis public policy and public safety, we cannot build Coolidge Dam, we cannot write construction codes for California, we cannot predict pollution in our air and water.

In speaking of the Earth Sciences, the authors state "Such models can never be verified. No doubt the same may be said of many biological, economic, and artificial intelligence models." Here the authors are offering a palliative. Surely, following their own philosophical definitions, we need not sweeten the medicine by saying "many" models can never be verified; the right words are "no models," if they tell it like it is, and as well include every other field of engineering and applied science. While everyone else in engineering attributes most of the advances in design capability during the last 10 years to computer modeling, the authors evaluate numerical modeling as virtually unusable anywhere, certainly not for radioactive waste disposal sites. (Then again, since the quantitative predictions of health and safety consequences of radiation are themselves based on numerical models, why worry?)

And if we stopped all nuclear energy production and all weapons production, we would still have tons of radioactive waste to dispose of. What exactly do the authors propose as a course of action? Just how do we "do" engineering in 1994 without numerical models?

For an article so long on philosophy and semantics, the authors somehow fail to address some elementary issues. They never define "model"! To their credit, they seem to understand the distinction between a numerical model and a computer code, but they missed the opportunity to make this important distinction for their readers. Also, they are not semantically careful when they say "[numerical] models require input parameters." Are they sure? Certainly computer codes require input parameters, but many practitioners would include the values of parameters in the definition of "model," as in "We intend to develop a hydrologic model of the lower Rio Grande basin."

Their biggest semantic problems revolve around the terms "verification" and "validation." There are indeed open-literature inconsistencies in the use of the terms "verification" and "validation" *as technical terms*, and practitioners constantly must define their terms. But it does no service to open up the Random House dictionary, as the authors do, and start lecturing practitioners because they are not living up to the most superficial definitions, extrapolated by the authors to legalistic absolutism. *All* terms must be taken in context, even in everyday conversation, and technical terms must be taken in their technical context, which I'll wager are defined (though not uniformly) with their limitations in each of the culprit reports cited. "Verification" and "validation" do not carry the absolute weight imputed by the authors, any more than their own term of choice, "confirmation." While a technical term should not be completely orthogonal to common usage, one cannot decide its technical meaning by studying its etymology and reading a common-use dictionary. What possible relevance to numerical modeling is there in quoting, as the authors do, dictionary definitions of "valid" in the context of logical arguments and legal contracts? Would a lawyer be expected to consult a software manual for background on "valid contracts"?

Perhaps the most laughable semantic aspect of the article is the authors' big punch line, their introduction of the term "confirmation." (To their credit, they did not find relevant the dictionary definition of "confirmation" as "a Christian rite admitting...") "To account for this discrepancy, *philosophers* [my emphasis] have developed a theory of confirmation, founded on the notion of science as a hypothetico-deductive [!] activity." The authors believe that its adoption will clarify things, and avoid the false absolute and affirmative quality that they claim is given by those nasty words "verification" and "validation." Their term "confirmation," we are told, suggests limitations. Alas, when we follow the bad example of the authors and refer to a dictionary or thesaurus (for what should have been defined in a technical context) we find that "confirmation" is synonymous with both "verification" and "validation"!

While criticizing modelers for circular definitions, they circle themselves, introducing terms that are no more clearly defined than the originals; instead of "verification" and "validation," they build their hopes on the undefined "confirmation," and float other undefined terms that do nothing to clarify the issue: "veracity," "weak verification," which they dismiss as an obvious oxymoron but then approvingly mention "probabilistic confirmation," and "certain degree of belief," "confirming," "confirming observations," "support," "corroboration," and of course, "hypothetico-deductive activity."

Ironically, when the authors attempt to invent technical definitions for "verification" and "validation" (which would be acceptable if others had not already done so) they come up with the opposite of common (but by no means universal) definitions: "...[T]he term valid might be useful for assertions about a generic computer code..." But this is what many people define as verification (Rechard et al, 1991). In this use of the terms, a computer code can be verified (meaning, it solved the equations "right"), but can never be validated in general. (How could it? Any code can be mis-applied.) Only a particular calculation or restricted class of calculations with the code can be validated (meaning, it solved the "right" equations for the problem.) The important point is not *which* common word (verification or validation) is given a more precise technical meaning in this specific technical context; the important point is the useful distinction in the activities. To quote older philosophies than those of Oreskes et al (1994): "The purpose of words is to convey ideas. When the ideas are grasped, the words are forgotten." (Merton, 1965)

The runner-up for semantic comedy is their answer to their rhetorical question, "Then what good are models?" One of their uses—"Models can also be used for sensitivity analysis..."

Wrong! *If* the world is really as unknowable as the authors claim, if models cannot be "validated," then they certainly cannot be useful for sensitivity analyses. The authors apparently do not appreciate the consequences of a real problem they discuss, that of nonuniqueness of models. As a matter of practical experience (if I may be so unfair), it is true that a relatively crude model can often be used with enhanced usefulness if it is used only differentially. Unfortunately, it is also true (again, in some real calculations) that a model that gives some kind of agreement with physical observation in an absolute sense can err in sensitivity calculations, and err unequivocally, as in predicting the wrong *direction* of sensitivity. (This should satisfy most anyone that it is "wrong" and not useful— or do I underestimate our philosophers?) And this practical difficulty occurs in more accessible areas than the Earth Sciences, e.g. aerodynamic turbulence modeling.

What the authors strain to introduce is the concept of probabilistic truth, or truth with uncertainty, or truth with accuracy tolerances or error bars, which is evidently new and difficult for them, but is taken for granted by most everyone else. What they fail to introduce is the hallmark of modeling for waste disposal sites, the concept of uncertainty estimation and probabilistic modeling. (It is this concept, plus a common-sense attitude towards accuracy, that allows rational inclusion of variations in parameters which the present authors think invalidates long-term repository modeling, namely climate changes and geological conditions.)

The authors' confusion over scientific method, engineering practice, and public policy definition is no more evident than in their footnotes 28 and 31. For the sake of brevity, consider only the latter. "They [waste repository modelers] propose that a model need only be 'adequate' for a 'given purpose', in this case compliance with federal regulations. But this definition begs the question of whether the regulations are adequate." But isn't it clear that the basic scientific principles, the engineering judgment, the numerical modeling, etc., is a rather distinct arena than regulations, which involve societal *values* of life and health probabilities?

There are indeed real problems (practical, non-academic, serious, uniquely difficult, perhaps even insoluble problems) associated with Earth Sciences, including many mentioned in the article, e.g., model calibration, scaling-up, long-time simulations. (The authors did not mention the difficult problem of establishing initial conditions.) There is indeed a fundamental difficulty with "validation" properly defined (assuring that the code solved the "right" equations) but the concept is still valuable so long as a common-sense engineering attitude is taken with regard to accuracy, uncertainty, and ultimate use. On the other hand, there is no fundamental difficulty, just work to be done, with *code* (not "model") "verification" properly defined (assuring that the code solved the equations "right"). But the realities are much better addressed in some of the references cited in Oreskes et al (1994) (e.g., their references 2, 4, 5, 7, 8, 14, 20) and elsewhere in the literature. When those respected practitioners talk about the real difficulties of numerical modeling, we all listen attentively. But the subject article contributes little new of value, and is actually pernicious in trivializing the real issues by association with pseudo-issues.

To end on a more positive note, I do agree heartily with the authors' criticism of the stated intention of some reports. "And in many cases, a positive result is presupposed. For example, the first step of validation has been defined by one group of scientists as developing 'a strategy for demonstrating [regulatory] compliance' (1,45). Such affirmative language is a roadblock to further scrutiny." And, I would add, a roadblock to credibility. I am sure that this language did not originate with the scientists involved but is DOE management boilerplate, which professional ethics demands should be replaced with something like "a strategy for determining [regulatory] compliance or lack thereof."

Patrick J. Roache, Ecodynamics Research Associates, Inc. P. O. Box 9229, Albuquerque, New Mexico 87119

REFERENCES

Oreskes, N., Shrader-Frechette, K., and Belitz, K. (1994), "Verification, Validation, and Confirmation of Numerical Models in the Earth Sciences," *Science*, Vol. 263, 4 February 1994, pp. 641–646.

Rechard, R. P., Roache, P. J., Blaine, R. L., Gilkey, A. P., and Rudeen, D. K. (1991), "Quality Assurance Procedures for Computer Software Supporting Performance Assessments of the Waste Isolation Pilot Plant," SAND90–1240/UC–721, Sandia National Laboratories, Albuquerque, New Mexico, 1991.

Merton, T. (1965), *The Way of Chuang Tzu*, "Means and Ends," New Directions, New York, 1965, p.154.

APPENDIX D. CFD PARODY: WILL THE WIND TUNNEL REPLACE THE COMPUTER?

The following anonymously authored parody of the tension between Computational Fluid Dynamics (CFD) and Experimental Fluid Dynamics (EFD) appeared in the mid 1980's, with the only indication of origin being "NASA Ames." I have slightly expanded the parody with my own response, 1997.

ORIGINAL

We all know of the importance of computers in today's aerospace engineering environment. The latest advances in cryogenically cooled semi-superconductor technology and microscopic germanium sub-wafer assembly has made desktop 100 MINS (**M**illions of **N**avier-**S**tokes solutions) machines commonplace in engineering use.

We are also aware, however, of the high cost of this aging technology. The most accurate aerodynamic prediction code available today, FLO-1234.5, is so complex and expensive that it has never been run. Many other codes, if run to completion, would require CPU time exceeding the average human lifespan. Most engineers attribute this situation to the time when the task of writing aerodynamic computer programs was automated and handed over to the computers. We now have codes too complex to be understood by any human being. The cost of computing has been rising exponentially over the years. Clearly, if these trends continue unabated, computational solutions will soon be beyond anyone's means.

Fortunately, there is an exciting new technology on the horizon which may someday replace the computer for aerodynamic design and analysis. Two workers at UNCAF (United Nations Computational Aerodynamics Facility) have recently made a startling discovery. They found that by building a small wooden model of an airplane and then blowing air past it in an enclosed tunnel, reasonably accurate predictions may be made of what the flow codes would compute. They refer to the method they have discovered as a "wind tunnel." At present, "wind tunnel" modeling is still in an early and relatively crude stage, and cannot be expected to precisely reproduce numerical results. For example, the continuous surface of a wood or metal airplane model will never exactly duplicate the discrete nature of a computational grid. Also, some factors, such as artificial viscosity, are neglected completely in wind tunnel modeling. It may be especially hard to accurately predict linearized potential flow in the tunnel. Nevertheless, in many cases, the wind tunnel agrees surprisingly well with the computer.

Constructing a wind tunnel model is much quicker and less labor-intensive than running all but the simplest computer programs. Shops such as Minicraft or Static Engineering complete even a highly detailed titanium model in a mere matter of months. Thus, many design iterations and trade-off studies can be conducted in a fraction of the time required via the computer. Advances in wind tunnel technology and model fabrication are expected to proceed at a rapid pace. Many promising new techniques, such as the chiseling of facets in Plaster-of-Paris models to more closely resemble computational panelings and grids, are already being suggested by researchers around the world. The future prospects of this amazing new wind tunnel technology are bounded only by the imagination.

But what, you may ask, will be the fate of the millions of computational aerodynamicists presently employed in the aerospace industry? Is the wind tunnel a threat to their job security? While it is true that some may lose their jobs, a brand-new demand will be created for those well-versed in state-of-the-art wind tunnel technology. Engineering graduate schools are already replacing course offerings in Finite Volume Methods and Grid Generation with curricula in Woodworking and Whittling. Clearly, the engineer will be freed from the tyranny and drudgery of computational methods, giving him more time to concentrate on creative tasks. It is doubtful, however, that the computer will ever be completely eliminated; the thought of an airplane designed solely from wind tunnel data without the aid of the computer seems too incredible to believe. While the wind tunnel may never fully replace the computer, it is almost certain to become the most useful engineering tool of the future.

NASA Ames 1984

RESPONSE

The limitations of the above description of an emerging technology cannot be permitted to pass without comment.

EFD can indeed come close to simulating geometric singularities like sharp corners. However, some physical rounding is inevitable due to machining practice, especially with wooden models. The more serious, and perhaps insurmountable obstacle is the inability of EFD physical models to simulate purely mathematical singularities such as polar coordinate origins, zero and negative jacobians, etc. Not only do no flow features attributable to the singularities appear, but the richness of the CFD simulations is lost. The EFD simulations (e.g., for an elliptically rounded leading edge of a flat plate) appear to give virtually the same flow solution no matter what the model material, no matter what coordinate system was used to define the geometry, and no matter what machining algorithm is used to produce the physical model. This poverty of the EFD solution space is in sharp and sad contrast to the rich multiplicity of solutions produced by CFD, depending on the coordinate system used, the time differencing, the iterative procedure, the choice of dependent variables, and of course the turbulence models, which now number in the hundreds (even thousands, if one includes the published variants of the parameter values).

Additionally, we have that most humane aspect of CFD, that which bonds the CFD community to the long heritage of western philosophical and aesthetic traditions, the underpinning of the *democratic values* implicit in CFD: the well-known yet mystical truth that each CFD practitioner, even though limited to using the same conceptual model, same code, and same parameter range, can still produce his or her own unique solution. (It is a curious historical fact that in the early days of CFD, this dependence of the solution on the human being who ran the code was incredibly considered a *disadvantage* of CFD. Apparently the old timers accepted the inhumane regimentation involved, which may be attributed to their psychological hangover from the robotic standardization enforced by the industrial revolution.)

On the other hand, some EFD practitioners have claimed (with a straight face) that their methodology not only *identically* conserves mass, momentum and energy, but *all the higher moments*. We obviously cannot accept such nonsense, which contradicts the Scheisskopf theorem, and we suggest that the guilty parties are compromising the credibility of the promising new field by making such outlandish claims.

APPENDIX E. A BIOGRAPHICAL SKETCH OF LEWIS FRY RICHARDSON

Lewis Fry Richardson (1881–1953) is a central figure in this book because of his papers on the "deferred approach to the limit", now more commonly known as "Richardson Extrapolation" (Richardson, 1910, 1927. See especially Chapter 5.) However, the present minor biographical sketch and tribute has little to do with error estimation, but much to do with personal admiration.

Considered for his scientific contributions and for all his humanity, Richardson is a singular figure of the 20th century. When I wrote the first edition of my CFD book in 1972, I opined that Richardson's 50-page paper presented to the Royal Society in 1910 "must be considered the cornerstone of modern numerical analysis of partial differential equations... He treated the iterative solution of Laplace's equation, the biharmonic equation, and others. He distinguished between steady-state problems 'according as the integral can or cannot be stepped out from a part of the boundary', i.e., between hyperbolic and elliptic problems, in modern terminology. He carefully treated numerical boundary conditions, including those at a sharp corner and those as infinity. He obtained error estimates and gave an accurate method of extrapolating answers toward the zero grid space limit, and further suggested checking the accuracy of numerical methods with exact solutions of simple geometries such as a cylinder. Finally, he was the first to actually apply these methods to a large-scale practical problem, that of determining stresses in a masonry dam." (Roache, 1972, p. 2.) Richardson had a hard time getting it published; one reviewer wanted Part A omitted and Part B condensed, while the second reviewer wanted Part B omitted and Part A condensed. (Ashford, 1985, p. 23.)

I also noted in a footnote (Roache, 1972, p. 2) that "Richardson presented what, in modern vocabulary, must be called a 'cost-effectiveness' study of the method, using human computers."

"So far I have paid piece rates for the operation [Laplacian] of about *n*/18 pence per coordinate point, *n* being the number of digits. The chief trouble to the computers has been the intermixture of plus and minus signs. As to the rate of working, one of the quickest boys averaged 2,000 operations [Laplacian] per week, for numbers of three digits, those done wrong being discounted." (Richardson, 1910, p. 325.)

O. Zienkiewicz, a dominant figure in Finite Element theory and applications, told me that he began his career as one of Richardson's later "computing boys." As I noted in (Roache, 1972) and even more applicable today and to the subject of the present book, "We may all be thankful that social conditions have changed since 1910. Many a computational fluid dynamicist would end up in the poorhouse if he were paid a certain fee per calculation, with 'those done wrong being discounted.'"

At that time, I knew nothing of Richardson's personal life, and I was *un*enthusiastic (see p. 177 of Roache, 1972) about Richardson Extrapolation! In Mandelbrot's book on Fractals (Mandelbrot, 1977) he includes a biographical sketch of Richardson because of his major conceptual contributions to this most *avant-garde* area of mathematics, fractals or fractal sets (terms coined subsequently by Mandelbrot). Fluid dynamicists recognize Richardson "for some of the most profound and most durable ideas regarding the nature of turbulence, a field into which he introduced a notion of self-similarity" (Mandelbrot, 1977, p. 30). His name is perpetuated in the Richardson Number, the fraction of turbulent energy from temperature gradients vs. velocity gradients. But I recently found out that the he also made major contributions to the field of experimental psychology, and the intellectual area in which Richardson is *most* widely known - his books and papers (Richardson, 1960a,b, 1961) being a canonical reference - is the sociometric area of quantitative peace studies! The processes in the theory of arms races are sometimes called "Richardson Processes." (See Boulding, 1978, p. 157.)

Richardson earned his B. A. in 1903 at Kings College, Cambridge with studies in physics, mathematics, chemistry, biology, and zoology. In his patent applications in 1912, he identified himself as "physicist." Later, after a career change, at the age of 47 he earned a degree in psychology.

His first scientific position was at the British Meteorological Office. He wrote the first book on numerical weather prediction (Richardson 1922, republished in 1965). His turbulence studies, which earned him election as a Fellow of the Royal Society of London, first elucidated the concept of energy cascade from low to high frequencies. The second section of his numerical weather prediction book was titled "Does the Wind Possess a Velocity?", an ostensibly foolish question that leads to the concept (though not the terminology) of fractals. As Mandelbrot (1977,

p. 270) states: "More important, at least to me, is that most of Richardson's arguments can very easily be translated into the language of the "fractal" vision of turbulence." His fractal concept is even more accessible in his (posthumously published) work on map metrics. Mandelbrot (1977) titles his own expository second chapter, "How Long is the Coast of Britain?", citing Richardson's work and his key (fractal) recognitions: that determination of coastline length from maps depends on the size of the calipers, i.e. on the measuring unit η; that, over a physically practical range of parameters, the imputed length does *not* converge as $\eta \to 0$; and that the characteristics of the coastline are consistent with the concept of fractal dimension. Mandelbrot's Plate 32 reproduces Richardson's empirical data, found among his papers after he died, on the rate of increase of the measured length of coastlines (Australia, South Africa, Germany, West Coast of Britain, and Portugal) as the caliper length η decreases. (By contrast, the measured "coastline" of a circle converges nicely as $\eta \to 0$.)

Richardson was a Quaker and a conscientious objector in the Second World War, but put his life on the line in an ambulance unit for three years. After the war, he resigned from the Meteorological Office because it was being merged into the Air Ministry. He then took up the study of the psychology of armed conflicts between nations, self-publishing a small book entitled *Mathematical Psychology of War*, and supporting himself by college teaching until he received an inheritance that allowed him to pursue his peace studies full time. From his study of the statistics and dynamics of war, he became the originator of the classical predator - prey equations.

His clearly expository papers have breadth and originality that are simply startling. Besides the numerical and meteorological work, his papers cover the following sampled diverse subjects: the measurement of mental "nature" [in the "Nature vs. Nurture" debate] and the study of adopted children; international voting power; thresholds when sensation is regarded as quantitative; the quantitative mental estimation of hue, brightness or saturation; the analogy between mental images and sparks; a quantitative view of pain; mathematical psychology of war; mental periodicities; the distribution of wars in time; chaos, international and inter-molecular; war-moods; variation of the frequency of fatal quarrels with magnitude; statistics of deadly quarrels; the problem of contiguity in population maps; the outcome of arms races (could an arms race end without fighting?); the nature of historical "fact" (is it possible to prove any general statements about historical fact?). He also had two patent applications for acoustic devices warning ships of possible collisions with large objects, above and below the water. He was inventive as an experimentalist, including measurements of turbulent dispersion, meteorology, and albedo. He tinkered with building his own analog computer in his retirement (Hunt, 1998, p. xxxiii).

Mandelbrot (1977) states that "Lewis Fry Richardson was a great scientist whose originality ... mixed with eccentricity, and who did not in his lifetime achieve the fame he deserves... His work is characterized by the conception and execution of experiments of classic simplicity, and by respect for the facts thus revealed. He never hesitated to use precise and refined concepts when he deemed them necessary." In regard to the claim of "experiments of classic simplicity", a famous Richardson anecdote is his (last) field experiment in turbulent dispersion; with H.

Stommel, he threw parsnips from a pier into a lake, observed their dispersion with an ingenious non-optical measuring instrument of his own design, and confirmed his famous 4/3 power law for turbulence.

He also had a delightful sense of humor. His *Mathematical Gazette* article (Richardson, 1925 in Ashford et al, 1993) on "How to Solve Differential Equations Approximately by Arithmetic" was a tutorial written for an audience of mathematicians in a day when numerical methods were not the fashion, and their lack of mathematical elegance was frowned upon. Richardson begins thus. "I. *The first obstacle may be one of sentiment*. It is said that in a certain grassy part of the world a man will walk a mile to catch a horse, whereon to ride a quarter of a mile to pay an afternoon call. Similarly, it is not quite respectable to arrive at a mathematical destination, under the gaze of a learned society, at the mere footpace of arithmetic. Even at the expense of considerable time and effort, one should be mounted on the swift steed of symbolic analysis. The following notes are written for those who desire to arrive by the easiest route, and who are not self-conscious about the respectability of their means of locomotion."

Richardson's (1922) book on numerical weather prediction, "arguably the first modern treatise in... dynamic meteorology" (Somerville, 1996, p. 61) is a classic, in part due to what he himself describes as his "fantasy." "After so much hard reasoning, may one play with a fantasy? Imagine a large hall like a theatre, ..." filled with 64000 human computers! This vision of massively parallel computation includes consideration of unbalanced processor loads (some people calculate faster than others), inter-processor communication band widths (posting of results on "numerous little 'night signs' so neighbouring computers can read them"), a hierarchical control structure ("each region is coordinated by an official of higher rank" and a conductor in a central pulpit who regulates the slide-rule wielding, calculator-punching processors by shining "a beam of rosy light upon any region that is running ahead of the rest, and a beam of blue light on those who are behind"), and information processing ("four senior clerks" who collect the weather prediction and are "despatching it by pneumatic carrier to a quiet room" whence it is "coded and telephoned to the radio transmitting station"). His "fantasy" includes humane considerations: "Outside are playing fields, houses, mountains and lakes, for it was thought that those who compute the weather should breathe of it freely."

His vision is charming, yet he was known for being caustic at times, and was never shy of controversy. He started a big one in 1922 with his experimental psychology paper entitled "Imagery, conation and cerebral conductance." His interest at this time was in the quantitative measurement of thought itself and of subjective sensation, including color (hue, brightness, saturation), loudness, touch, and pain. His ideas sharply divided the scientific community, many of whom thought the entire concept was nonsense. In 1932, the British Association for the Advancement of Science appointed a committee (including Richardson) to resolve the problem. Six years later, they had come to no conclusion! After eight years, they had made "some progress" but were still unable to agree on "certain important points." (Ashford, 1985, pp. 130–133, 147–149.) His concepts are now well accepted. To the last, he was disputatious; in a letter written three days before he died to a friend and fellow war researcher Quincy Wright,

Richardson chided him harshly for clinging to the 19th century concept of "balance of power" and ignoring the concept of stability. (Ashford, 1985, p. 235.)

Except for his books, his entire published works (including journal letters and posthumously published work) are available in a two-volume set (Ashford et al, 1993). His work has the quality of perpetually renewing in interest. Mandelbrot (1977, p. 270) notes that his turbulence papers "remain interesting; each fresh glance at them seems to show some angle that has passed unnoticed." I re-read his classic 1910 paper 2 decades after I had already opined (in 1972) that it was the foundation of my own field of CFD, and was astounded at how much I had overlooked or failed to appreciate in several previous unhurried readings. A short and readable biography was written by Hunt (1998) in the 1998 *Annual Review of Fluid Mechanics*. The full biography entitled "Prophet or Professor? The Life and Work of Lewis Fry Richardson" was written by Ashford (1985); eleven years later, the mathematician and humanist Philip J. Davis still considered the book worth reviewing (and his life reflecting upon) in a SIAM NEWS review titled "Weather, War, and Mathematics" (Davis, 1996). Davis also reviewed a newer popular science book on climate change by Somerville (1996), who also considers Richardson "profoundly original" (p. 57) and "a special case in the spectrum of humanity" who "makes ordinary obsessive scientists...look normal by contrast" (p. 151).

Richardson's other activities and interests included participation in the simplified spelling movement, rational language movements (Ido and Esperanto), Eugenics Societies (when eugenics was still politically correct and in fact espoused by most leading liberals), hiking and scouting (he was a serious naturalist), writing one unpublished play (he also used Socratic dialogue in his treatise on arms races), and of course religion and philosophy. One might think that his Quaker religion defined him, yet he did not fail to reflect on his religious inheritance, and he rejected much of it; his religious philosophy was far from naive (e.g., see Ashford, 1985, p. 69).

Davis (1996) gave this admiring observation.

Richardson was a man who walked independently, who openly admitted his failures ... even publishing them. He was an unorthodox scientist, never in the mainstream. He never played it safe.

REFERENCES

Ashford, O. M. (1985), *Prophet or Professor? The Life and Work of Lewis Fry Richardson*, A. Hilger, Boston.

Ashford, O. M. et al (1993), Eds., *The Collected Papers of Lewis Fry Richardson*, (2 Volumes), Cambridge University Press, Cambridge.

Boulding, K. E. (1978), *Ecodynamics: A New Theory of Societal Evolution*, Sage Publications Ltd. London.

Davis, P. J. (1996), "Weather, War, and Mathematics", Book Review, *SIAM News*, November 1996, p. 4.

Hunt, J. C. R. (1998), "Lewis Fry Richardson and His Contributions to Mathematics, Meteorology, and Models of Conflict", *Annual Reviews of Fluid Mechanics*, Volume 30, 1998, pp. xii–xxxvi.

Mandelbrot, B. B. (1977), *Fractals: Form, Chance, and Dimension*, W. H. Freeman and Company, San Francisco.

Richardson, L. F. (1910), "The Approximate Arithmetical Solution by Finite Differences of Physical Problems Involving Differential Equations, with an Application to the Stresses in a Masonry Dam", *Transactions of the Royal Society of London, Series A*, Vol. 210, pp. 307–357.

Richardson, L. F. (1922), *Weather Prediction by Numerical Process*, Cambridge University Press. (Re-published in 1965.)

Richardson, L. F. (1925), "How to Solve Differential Equations Approximately by Arithmetic", *The Mathematical Gazette*, London, Vol. XII, No. 177, July 1925, pp. 415–421.

Richardson, L. F. (1927), "The Deferred Approach to the Limit", *Transactions of the Royal Society of London, Series A.*, Vol. 226, pp. 299–361. "Part I. Single Lattice." "Part II. Interpenetrating Lattices" by J. A. Gaunt.

Richardson, L. F. (1960a), *Arms and Insecurity: A Mathematical Study of the Causes and Origins of War*, Ed. N. Rashevsky and E. Trucco, Boxwood Press, Pittsburgh (now, Pacific Grove).

Richardson, L. F. (1960b), *Statistics of Deadly Quarrels*, Q. Wright and C. C. Kienau, eds., Boxwood Press, Pittsburgh (now, Pacific Grove).

Richardson, L. F. (1961), "The Problem of Contiguity: An Appendix of Statistics of Deadly Quarrels", *General Systems Yearbook*, Vol. 6, 1961, pp. 139–187.

Roache, Patrick J. (1972), *Computational Fluid Dynamics*, First Edition, Hermosa Publishers, P. O. Box 9110, Albuquerque, New Mexico 87119-9110, USA. http://www.hermosa-pub.com/hermosa. Revised Printing 1976. See also Roache (1998).

Somerville, R. C. J. (1996), *The Forgiving Air: Understanding Environmental Change*, University of California Press, Berkeley.

INDEX

B

C

D

J

K

L

M

N

O

P

Q

R

T

U

ORDER FORM

Postal Orders: Just return this Order Form with your cheque or credit card number to
Hermosa Publishers
P. O. Box 9110
Albuquerque, New Mexico 87119-9110
USA

FAX Orders: Just FAX this Order Form with your credit card number to
(USA) (505) 866-5323

Web Orders: For Table of Contents, errata, other information, and to **Order by credit card (secure)** visit

http://www.hermosa-pub.com/hermosa

VERIFICATION AND VALIDATION IN COMPUTATIONAL SCIENCE AND ENGINEERING
by ***Patrick J. Roache***

- Applicable to commercial codes and to all problems modeled by partial differential equations.
- Examples include CFD, Groundwater Modeling, Grid Generation, Electrodynamics, etc.
- Useable formulas & practical guidance for error estimation/banding & experimental Validation.
- 464 pages, Hard Cover, subject index.

FUNDAMENTALS OF COMPUTATIONAL FLUID DYNAMICS
by ***Patrick J. Roache***

- The successor to the original pioneering text *Computational Fluid Dynamics*,
- Translated into Russian, Japanese and Chinese, and cited over 2,066 times through 1995.
- **Part I** - original text with corrections, minor revisions, and updates. **Part II** - more recent topics.
- 650+ pages, Hard Cover, subject index.

Your Name: ______________________________
Company: ______________________________
Address (line 1): ______________________________
Address:(line 2): ______________________________
City: ______________________________
State or Province: ____________ ZIP (Postal Code) ____________ Country: ____________

Verification and Validation in Computational Science and Engineering
Number of copies ordered [___]× US$85.00 = US$ ___.___
Fundamentals of Computational Fluid Dynamics
Number of copies ordered [___]× US$75.00 = US$ ___.___
Shipping and handling charge per order (We pay any sales tax) US$ 3.50
Moneyback guarantee US$ ___.___ Total

Payment by enclosed cheque [_] or credit card: VISA [_] MasterCard [_] AMEX [_] Discover [_]
Card Number: ______________________________
Name on Card: ______________________________ Exp. Date: ____________